P9-AEY-698

Arabidopsis

Springer-Verlag

New York
Berlin
Heidelberg
Barcelona
Budapest
Hong Kong
London
Milan
Paris
Santa Clara
Singapore
Tokyo

John Bowman

Editor

Arabidopsis

An Atlas of Morphology
and Development

With 167 Figures

Springer-Verlag

John L. Bowman
University of California, Davis
Department of Plant Biology
Davis, CA 95616
USA

Library of Congress Cataloging-in-Publication Data
Arabidopsis: an atlas of morphology and development / edited by John
    Bowman. — [1st ed.]
        p.   cm.
    Includes bibliographical references (p.   ) and index.
    ISBN 0-387-94089-8. — ISBN 3-540-94089-8 (Berlin)
    1. Arabidopsis — Morphology — Atlases.   2. Arabidopsis — Development —
Atlases.   I. Bowman, John
QK495.C9A69   1993
583´.123 — dc20   93-5262

Cover art credits:

Illustration of *Arabidopsis* courtesy of Stephen H. Howell, Boyce Thompson Institute for Plant Research, Ithaca, New York.

Digitized picture shows luciferase expression in an *Arabidopsis* plant that has been transformed with the bacterial luciferase (*Lux* $\alpha\beta$) gene. Courtesy of G.P. Rédei, C. Koncz, and W.H.R. Langridge.

Printed on acid-free paper.

© 1994 Springer-Verlag New York, Inc.
All rights reserved. This work may not be translated or copied in whole or in part without the written permission of the publisher (Springer-Verlag New York, Inc., 175 Fifth Avenue, New York, NY 10010, USA), except for brief excerpts in connection with reviews or scholarly analysis. Use in connection with any form of information storage and retrieval, electronic adaptation, computer software, or by similar or dissimilar methodology now known or hereafter developed is forbidden.
The use of general descriptive names, trade names, trademarks, etc., in this publication, even if the former are not especially identified, is not to be taken as a sign that such names, as understood by the Trade Marks and Merchandise Marks Act, may accordingly be used freely by anyone.

Acquiring Editor: Robert C. Garber
Production coordinated by Chernow Editorial Services, Inc. and managed by
    Theresa Kornak; manufacturing supervised by Jacqui Ashri.
Typeset by Asco Trade Typesetting Ltd., Hong Kong.
Printed and bound by Edwards Brothers, Inc., Ann Arbor, MI.
Printed in the United States of America.

9 8 7 6 5 4 3 2

ISBN 0-387-94089-8 Springer-Verlag New York Berlin Heidelberg
ISBN 3-540-94089-8 Springer-Verlag Berlin Heidelberg New York   SPIN 10560913

Preface

The recent application of molecular genetics to problems of developmental biology has provided us with greater insight into the molecular mechanisms by which cells determine their developmental fate. This is particularly evident in the recent progress in understanding of developmental processes in model animal systems such as *Drosophila melanogaster* and *Caenorhabditis elegans*. Despite the use of plants in some of the earliest genetics experiments, the elucidation of the molecular bases of plant development has lagged behind that of animal development. However, the emergence of model systems such as *Arabidopsis thaliana*, amenable to developmental genetics, has led to the beginning of the unraveling of the mysteries behind plant morphogenesis.

This atlas of the morphology and development of the weed *Arabidopsis* is intended to be a reference book, both for scientists already familiar with plant anatomy and for those utilizing *Arabidopsis* who have come from other fields. The primary concentration is on descriptions rather than interpretations, as interpretations evolve and change relatively rapidly, whereas the evolution of plant form takes place on a much longer time scale. Molecular genetics and the use of mutants to probe wild-type gene function rely on the wild-type being well characterized. With this in mind, an attempt was made to present detailed descriptions of wild-type structure and development, to provide a foundation for comparison with the selected mutants in the atlas. More importantly, it is hoped that the atlas will serve as a valuable resource in the characterization of new mutants.

The atlas is arranged in sections devoted to various aspects of the growth and development of the plant. However, because development is a continuous process and is not neatly broken into discrete chapters, it is inevitable that there is some overlap between the sections, especially in the case of pleiotropic mutants. In the process of compiling the atlas, it has become obvious that there are gaps in the description of the morphology and development of wild-type *Arabidopsis*. If these gaps are real, rather than just holes in the editor's knowledge of the plant, perhaps they will stimulate ideas for future directions.

I would like to thank the many people who have contributed to this book, since they deserve the credit for illustrating a common weed in such a beautiful manner, and I hope the atlas will be as illuminating to the reader as it has been to the editor. I have attempted to acknowledge the contributors on every plate, even though their material may not be contiguous in the book. Special thanks go to Gary Mansfield, Jeff Dangl, and Liam Dolan who took the time to write enlightening introductions to some of the sections. I would also like to thank David Smyth for supporting me while working on this time-consuming project in his laboratory and for his helpful discussions about the book's organization. Finally, I am grateful to Pene Brockie for her time spent compiling the atlas.

Contents

Contributors

JOHN ALVAREZ
Department of Genetics and Developmental Biology, Monash University, Clayton, Victoria 3168, Australia

RICARDO AZPIROZ
Department of Plant Sciences, College of Agriculture, University of Arizona, Tucson, AZ 86721, USA

TOBIAS I. BASKIN
Division of Biological Sciences, University of Missouri, Columbia, MO 65211, USA

FRIEDRICH J. BEHRINGER
Department of Biology, The Pennsylvania State University, University Park, PA 16802, USA

PHILIP N. BENFEY
Department of Biology, New York University, New York, NY 10003, USA

GERD BOSSINGER
Department of Genetics and Developmental Biology, Monash University, Clayton, Victoria 3168, Australia

JOHN L. BOWMAN
Department of Genetics and Developmental Biology, Monash University, Clayton, Victoria 3168, Australia

L. G. BRIARTY
Department of Life Science, University of Nottingham, University Park, Nottingham NG7 2RD, UK

JOSEPH D. CALLOS
Department of Biology, The Pennsylvania State University, University Park, PA 16802, USA

ABED CHAUDHURY
CSIRO Division of Plant Industry, GPO Box 1600, Canberra, ACT 2601, Australia

JOANNE CHORY
Plant Biology Lab, Salk Institute, La Jolla, CA 92037, USA

STEVE E. CLARK
Division of Biology, Caltech, Pasadena, CA 91125, USA

STUART CRAIG
CSIRO Division of Plant Industry, GPO Box 1600, Canberra, ACT 2601, Australia

WILLIAM CRONE
Department of Botany and Plant Sciences, University of California, Riverside, Riverside, CA 92521, USA

JEFF DANGL
Max Delbrück Laboratory, Max Planck Institut für Zuchtungsforschung, 50829 Köln 30, Germany

ARTHUR RALPH DAVIS
Biology Department, University of Saskatchewan, Saskatoon, Saskatchewan S7N 0W0, Canada

J. DAWSON
Department of Life Science, University of Nottingham, University Park, Nottingham NG7 2RD, UK

THOMAS DEBENER
Max Delbrück Laboratory, Max Planck Institut für Zuchtungsforschung, 50829 Köln 30, Germany

LIAM DOLAN
Department of Cell Biology, John Innes Institute, Norwich NR4 7UH, UK

JEFFERY J. ESCH
School of Biological Sciences, University of Nebraska-Lincoln, Lincoln, NE 68588-0118, USA

MARK A. ESTELLE
Department of Biology, Indiana University, Bloomington, IN 47405, USA

KENNETH A. FELDMANN
Department of Plant Sciences, College of Agriculture, University of Arizona, Tucson, AZ 86721, USA

ROBERT L. FISCHER
Department of Plant Biology, University of California, Berkeley, CA 94720, USA

W. GOLINOWSKI
Department of Botany, Agricultural University of Warsaw, Warsaw, Poland

MEGAN GRIFFITH
Department of Genetics and Developmental Biology, Monash University, Clayton, Victoria 3168, Australia

FLORIAN M. W. GRUNDLER
Institut für Phytopathologie, Universität Kiel, Kiel, Germany

GEORGE W. HAUGHN
Biology Department, University of Saskatchewan, Saskatoon, Saskatchewan S7N 0W0, Canada

JEFFREY P. HILL
Department of Botany and Plant Sciences, University of California, Riverside, Riverside, CA 92521, USA

MAARTEN KOORNNEEF
Department Genetics, Agricultural University, 6703 HA Wageningen, The Netherlands

RACHEL M. LEECH
Department of Biology, University of York, Heslington, York YO1 5DD, UK

Contributors

HILTRUD LIEDGENS
Max Delbrück Laboratory, Max Planck Institut für Zuchtungsforschung, 50829 Köln 20, Germany

BRUCE M. LINK
Department of Biology, The Pennsylvania State University, University Park, PA 16802, USA

PAUL LINSTEAD
Department of Cell Biology, John Innes Institute, Norwich NR4 7UH, UK

ELIZABETH M. LORD
Department of Botany and Plant Sciences, University of California, Riverside, Riverside, CA 92521, USA

S. GARY MANSFIELD
Department of Cell Biology, Duke University Medical Center, Durham, NC 27710, USA

M. DAVID MARKS
School of Biological Sciences, University of Nebraska-Lincoln, Lincoln, NE 68588-0118, USA

JOANNE L. MARRISON
Department of Biology, University of York, Heslington, York YO1 5DD, UK

BRIGITTE MAUCH-MANI
Institute für Pflanzenbiologie, CH 8008 Zürich, Switzerland

JUNE I. MEDFORD
Department of Biology, The Pennsylvania State University, University Park, PA 16802, USA

ELLIOT M. MEYEROWITZ
Division of Biology, Caltech, Pasadena, CA 91125 USA

ZORA MODRUSAN
Biology Department, University of Saskatchewan, Saskatoon, Saskatchewan S7N 0W0, Canada

B. J. MULLIGAN
Department of Life Science, University of Nottingham, University Park, Nottingham NG7 2RD, UK

KIYOTAKA OKADA
Division I of Gene Expression and Regulation, National Institute for Basic Biology, 38 Nishigonaka, Myodaijicho, Okazaki 444, Japan

KITTY PLASKITT
Department of Cell Biology, John Innes Institute, Norwich NR4 7UH, UK

DANIEL S. POOLE
Plant Biology Lab, Salk Institute, La Jolla, CA 92037, USA

KEVIN A. PYKE
Department of Biology, University of York, Heslington, York YO1 5DD, UK

LEONORE REISER
Department of Plant Biology, University of California, Berkeley, CA 94720, USA

KEITH ROBERTS
Department of Cell Biology, John Innes Institute, Norwich NR4 7UH, UK

HAJIME SAKAI
Division of Biology, California Institute of Technology, Pasadena, CA 91125, USA

JOHN W. SCHIEFELBEIN
Department of Biology, University of Michigan, Ann Arbor, MI 48109, USA

YOSHIRO SHIMURA
Division I of Gene Expression and Regulation, National Institute for Basic Biology, 38 Nishigonaka, Myodaijicho, Okazaki 444, Japan and Department of Biophysics, Faculty of Science, Kyoto University, Kitashirakawa-Oiwakecho, Sakyoku, Kyoto 606, Japan

ALAN J. SLUSARENKO
Institut für Pflanzenbiologie, CH 8008 Zürich, Switzerland

DAVID R. SMYTH
Department Genetics and Developmental Biology, Monash University, Clayton, Victoria 3168, Australia

DANIEL STEWART
Department of Biology, The Pennsylvania State University, University Park, PA 16802, USA

JENNIFER VASINDA
Department of Biology, The Pennsylvania State University, University Park, PA 16802, USA

MARY C. WEBB
Plant Cell Biology Group, Research School of Biosciences, Australia National University, Canberra, ACT 2601, Australia

DETLEF WEIGEL
Division of Biology, California Institute of Technology, Pasadena, CA 91125, USA

MARK WILKINSON
Biology Department, University of Saskatchewan, Saskatoon, Saskatchewan S7N 0W0, Canada

RICHARD WILLIAMSON
Plant Cell Biology Group and Plant Science Center, Research School of Biological Sciences, Australia National University, Canberra, ACT 2601, Australia

ZOE A. WILSON
Department of Life Science, University of Nottingham, University Park, Nottingham NG7 2RD, UK

U. WYSS
Institut für Phytopathologie, Universität Kiel, Kiel, Germany

Introduction

Genetics and Development

Over two centuries ago, Goethe (1790) noted the value of mutants, or in his term, *"retrograde* metamorphosis," in elucidating the mechanisms by which living organisms develop: "From our acquaintance with this abnormal metamorphosis, we are enabled to unveil the secrets that normal metamorphosis conceals from us, and to see distinctly what, from the regular course of development, we can only infer." However, it was not until early in the twentieth century, following the rediscovery of Mendel's work, that the disciplines of genetics and developmental biology were united. This was in part due to Thomas Hunt Morgan's conclusion that the problems of experimental embryology were intractable, and his move into genetics in the hope that it would illuminate the problems of development; this has been referred to as Morgan's Deviation. In the next thirty years much progress was made in the field of genetics. However, due to the lack of knowledge of the molecular basis of inheritance, there remained a gulf in understanding between the nature of the gene and development: "Between the characters that are used by the geneticist and the genes that his theory postulates lies the whole field of embryonic development, where the properties implicit in the genes become explicit in . . . the cells." (Morgan, 1934).

The gap between gene and metamorphosis has been partially bridged with the advent of molecular genetics, as the ability to purify and sequence DNA has allowed the relationship to be examined at the molecular level. In large part, molecular genetic studies rely on well-characterized organisms that are easily manipulated. In the domain of flowering plants, the small cruciferous weed *Arabidopsis thaliana* has become a model system for many molecular genetic investigations of plant development, including pattern formation, physiology, biochemistry, and plant–pathogen interactions. However, before mutants can "unveil the secrets" of normal development, it is crucial that we have a detailed description of wild-type development. This book is intended to provide a foundation of knowledge about the normal development of *Arabidopsis*, a baseline with which the abnormal morphogenesis of mutants can be compared. As a first step, a selection of already well-characterized mutants is also presented, demonstrating that the study of the morphology of the mutant can illuminate wild-type gene function.

The Plant: Taxonomy, Habitat, and Variation

Arabidopsis thaliana is a small herbaceous annual of the Brassicaceae. The family Brassicaceae is large (ca. 350 genera and 3,000 species) and unusually homogeneous, making delineations above the generic level difficult (Hedge, 1976). For this reason, the precise taxonomic relationship of *Arabidopsis* to the other members of the Brassicaceae is uncertain, with different classifications of the crucifers according the genus *Arabidopsis* different positions (Schulz, 1936; Janchen, 1942; Berger, 1965; see Rédei, 1970; Hedge, 1976 for reviews).

Although the exact geographical origin of *Arabidopsis thaliana* is unknown, it is thought to have arisen in Eurasia. Some authors have suggested the Central Asian highlands of the Western Himalayas as the source (Berger, 1965; Rédei, 1970). It is found in many different ecological and geographical regions, albeit predominantly in the moderate temperate climates of the northern hemisphere. *A. thaliana* is common through Europe and western Asia, and can be found in northern Africa and east Asia (Rédei, 1970). Although *A. thaliana* is common in temperate North America, it is likely not native as its distribution maps early settlers' shipping routes (Rédei, 1992). *A. thaliana* is also naturalized in localized areas of Australia (Rédei, 1970).

The habitat of *Arabidopsis* ranges from the high mountains of equatorial Africa to highlands of the Himalayas to most of temperate Europe. Consistent with its broad ecological and geographical distribution, a large amount of naturally occurring variation exists in *Arabidopsis thaliana*. Wild populations (ecotypes) have been collected primarily from Europe and western Asia (e.g., the former Soviet Union, Kashmir), but also from northern Africa, Japan, and North America (including one naturalized population that was found in Yosemite National Park, USA) (Röbbelen, 1965b; Kranz and Kirchheim, 1987). Most of the ecotypes are winter annuals, with seed germination in the autumn, survival through the winter in the vegetative rosette phase, and transition to flowering in the spring. Among the wild ecotypes, the major source of phenotypic variation is in the response to daylength and to vernalization with respect to the transition to flowering, although leaf characteristics vary as well (Rédei, 1970; Lawrence, 1976). However, it must also be noted that most ecotypes collected from the wild are heteromorphic for several characters including flowering time (Napp-Zinn, 1964; Effmertová, 1967; Rédei, 1970; Lawrence, 1976).

Most of the commonly used laboratory strains have a single seed as the original source and have been inbred for many generations. They are thus homozygous at most loci. Some common laboratory strains, such as Landsberg *erecta* (Rédei, 1970), are not true wild-type in that they have been selected to be homozygous for mutations that result in a growth habit favorable for laboratory/greenhouse conditions. An estimate of DNA sequence divergence between ecotypes has been obtained using restriction fragment length polymorphism (RFLP) analysis. A survey of a large number of RFLPs between three inbred laboratory ecotypes (Landsberg *erecta*, Columbia, and Niederzenz) suggests that these ecotypes differ from between 1% and 1.5% in their low-copy-number genomic DNA nucleotides (Chang et al., 1988).

The Plant: A Brief History

Arabidopsis thaliana has been known to botanists for at least four centuries and has been utilized in experimental research for nearly a century (for review see Rédei, 1992). In 1907, when the relationship between chromosomes and genetics was just coming to light, Laibach reported the continuity of the *Arabidopsis* chromosomes (2n = 10) during interphase, a primary requirement for the hereditary material. Also noted was the small size of the *Arabidopsis* chromosomes compared to other closely related species such as *Brassica* (Laibach,

1907), a feature that hindered early cytological studies but has facilitated recent molecular genetic studies. Despite the small size of the chromosomes, cytological studies utilizing trisomics have correlated karyotype with linkage groups (Steinitz-Sears, 1963; Schweizer et al., 1988), and major and minor nucleolar organizing regions (rDNA) have been identified cytologically (Sears and Lee-Chen, 1970; Schweizer et al., 1988).

The concept of *Arabidopsis* as a model plant system was initiated in 1943 when Laibach outlined the advantages of using *Arabidopsis* in genetic experiments. Several features make *Arabidopsis* amenable to experimental classical genetics: (i) small size; (ii) rapid generation time (5–6 weeks under optimum growth conditions); (iii) ability to grow well in controlled conditions (either on soil or defined media); (iv) ease of cross- and self-fertilization; (v) fecundity (up to 10,000 seeds per plant); (vi) small chromosome number; and (vii) ease of mutagenesis. During the next few decades *Arabidopsis* was used in a variety of mutagenesis studies (see Rédei, 1970; Rédei, 1975; Estelle and Somerville, 1986; Meyerowitz, 1987; Bowman et al., 1988; Haughn and Somerville, 1988; Meyerowitz, 1989; Griffing and Scholl, 1991; Rédei and Koncz, 1992; Koornneef and Stam, 1992, for reviews), including the first isolation of auxotrophs in plants (Langridge, 1955; Li and Rédei, 1969b). More recently, *Arabidopsis* has been used as a model to examine the genetic effects of radiation from both the cosmos (e.g., Kranz, 1986) and sources a bit closer to home, such as the area surrounding the Chernobyl Nuclear Power Station (Abramov et al., 1990).

The blossoming of *Arabidopsis* as a model system began with the discovery that the size of its nuclear genome is the smallest known among flowering plants (Sparrow et al., 1972; Leutwiler et al., 1984), and that it contained a very low level of dispersed repetitive DNA (Pruitt and Meyerowitz, 1986), making *Arabidopsis* suitable for molecular as well as classical genetics (Meyerowitz and Pruitt, 1985). The size of the nuclear genome is estimated to be in the range of 60,000 to 100,000 kilobases, roughly the same as those of *Caenorhabditis elegans* and *Drosophila melanogaster*, and only 5–7 times larger than that of *Saccharomyces cerevisiae* (Leutwiler et al., 1984; Galbraith et al., 1991). The advantages of using *Arabidopsis* in molecular genetic experiments have been reviewed many times and will not be detailed here (Meyerowitz and Pruitt, 1985; Meyerowitz, 1987; Bowman et al., 1988; Meyerowitz, 1989; Griffing and Scholl, 1991).

Many biochemical and morphological mutations have been mapped on the five linkage groups of the genetic map (Koornneef et al., 1983; Koornneef, 1990b). In addition, linkage maps of RFLP and RAPD (random amplified polymorphic DNA) markers have been constructed (Chang et al., 1988; Nam et al., 1989; Reiter et al., 1992) as a first step towards the physical mapping of the *Arabidopsis* genome (e.g., van Montagu et al., 1992).

Thus, the tools are available to take the next step in attempting to bridge the gap between genes and morphogenesis in flowering plants. Plants have evolved multicellularity independently of animals, and are unique in several aspects of their developmental processes. In light of these fundamental differences, it will be of interest to see whether the plant and animal kingdoms have evolved similar mechanisms to solve the problems of development. The application of molecular genetics to *Arabidopsis*, as well as to other model flowering plant species (e.g., maize, *Antirrhinum majus*), should contribute to the closing of the gap outlined over half a century ago.

JOHN L. BOWMAN

1
Vegetative Development

Introduction

Arabidopsis thaliana is a rosette annual with separate vegetative and reproductive growth phases. Vegetative development is characterized by the production of rosette leaves in a phyllotactic spiral with little internode elongation between successive leaves. The number of rosette leaves formed prior to the transition to flowering (leaf number) varies with growth conditions, with low growth temperatures and short days retarding the transition to flowering. An increase in leaf number is closely correlated with a delay in the transition to flowering (Rédei, 1970; Napp-Zinn, 1985). Vegetative growth also varies with genetic background, with different ecotypes having widely varying leaf number (from a low of 5–8 for early-flowering ecotypes to more than 30 for late-flowering ecotypes, when grown in long days). Likewise, leaf characteristics such as shape, stem attachment, hairiness, type of margin, and color also vary between ecotypes (Rédei, 1970).

Vegetative Apical Meristem

The structure of the apical meristem of *Arabidopsis* is typical of that of the Brassicaceae, with distinct morphological and functional zonation (Vaughan 1952; Vaughan, 1955; Brown et al., 1964; Bernier, 1964; Miksche and Brown, 1965; Bernier et al., 1970; Besnard-Wibaut, 1970; Besnard-Wibaut, 1977; Orr, 1978).

In terms of the tunica-corpus concept (Schmidt, 1924; Satina et al., 1940), the mature vegetative meristem of *Arabidopsis thaliana* has a tunica of two layers of cells (L1 and L2), and a corpus represented by L3 cells (Vaughan, 1955). The L1 cells give rise to the epidermal cells, whereas the L2 and L3 cells give rise to the internal portions of the plant. Cell divisions in the two surface layers (L1 and L2) are nearly exclusively anticlinal, except when leaf primordia are initiated by periclinal divisions in the L2 (Vaughan, 1955).

The adult (see below) apical vegetative meristem of *A. thaliana* can be subdivided into distinct zones that are characterized by cell size, cell division rates, and density of histological staining (Vaughan, 1955). In the center of the adult vegetative meristem, which is flat to slightly convex, is a lens-shaped region referred to as the central initiation zone characterized by a relatively low cell division rate. Surrounding the central initiation zone is the doughnut-shaped flank, or peripheral, meristem characterized by a relatively high cell division rate. The cells of the flank meristem stain more deeply than those of the central initiation zone, and like those of the central initiation zone, exhibit no obvious vacuolation. Below the central initiation zone lies the file, or rib, meristem, whose lightly staining cells divide almost exclusively transversely. It must be noted that this zonation pattern is established sometime after vegetative growth begins, and it is difficult to recognize in younger, flat vegetative meristems (Miksche and Brown, 1965).

Distinct roles can be assigned to each of the zones of the vegetative meristem (Vaughan, 1955). Leaf primordia are derived from the flank meristem, their initiation brought about by periclinal divisions in the L2. The procambial strands of the leaf primordia and the cortex are established by the flank meristem as well.

The file meristem gives rise to the pith of the stem. The upper few cell layers of the file meristem have been referred to as cambium-like and they are flattened in shape. Below this, the cells which are derived from the file meristem are highly vacuolate and extended in a lateral direction, with little vertical extension.

The central initiation zone is a self-perpetuating region responsible for maintaining both the flank and file meristems, as these zones become depleted due to the production of leaf primordia and pith, respectively. Although the cells of the central initiation region act as self-renewing "stem" cells, they are thought not to be permanent "stem" cells (Ruth et al., 1985; Furner and Pumfrey, 1992).

Cell Fate Studies

The fates, both somatic and germinal, of the cells in the shoot apical meristem of the mature dry seed have been examined in several studies utilizing irradiation to induce clonal mutant sectors (Müller, 1965; Li and Rédei, 1969a; Grinikh et al., 1974; Grinikh and Shevchenko, 1976; Irish and Sussex, 1992; Furner and Pumphrey, 1992). The shoot apical meristem of the dry seed, which will give rise to all aerial portions of the plant (excluding the hypocotyl and cotyledons), comprises approximately 60–110 cells (Irish and Sussex, 1992; Medford et al., 1992). The genetically effective cell number (the number of cells of the seed shoot apical meristem that will give rise to the gametes) is estimated to be about two or three (Müller, 1965; Li and Rédei, 1969a; Grinikh et al., 1974; Grinikh and Shevchenko, 1976).

Studies examining the contribution of cells of the seed shoot apical meristem to the somatic tissues have indicated that there is no strictly defined pattern of cell lineage, but rather probable fates that are only somewhat position-dependent (Irish and Sussex, 1992; Furner and Pumphrey, 1992). Single cells of the seed shoot apical meristem can give rise to variable portions of the plant, ranging from a sector on a single leaf to portions of several leaves, as well as portions of the inflorescence (Irish and Sussex, 1992; Furner and Pumphrey, 1992). Axillary meristems are clonally related to the subtending leaf (Irish and Sussex, 1992; Furner and Pumphrey, 1992). Sectors within leaves generally extend from the base to the tip. There is a general pattern of acropetal variation in sector size and frequency in plants grown from irradiated seeds, with frequent and relatively small basal sectors, and sparse and relatively large apical sectors. This is consistent with the dynamic view of meristem structure and maintenance as described above, with nonpermanent "stem" cells of the central initiation zone continually replenishing the flank and file meristems. In summary, there are few if any restrictions on the fates of the cells of the seed shoot apical meristem.

Phyllotaxy and Heteroblasty

Vegetative growth in *Arabidopsis* exhibits changes in both phyllotaxy and leaf shape, correlated with the transition of the vegetative meristem from juvenile to adult stages (Röbbelen, 1957a; Medford et al., 1992). The relatively flat, bilaterally symmetric, juvenile vegetative meristem produces small, round, and entire leaves in a decussate phyllotaxy. In contrast, the slightly convex, radially symmetric, adult vegetative meristem produces larger serrate and increasingly spatulate leaves in a spiral phyllotaxy (Figure 1). Adult leaves are characterized by a high density of stellate trichomes on their adaxial (upper) surface, and a lower density on their abaxial (lower) surface, whereas juvenile leaves have a low density on their adaxial surface and lack trichomes on their abaxial surface.

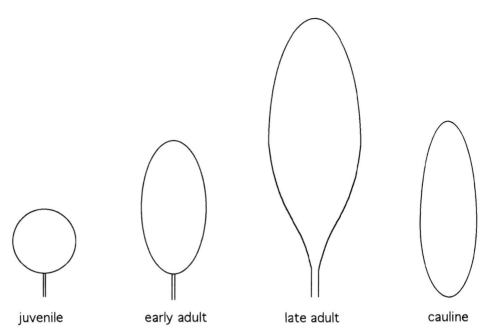

juvenile   early adult   late adult   cauline

Figure 1. Leaf heteroblasty. The juvenile vegetative meristem produces small round leaves while the adult vegetative meristem produces larger increasingly spatulate leaves. The last leaves produced before the transition to flowering are the cauline (stem) leaves. Cauline leaves are small, lanceolate, and, in contrast to the earlier produced leaves, lack a petiole.

For plants of the Wassilewskija and Landsberg *erecta* ecotypes grown at 20–25°C in long-day photoperiods, the transition from the juvenile to the adult vegetative meristem generally occurs between leaf four and five (Medford et al., 1992), but it may be delayed in other growth conditions. Leaf primordia are initiated from the adult vegetative meristem in a spiral phyllotaxy, with successive leaf primordia separated by an angle of approximately 137.5°. The direction of spiral phyllotaxy in the adult meristem may be either clockwise or counter-clockwise, presumably reflecting a stochastic event in the initiation and development of the juvenile leaves. It is not known if the transition from juvenile to adult vegetative meristem must occur before the transition to an inflorescence meristem can occur.

Leaf Development

Mature *Arabidopsis* leaves are typical of mesophytic dicotyledons with a dorsiventral structure (Pyke et al., 1991). The mesophyll tissue consists of an upper, adaxial layer of elongate palisade cells, with lower, spongy, mesophyll located abaxially. Trichomes and stomata are interspersed throughout the epidermis, with trichomes more abundant on the adaxial surface, and stomata more prevalent on the abaxial surface. The vascular system within the leaf is reticulate. *Arabidopsis* leaves are accompanied by flanking stipules, which degenerate before full expansion of the leaf (Medford et al., 1992).

Leaf primordia are initiated by periclinal divisions in the L2 of the flank meristem (Vaughan, 1955; Medford et al., 1992). The procambial strands of the leaf primordia, also established by the flank meristem, develop acropetally, from the base of the leaf to the tip. Soon after inception, vacuolation takes abaxial side to the adaxial side of the leaf, and stipules are conspicuous during the early stages of leaf primordia development. Differentiation of the epidermis, vacuolation of cells, and initiation of trichomes begin soon after inception of the leaf primordium, prior to differentiation of the ground tissue. Cells throughout the mesophyll remain meristematic until the unfolding of the leaf primordium. In contrast to the vascular tissue, the differentiation of the mesophyll occurs in a basipetal manner, starting at the leaf tip and proceeding towards the leaf base (Pyke et al., 1991). Thus, individual developing leaves exhibit a gradient of cells at varying stages of development, with fully differentiated cells at the tip, and meristematic cells toward the base. Expansion of the lamina is due primarily to cell expansion late in development throughout the leaf. Cell expansion results in a large increase in mesophyll volume and intercellular air space, with an accompanied increase in the vascular system.

Mutants

Mutations Affecting the Vegetative Meristem

Several mutations have been isolated that result in alterations in the structure and proper functioning of the vegetative meristem. The zonation of the vegetative meristem is disrupted in *forever young* (*fey*) and *disrupted* (*dip*) mutants (Medford et al., 1992). The cells of the file, or rib, meristem display both intra- and intercellular abnormalities in *schizoid* (*shz*) mutants (Medford et al., 1992). Recessive mutations at several loci lead to fasciation due to enlargement and lateral extension of the shoot apical meristem. In some cases, such as *fasciata1* (*fas1*) and *fasciata2* (*fas2*), abnormalities in leaf shape accompany the fasciation (Leyser and Furner, 1992). However, mutations at other loci, such as *CLAVATA1* (*CLV1*) (Koornneef et al., 1983; Leyser and Furner, 1992), *CLAVATA2* (*CLV2*) (Koornneef et al., 1983), *CLAVATA3* (*CLV3*) (J. Alvarez, M. Griffith, S. Clark, and D. Smyth, unpublished), and *FULLY FASCIATED* (*FUF*) (Medford et al., 1992), do not cause major alterations in organ shape. Generally, mutations that result in fasciation also cause alterations in leaf phyllotaxy.

Another mutation that affects the maintenance of the vegetative meristem is *embryonic flower* (*emf*) (Sung et al., 1992). The seed shoot apical meristem in homozygous *emf* mutants is disorganized, lacking a clear tunica-corpus organization, and develops directly into an abnormal inflorescence meristem rather than a vegetative meristem.

Since the inflorescence meristem is derived directly from the vegetative meristem in *Arabidopsis*, it is not surprising that each of the mutants described above also alters the proper maintenance and functioning of inflorescence and floral meristems as well.

Leaf Shape Mutants

Numerous mutations have been isolated that alter leaf-shape in several different ecotypes (e.g., Kranz and Kirchheim, 1987). However, most of these are not specific to leaf development, with pleiotropic alterations in plant form resulting as well. None of these has been characterized extensively. One of these mutations, *erecta*, has been extensively utilized in experiments since its pleiotropic phenotype (compact rosettes, shortened petioles, and reduced internode elongation) results in plants that are easily managed in laboratory and greenhouse conditions. For a discussion of the origin of the commonly used laboratory ecotypes, Landsberg *erecta* and Columbia, see Rédei (1992).

Biochemical and Physiological Mutants

Due to their immobile lifestyle, plants must adapt to their environment by utilizing plastic modes of growth and development. The basis of the plasticity is the ability to regulate cell division, cell elongation, and differentiation in response to changing environmental conditions. Of the many external cues to which plants respond, light, gravity, daylength, temperature, and water are some of the more conspicuous. Thus, while some patterns of plant growth are genetically fixed, much of plant form can be plastic as the result of environmental stimuli activating specific receptors. In some cases, the action of the receptors may be ultimately mediated through phytohormones, which play an important role in regulating plant growth and development.

The following short descriptions highlight some of the mutants that have been isolated in *Arabidopsis*, concentrating on those that lead to morphological and developmental abnormalities, often due to a failure to respond appropriately to environmental conditions. However, this list is continually expanding and is by no means exhaustive. Mutants have been selected that are defective in a variety of biochemical or phytohormonal pathways. Frequently these are isolated by screening for plants that behave inappropriately under specific environmental conditions, or for plants resistant to either normally toxic compounds or to high concentrations of substrate analogues. However, in some cases, mass nonselective screening of plants has been used to identify mutants in specific biochemical pathways. Several reviews discuss various aspects of biochemical and physiological mutants in more detail (Rédei, 1970; Rédei, 1975; Estelle and Somerville, 1986; Meyerowitz, 1987; Bowman et al., 1988; Meyerowitz, 1989; Klee and Estelle, 1991; Rédei and Koncz, 1992).

Light Perception Mutants

Light is one of the many environmental stimuli to which plants respond both by positioning their growing point in the most favorable photosynthetic environment and activating photomorphogenesis. Three distinct classes of mutants affecting light perception have been identified. Mutants of the first class, such as the recessive *hy* mutants, exhibit characteristics of dark grown seedlings, i.e., an elongated stem, apical hook, and reduced greening, when grown in the light (Koornneef et al., 1980; Parks et al., 1989; Chory et al., 1989a; Somers et al., 1991; Parks and Quail, 1993). The *hy* mutants are generally deficient in either phytochrome apoproteins or phytochrome chromophores. Those of the second class, such as the recessive *de-etiolated* (*det*) and *constitutive photomorphogenic* (*cop*) mutants, exhibit the converse phenotype, characteristics of light grown plants (i.e., a short stem, lack of apical hook, and greening), when grown in the dark (Chory et al., 1989b; Chory et al., 1991; Deng et al., 1991; Wei and Deng, 1992). Mutants of these first two classes can be considered photomorphogenetic mutants in that they uncouple to some extent the developmental events of photomorphogenesis (e.g., chloroplast differentiation, cotyledon expansion, and leaf development) from the light signals.

The third class of light perception mutants is composed of those that exhibit altered shoot phototropism (Konjević et al., 1989; Khurana and Poff, 1989; Khurana et al., 1989; Okada and Shimura, 1992a). One of these mutants completely lacks a phototropic response of the shoot to light at any fluence rate, while others display reduced phototropic responses, depending on the fluence and duration of the light stimulus (Khurana and Poff, 1989; Khurana et al., 1989). In addition, some of the mutants that exhibit a reduced phototropic response also display reduced root gravitropism (Khurana and Poff, 1989; Khurana et al., 1989).

Mutants have also been selected for the absence of root phototropism (wild-type roots display negative phototropism; Okada and Shimura, 1992a). These mutants display normal shoot phototropism and root gravitropism (Okada and Shimura, 1992a).

Hormone Mutants

Mutations that alter the production or perception of hormones have been identified for each of the five classes of known phytohormones: gibberellins, auxins, cytokinins, abscisic acid (ABA), and ethylene (reviewed in Klee and Estelle, 1991).

For gibberellins, mutants defective in both production and sensitivity have been isolated (Koornneef and van der Veen, 1980; Koornneef et al., 1985). Both classes are dwarfs. The mutants with altered gibberellin production can be rescued to a wild-type phenotype by exogenously applied gibberellins, while those altered in perception are insensitive to exogenous gibberellins. Likewise, both ABA deficient and ABA insensitive mutants have been isolated (Koornneef et al., 1982b; Koornneef et al., 1984; Koornneef et al, 1989a; Nambara et al., 1992). Both classes of ABA mutants are characterized by a reduction in seed dormancy and symptoms of withering in mature plants. Differential effects with respect to seed dormancy and water relations are observed with mutations at different loci.

Wild-type root growth is impaired in the presence of low concentrations of cytokinin, with primary root growth inhibited, but root hair elongation stimulated. Selection for wild-type root growth in the presence of low concentrations of cytokinin has allowed the identification of *cytokinin resistant* (*ckr*) mutants altered in their response to cytokinins (Su and Howell, 1992). Root growth of *ckr* mutants in the absence of exogenous cytokinins differs from wild-type root growth in that the roots of the mutants are slightly longer, while the root hairs are shorter.

Selection on the basis of alterations in the response of dark-grown seedlings to ethylene has been used to isolate mutants altered in ethylene perception and synthesis (Bleecker et al., 1988; Guzmán and Ecker, 1990; Harpham et al., 1991). When grown in the dark in the presence of ethylene, wild-type seedlings exhibit an inhibition of stem elongation, a thickening of the stem, and an absence of normal geotropic response such that the apical hook of the seedling remains present. This response is likely an ecological adaptation to penetrate the soil surface after subsurface germination. The identified mutants fall into three classes, those that are insensitive to ethylene, those that over-produce ethylene, and those that have reduced ethylene synthesis. Mutants insensitive to ethylene generally lack all three characteristics of the response; seedlings have elongated, thin stems and no apical hook even when grown in a high concentration of exogenous ethylene. Mutants that over-produce ethylene exhibit a constitutive triple response, and mutants that have reduced ethylene synthesis display an altered triple response, such as the lack of an apical hook.

Mutants affected in auxin perception representing five loci have been isolated by screening for resistance to artificial auxin, the herbicide 2,4-dichlorophenoxyacetic acid (2,4-D). Mutations at three of the loci, *auxin resis-*

tant (*aux1*) (Maher and Martindale, 1980; Mirza et al., 1984; Mirza, 1987b), *auxin resistant* (*axr1*) (Estelle and Somerville, 1987, Lincoln et al., 1990), and *agravitropic roots* (*agr1*) (Bell and Maher, 1990) are recessive, while those at *dwarf* (*dwf*) (Mirza et al., 1984) and *axr2* (Wilson et al., 1990) are dominant, with *dwf* being lethal when homozygous. Several of these mutations exhibit resistance to high concentrations of other phytohormones as well. For example, *axr2* mutants are resistant to ethylene and abscisic acid as well as auxin (Wilson et al., 1990), *aux1* mutants are resistant to both auxin and ethylene (Pickett et al., 1990), and *axr1* mutants are resistant to both auxin and cytokinin (Estelle, 1992).

The auxin resistant mutants exhibit a spectrum of phenotypes, with the common theme that each has agravitropic roots (Mirza et al., 1984; Olsen et al., 1984; Mirza, 1987b; Bell and Maher, 1990; Lincoln et al., 1990; Maher and Bell, 1990; Wilson et al., 1990; Okada and Shimura, 1992a). In the case of *aux1* and *agr1* mutants, root agravitropism is the primary defect. In contrast, *dwf*, *axr1* and *axr2* plants exhibit more pleiotropic defects with the aerial part of the plant affected as well (Mirza et al., 1984; Estelle and Somerville, 1987; Lincoln et al., 1990; Wilson et al., 1990; Timpte et al., 1992). One theory of gravitropism is that amyloplasts with starch grains act as statoliths and form an integral part of gravity perception. This hypothesis has been investigated utilizing agravitropic mutants and mutants deficient in chloroplast phosphoglucomutase activity, which lack plastid starch (Caspar et al., 1985). These studies have proved inconclusive. Although the ultrastructure of *aux1* and *dwf* root caps is indistinguishable from wild type, a difference was found in the rate of redistribution of amyloplasts in *aux1* mutants following inversion (Olsen et al., 1984). However, measurements of the gravitropic responsiveness and the amount of plastid starch in the starch-deficient mutant have been controversial (Caspar and Pickard, 1989; Kiss et al., 1989; Moore, 1989; Sæther and Iverson, 1991). Further experiments with other starch-deficient mutants, perhaps in multiple mutant combinations (Lin et al., 1988), are needed to clarify these uncertainties.

The aromatic amino acid biosynthetic pathway is the source not only of tryptophan, but many other secondary products including auxins such as indole acetic acid. No mutants have been isolated that fail to produce auxin. However, increased levels of endogenous auxin occur in *tryptophan auxotrophy2* (*trp2*) mutants, which were selected on the basis of resistance to 5-methylanthranilic acid plus tryptophan (Last and Fink, 1988; Last et al., 1991; J. Normanly, J. Cohen, and G. Fink quoted in Estelle, 1992). The *trp2* mutants are conditional tryptophan auxotrophs, with exogenous tryptophan required for growth in high light conditions (Last et al., 1991). All *trp2* mutants fail to develop true leaves without exogenous tryptophan, and when supplied with tryptophan, are bushy plants with rounded, light green leaves, a phenotype consistent with overproduction of auxins. The *TRP2* gene encodes a tryptophan synthase β subunit, part of the enzyme complex responsible for the last step in tryptophan biosynthesis, the conversion of indole to tryptophan. Under low-light growth conditions, the requirement for tryptophan is lost in *trp2* mutants, probably due to a second, less highly expressed tryptophan synthase β subunit gene. Mutants at a second locus, *TRP1*, also result in tryptophan auxotrophy, and can be detected by their blue fluorescence in ultraviolet light caused by their accumulation of anthranilate, a tryptophan biosynthesis intermediate (Last and Fink, 1988). They lack phosphoribosylanthranilate transferase activity, an early step in tryptophan and indole biosynthesis (Last and Fink, 1988; Rose et al., 1992). *trp1* mutants are small and bushy, consistent with a defect in auxin synthesis. The phenotypes, including the accumulation of high levels of endogenous auxins, of the tryptophan auxotrophs suggest that in *Arabidopsis* auxin is primarily derived from indole, not tryptophan (Last and Fink, 1988; Last et al., 1991; Rose et al., 1992). The high levels of endogenous auxins in the mutants may reflect an accumulation of the tryptophan (and auxin) intermediate, indole.

Mutants of Lipid Metabolism

Mutants affecting the formation of cuticular lipids have been selected on the basis that their stems and siliques are a brighter green than wild type (Koornneef et al., 1989b). These *eceriferum* (*cer*) mutants are characterized by altered epicuticular wax deposition.

Mutants affecting glycerolipid biosynthesis in leaf cells have been isolated by measuring fatty acid composition in leaf extracts by gas chromatography (Browse et al., 1985; reviewed in Somerville and Browse, 1991). These mutants do not display any distinguishable gross morphological aberration when grown under normal growth conditions, except a slight reduction in leaf pigmentation. Additional mutants in which the lipid composition in seeds is affected have been isolated by applying techniques similar to those used for leaves to extracts of seeds (Lemieux et al., 1990). Mutations at seven loci have been analyzed, with most of the mutations resulting in the loss or reduction of an unsaturated fatty acid, and an accumulation of the corresponding precursor. Six of the *FATTY ACID DEFICIENT* (*FAD*) loci, *FAD2, FAD3, FAD4* (*FADA*), *FAD5* (*FADB*), *FAD6* (*FADC*), and *FAD7* (*FADD*), appear to encode desaturases, some of which are specific for the length of the fatty acid, while others have a more general effect (Somerville and Browse, 1991 and references therein). There are two sites of glycerolipid synthesis in the plant cell, the plastid and the endoplasmic reticulum. Some of the mutants are specific to the plastid pathway of glycerolipid synthesis, while others specifically affect the endoplasmic reticulum pathway (Somerville and Browse, 1991 and references therein). Both genetic and biochemical studies indicate that there is considerable exchange of lipids between the two pathways of glycerolipid synthesis (Somerville and Browse, 1991; Miquel and Browse, 1992). Mutants at the *ACYL TRANSFERASE* (*ACT1*) locus are deficient in plastid glycerol-3-phosphate acyltransferase activity (Kunst et al., 1989a), resulting in most of the fatty acids synthesized in the plastid to be shunted to the endoplasmic reticulum pathway of glycerolipid synthesis. The *FAD3* gene has been cloned by chromosome walking (Arondel et al., 1992).

Although the known mutants affecting glycerolipid biosynthesis exhibit few morphological alterations under normal growth conditions, phenotypic effects are observed when some mutants, such as *fad2* and *fad3*, and possibly others, are grown at either elevated or reduced temperatures (reviewed in Somerville and Browse, 1991). Specifically, the mutants become chlorotic at reduced growth temperatures and display enhanced thermal stability of chloroplasts at elevated growth temperatures. Another chilling-sensitive mutant, *chs1*, has been shown to have altered steryl-ester metabolism (Hugly et al., 1990). In addition, some of the mutants altering plastid lipid composition, such as *fad6* and *fad7*, exhibit abnormal chloroplast morphology and altered chloroplast numbers even at normal growth temperatures (Hugly et al., 1989; Somerville and Browse, 1991). For example, in *fad7* mutants, the cross-sectional area of the chloroplast is reduced by 50%, but there is a corresponding increase in chloroplast number such that the total amount of chloroplast membrane is close to that of wild type (Somerville and Browse, 1991).

Mutants of Photorespiration and Photosynthesis

Selection on the basis of the ability to grow in elevated carbon dioxide levels, but not in atmospheric CO_2 levels, has led to the isolation of an extensive collection of mutants with defects in photorespiratory metabolism (for reviews see Somerville, 1986; Estelle and Somerville, 1986). Since CO_2 at high concentration competitively inhibits ribulose bisphosphate (RuBP) oxygenase activity, the first enzyme involved in the photorespiration pathway, mutants in which the pathway is blocked do not grow when CO_2 concentration is low, either due to accumulation of lethal intermediates, or failure to recycle carbon into the Calvin cycle, through which CO_2 is fixed. However, in high CO_2 environments, where photorespiration is reduced, the mutants grow normally. While many of the mutations cause loss of enzyme activities in the photorespiratory pathway (for review see Somerville, 1986), others affect nitrogen recycling or metabolite transport into the chloroplast (Somerville and Ogren, 1983). In one case, the mutant phenotype is due to failure to activate RuBP carboxylase/oxygenase (Somerville et al., 1982), and surprisingly, the defect is not in RuBP carboxylase/oxygenase, but rather in the chloroplast protein, RuBP carboxylase/oxygenase activase. Since RuBP carboxylase/oxygenase is the first enzyme required in the Calvin cycle, photosynthesis as well as photorespiration is reduced in this mutant.

Nonconditional photosynthetic mutations, such as those causing alterations in chlorophyll or primary photosynthetic enzymes, result in seedling lethality and are discussed briefly under pigment mutants in the embryogenesis section.

Mutants of Nitrate Metabolism

Selection on the basis of resistance to chlorate has led to the isolation of mutants with altered nitrate metabolism (Braaksma and Feenstra, 1973; Braaksma and Feenstra, 1982; LaBrie et al., 1992; Crawford et al., 1992). After uptake, chlorate is converted to toxic chlorite by nitrate reductase. Eight *chlorate resistant* (*chl*) complementation groups have been identified, with all but one, *chl1*, having reduced nitrate reductase activity. The *CHL3* locus encodes a nitrate reductase enzyme (Cheng et al., 1988; Crawford et al., 1988; Wilkinson and Crawford, 1991), with at least four other loci (*CHL2*, *CHL4*, *CHL6*, *CHL7*) involved with the molybdenum pterin cofactor of the nitrate reductase enzyme (Braaksma and Feenstra, 1982; LaBrie et al., 1992). A second structural gene has been identified, but no corresponding mutant has been isolated (Cheng et al., 1988). Mutations at the *CHL1* locus have normal nitrate reductase activity, but they exhibit reduced chlorate uptake (Braaksma and Feenstra, 1982).

J.L. Bowman

Plate 1.1
The seedling

In typical winter annual species, seeds are dispersed in the spring, remain dormant during the summer, and germinate in the autumn. The winter is spent in vegetative growth and the longer days of spring induce the transition to flowering. The winter annual ecotypes of *Arabidopsis* follow this pattern of ecological adaptation (Baskin and Baskin, 1972). Following dispersal, *Arabidopsis* seeds remain dormant in the warm dry conditions of summer, the extent of dormancy varying among ecotypes, with germination encouraged by the cooler days of autumn. The ecological adaptation of seed dormancy can be overcome by exposing recently shed seeds to cool temperatures. The frequency of germination increases both with increased time of cold exposure and increased periods of dormancy prior to cold exposure. For example, nearly 100% germination occurs with a dormancy period of a few weeks followed by a cold treatment of a couple days.

(**A**) Two-day-old seedling. Seed was imbibed for 24 hours at 4°C on wet filter paper and then transferred to continuous fluorescent light at 25°C for two days. At this stage the cotyledons have not yet fully expanded. When germinated in the dark (e.g., below the soil surface) the apex of the seedling forms a hypocotyl hook (not shown), with the apical part of the hypocotyl curving such that the shoot apex is pointing downward. This presumably protects the shoot apex while the seedling breaks through the soil.

(**B**) Germinating seed. The seed coat, which has expanded due to imbibition, is often temporarily retained at the root–hypocotyl transition zone (Mirza, 1987a).

(**C**) Vascular pattern of expanded cotyledons. The exact pattern of vasculature varies somewhat between cotyledons but the extent of branching is essentially constant, with a central vein and two or three interconnecting side branches. Lignification of the xylem, in the cotyledons as well as the hypocotyl and root, begins before the cotyledons are fully expanded, and spreads acropetally within the cotyledons (Dharmawardhana et al., 1992).

(**D**) Adaxial (upper) surface of an expanded cotyledon. Numerous stomata are interspersed among the jigsaw-shaped cells. In contrast to *Arabidopsis* leaves, cotyledons lack both trichomes and flanking stipules.

(**E**) Shoot apex of 2-day-old seedling with one cotyledon removed. The first two leaf primordia are conspicuous.

(**F**) Hypocotyl of 2-day-old seedling.

(**G**) Transition between the hypocotyl and root of 2-day-old seedling.

(**H**) Primary root with numerous root hairs of 2-day-old seedling.

(**I**) Root tip of 2-day-old seedling.

Bar = 20 μm in **E**; 50 μm in **F**, **G**, **H**, **I**; 100 μm in **B**, **D**; 500 μm in **A**.

J.L. Bowman

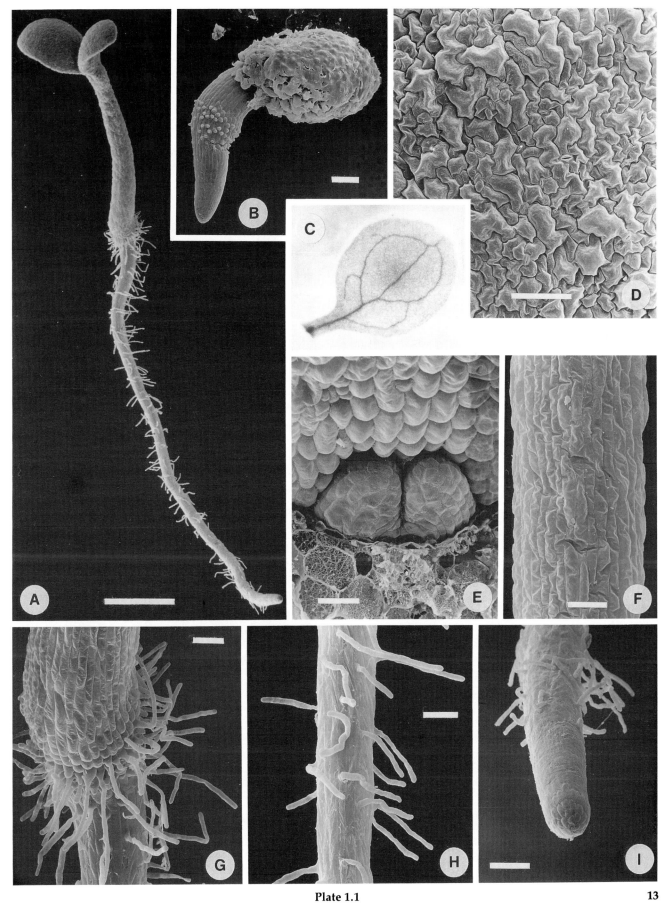

Plate 1.1

13

Plate 1.2
**The seed shoot meristem
and young juvenile shoot
meristem**

The vegetative shoot apical meristem consists of a small group of undifferentiated cells located at the distal end of the shoot, and is surrounded by the youngest leaf primordia (Cutter, 1965). The shoot apical meristem gives rise to the entire above ground portion of the plant except the hypocotyl and cotyledons (Sussex, 1989). The shoot apical meristem has several functions: it initiates tissues, initiates organs, communicates signals to other parts of the plant, and maintains itself as a formative region (Steeves and Sussex, 1989; Sachs, 1991; Medford, 1992).

In the following descriptions of wild-type development, plants were grown at a constant temperature of 22°C and in a 16 hour/8 hour light/dark cycle (Medford et al., 1992).

The Shoot Apical Meristem in the Mature Seed

(A–C) In the mature embryo, the wild-type shoot apical meristem consists of a rectangular group of approximately 60–80 cells in total, with a thickness of about four cells in the center and two cells at the periphery. Although no zonation is apparent at this time, bulges at the periphery of the meristem suggest that the first two leaf primordia have been initiated simultaneously or very nearly so. Sections are through a young shoot apical meristem.

(A) Diagram of the *Arabidopsis* embryo in the mature seed.

(B) Section through the shoot apical meristem (m) in the plane parallel to the cotyledons shown in A.

(C) Section through the shoot apical meristem (m) in a plane perpendicular to the cotyledons showing the emergence of the first pair of leaf primordia.

Characteristics of the Juvenile Shoot Meristem: The Shoot Apex at 4 Days

(D) and (E) By day 4 postgermination, the first two leaf primordia develop in a plane 90° from that of the cotyledons. Further, the primordia form as radially symmetric structures on opposite sides of the rectangular shoot apical meristem. At this stage, the leaf primordia are approximately 35 μm in height, and stipules, paired leafy appendages that develop at the base of leaves, have not yet been delineated. Twelve hours later, the first two leaf primordia have enlarged to 65 μm, trichomes begin to emerge at their distal end, and stipules have started to emerge on each side of the leaf base.

(D) Vertical view of the shoot apex showing that the first two leaf primordia (p) are formed as radially symmetrical structures on opposite sides of a rectangular shoot apical meristem (m).

(E) Side view showing the emerging stipules (es) at the base of the leaf, and trichomes (et) on the distal end of the leaf. By this stage it appears that one of the leaf primordia is larger and developmentally more advanced than the other.

Bar = 10 μm.

J.D. Callos, F.J. Behringer, J. Vasinda, D. Stewart, B.M. Link, and J.I. Medford

B, D, and **E** reproduced from Medford et al. (1992) with permission from American Society of Plant Physiologists.

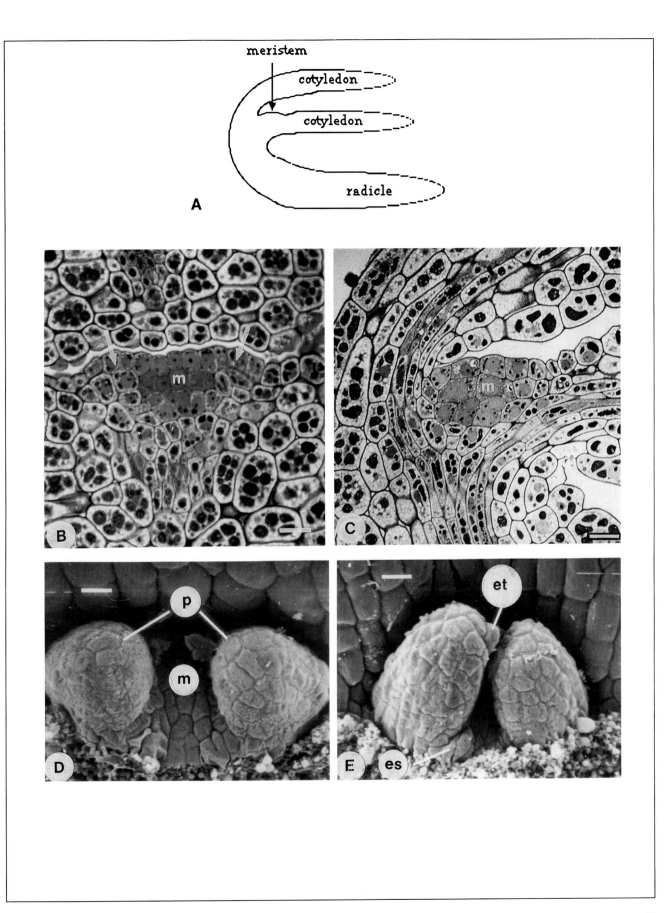

Plate 1.2

15

Plate 1.3
The juvenile and adult
shoot apical meristem

Characteristics of the Juvenile Shoot Apical Meristem: The Shoot Apex at 5–6 Days

(A–C) Between 5 and 6 days postgermination, the primordia for leaves 1 and 2 have enlarged to approximately 105 μm and have begun to adopt a dorsiventral symmetry. Trichomes become conspicuous on the adaxial surface at the distal end of each primordium. Stipules have enlarged and begun to differentiate into club-like structures. The meristem is still flat, but has begun to take on a trapezoidal shape, perhaps in anticipation of leaf primordium 3, which is initiated slightly in advance of leaf primordium 4. Although leaf primordium 4 is not visible at this time, periclinal cell divisions in L2 suggest that it has been initiated.

(A) Side view of 5- to 6-day-old shoot apex. The leaf primordia have developed as dorsiventral structures. Trichomes (t) and stipules (s) are prominent.

(B) Vertical view of 5- to 6-day-old shoot apex showing that the shoot apical meristem (m) has changed from a rectangular-shaped structure to a trapezoidal-shaped structure (l, leaf trace; s, stipule).

(C) Section through a 5- to 6-day-old shoot apical meristem in a plane perpendicular to the first pair of leaf primordia. The shoot apical meristem is still relatively flat and leaf primordium 3 (p) can be seen as a slight bulge. On the opposite side, a periclinal division (*arrow*) in the L2 indicates that leaf primordium 4 has been initiated. Stipules (s) of the first two leaves are prominent.

Characteristics of the Mature Shoot Apical Meristem

(D–F) At day 7 postgermination, the shoot apical meristem starts to enlarge and begins to adopt a radial symmetry. The vegetative shoot apical meristem changes from a relatively flattened, bilaterally symmetric structure to a dome-shaped, radially symmetric structure consisting of approximately 450 cells. Leaves 3 and 4 contribute to the basal rosette in an intriguing pattern. The primordia for leaves 3 and 4 are initiated 180° from one another, yet the developing leaves emerge at an angle different from 180°. Leaf primordium 5 forms at a position between the positions of leaf primordium 2 and 3, initiating a spiral phyllotactic pattern. The spiral pattern may be either clockwise or counterclockwise. In *Arabidopsis*, the average angle between leaves initiated from the adult vegetative shoot apical meristem was calculated to be 136.4° (Callos and Medford, unpublished). The phyllotactic spiral initiated in vegetative development continues through floral development. The structure of the mature vegetative apical meristem can be described as consisting of three distinct layers as well as three distinct zones, with respect to cell division patterns and histochemical staining (Vaughan, 1955; see Vegetative Development: Introduction).

(D) Longitudinal section through the vegetative shoot apical meristem showing that the meristem has changed from a flattened, bilaterally symmetric structure to a dome-shaped, radially symmetric structure (p, leaf primordia).

(E) Cross-section through the shoot apex showing the angle on emergence between leaf primordia 3 and 4 is different from the angle at initiation (c, cotyledon; m, meristem). Leaves are numbered from oldest (1) to youngest (5).

(F) Scanning electron micrograph showing continuation of phyllotactic spiral (p, leaf primordia; s, stipule; m, meristem).

Bar = 10 μm in **A**, **B**, **C**; 25 μm in **D**, **E**, **F**.

J.D. Callos, F.J. Behringer, J. Vasinda, D. Stewart, B.M. Link, and J.I. Medford

Reproduced from Medford et al. (1992) with permission from American Society of Plant Physiologists.

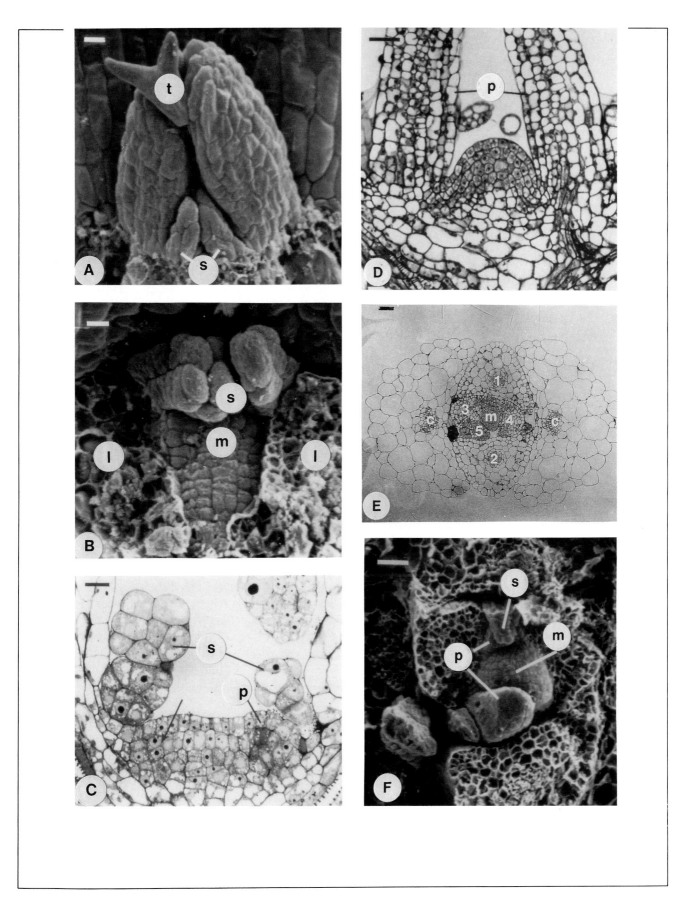

Plate 1.3

17

Plate 1.4
Apical meristem mutants

clavata (*clv*)

Mutations in three loci: *CLAVATA1* (*CLV1*) (McKelvie, 1962; Leyser and Furner 1992), *CLAVATA2* (*CLV2*) (Koornneef et al., 1983), and *CLAVATA3* (*CLV3*), (Griffith, Alvarez, and Smyth unpublished) all affect the development of the vegetative, inflorescence, and flower meristems. The mutant phenotypes appear to result from an increase in the size of meristems.

Three days after germination, the mutant shoot apical meristem is already larger than the wild type meristem. This may be because there is an increased rate of cell division in the meristem, or because an increased number of meristematic cells is laid down during embryogenesis, or both. An increased number of rosette leaves and cauline leaves (with axillary inflorescence meristems) is seen in some mutants, for example *clv1-1* and *clv3-2*. The expansion in size of the shoot apex continues as the vegetative meristem changes over to an inflorescence meristem. The shoot apical meristem continues to broaden and become more distorted relative to wild type so that fasciation of the shoot apex often occurs (Leyser and Furner, 1992). An increased number of floral organs is produced in *clv* mutant flowers, presumably because flower meristems also appear to be larger than in wild-type plants. The gynoecium usually has four carpels instead of two, and club-shaped siliques, after which the phenotype was named, are usually produced.

The spiral phyllotaxy of leaf and flower development is also altered to varying degrees. When grown at 24°C the leaves and flowers of the *clv2-1* mutant develop as in wild type with a spacing of approximately 137.5° between each primordium. In contrast, *clv1-1* and *clv3-2* plants almost always have altered phyllotaxy. The angle between successive leaves varies widely (between 45° and 180°) and sometimes there is complete loss of directional spiral phyllotaxy (Leyser and Furner, 1992).

Early Development of the Apical Meristem of Wild-Type *A. thaliana*

(**A–C**) Wild-type apical meristem (Landsberg *erecta*) after 3 days (**A**) and 6 days (**B–C**) growth at 20°C under continuous light. After 3 days of growth (**A**), the average number of epidermal cells across the wild-type apex (*arrow*) between the first two rosette leaves (l) was 3.3 ± 0.53. One of the two cotyledons (c) has been removed. After 6 days of growth (**B**) it was necessary to remove 4 of the older leaves (l) to expose the apical meristem (*arrow*) and the youngest leaf primordium. At this age the average number of leaves and distinct leaf primordia initiated from the wild-type apical meristem was 5.3 ± 0.65. Stipules are visible on each side of the base of petioles in **B**. A different 6-day-old apex is shown in side view in **C**.

Early Development of the Apical Meristem of *clv* Mutants

Apical meristems of *clv2-1* (**D–F**), *clv1-1* (**G–I**), and *clv3-2* (**J–L**) mutants (all in Landsberg *erecta* background) grown at 20°C under continuous light. The *clv* apices are wider and more dome-shaped than in wild type, and some mutants already have more leaves after 6 days of growth.

(*Text continued on p. 21*)

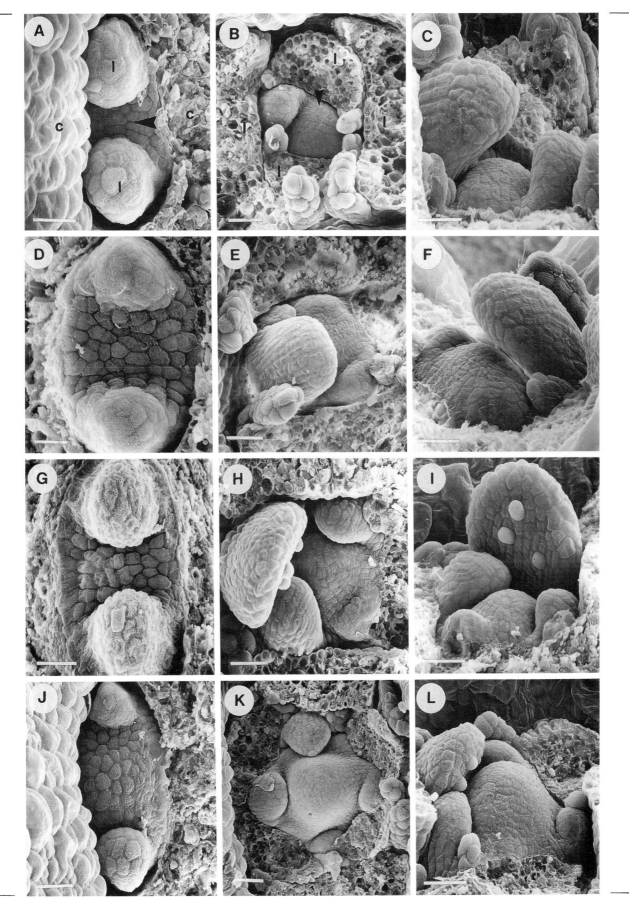

Plate 1.4

Plate 1.4
Apical meristem mutants
(*continued*)

(D–F) Apical meristems of *clv2-1* mutant plants after 3 days (**D**) and 6 days (**E** and **F**) of growth. The meristem is clearly larger than in wild type. The average number of epidermal cells separating the first 2 rosette leaves at 3 days was 6.2 ± 0.84 compared with 3.3 ± 0.53 in wild type. However at 6 days the average number of leaves and distinct leaf primordia initiated from the apical meristem was observed to be 5.4 ± 0.50, close to that in wild type. In **E**, 3 leaves have been removed to expose the apical meristem and 2 developing leaf primordia. **F** shows the same apex in side view.

(G–I) Apical meristems of *clv1-1* mutant plants after 3 days (**G**) and 6 days (**H, I**) of growth. The average number of epidermal cells separating the first 2 rosette leaves at 3 days was 5.5 ± 1.2, more than in wild type. At 6 days the average number of leaves and distinct leaf primordia initiated from the apical meristem was 7.1 ± 0.60, slightly more than wild type. In **H**, 3 leaves have been removed to expose the apical meristem and 4 developing leaf primordia. **I** is a side view of the apex shown vertically in **H**.

(J–L) Apical meristems of *clv3-2* mutant plants after 3 days (**J**) and 6 days (**K, L**) of growth. The average number of epidermal cells separating the first 2 rosette leaves at 3 days was 8.4 ± 0.73, nearly 3 times that in wild type. At 6 days, the average number of leaves and distinct leaf primordia initiated from the apical meristem was observed to be 8.1 ± 1.73, nearly 3 more than in wild type. In **K**, 6 leaves have been removed, exposing the apical meristem. Four more developing leaf primordia are evident. **L** shows in side view another 6-day-old apex of *clv3-2*.

Bar = 10 μm in **D, J**; 15 μm in **G**; 20 μm in **A**; 25 μm in **B, C, E, F, H, I, K, L**.

M. Griffith

Plate 1.5
Apical meristem mutants

forever young (*fey*)

The recessive mutant *fey* was generated by T-DNA insertional mutagenesis (Feldmann, 1991). The lesion in *fey* plants is specific to the shoot apex, as cotyledons, hypocotyl, and roots are unaffected (Medford et al., 1992). Plants containing *fey* have abnormalities in the organization of the apical meristem and developing leaf primordia such that leaf primordia are usually initiated with greater than the normal number of cells. These aberrations result in an irregular pattern of leaf formation and usually lead to premature death of the shoot. At low temperatures, the shoot apical meristem reorganizes and forms multiple meristems.

(**A**) *fey* mutant at 16 days.

(**B**) Median section through a *fey* apex at 8 days. The third leaf primordium (p) is initiated with more cells than in wild type.

(**C**) Cross section through a *fey* shoot apex at 30 days. Leaf primordia are are initiated in a random pattern.

(**D**) *fey* mutant at 27 days. When *fey* plants are grown at low temperature, the shoot apical meristem reorganizes and gives rise to multiple shoots.

Bar = 25 μm in **C**; 50 μm in **B**; 100 μm in **D**; 1 cm in **A**.

J.D. Callos, F.J. Behringer, J. Vasinda, D. Stewart, B.M. Link, and J.I. Medford

A and **B** reproduced from Medford et al. (1992) with permission from American Society of Plant Physiologists.

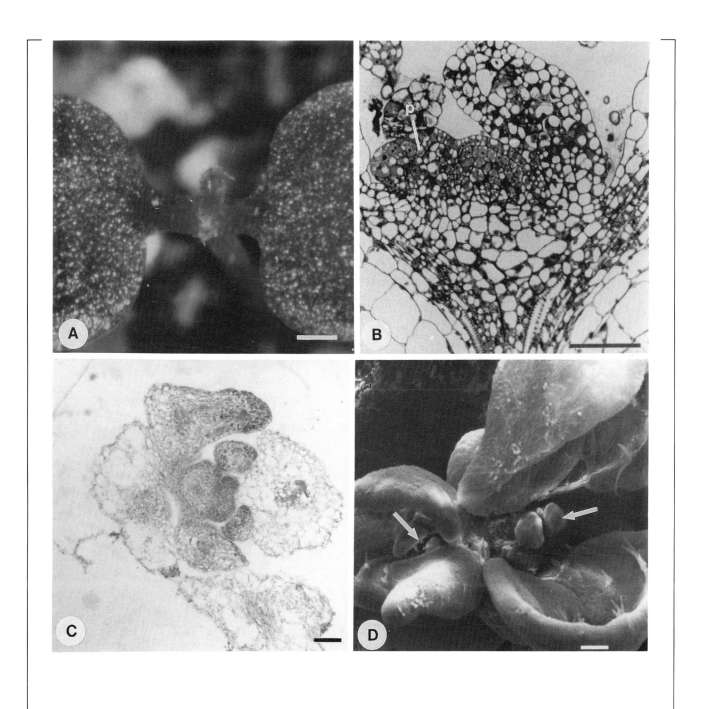

Plate 1.5

23

Plate 1.6
Apical meristem mutants

schizoid (*shz*)

The recessive mutation *shz* affects the cells in the rib zone of the shoot apical meristem. In a wild-type plant, the rib zone contributes to the growth of the stem and cortical tissues. The effects of *shz* are apparent when the stem is forming just prior to bolting. In *shz* plants, the cells in the stem appear necrotic, resulting in the premature release of axillary buds from dormancy. The phenotype is not observed in axillary meristems, implying that the *SHZ* gene product is required only early in development, or that differences exist between the apical shoot meristem and other shoot meristems (Shannon and Meeks-Wagner, 1991; Medford et al., 1992).

(**A**) Longitudinal section through a *shz* shoot apex at 12 days showing the disorganization of cells in the rib zone (*arrow*).

(**B**) Longitudinal section through the shoot apex of a *shz* plant at 16 days showing necrosis in the stem and premature release from dormancy of axillary buds (ax).

(**C**) *shz* plant (*right*) in comparison with wild type (*left*) showing elongation of the rosette axillary shoots (*arrows*).

disrupted (*dip*)

Plants homozygous for the recessive mutation *dip* display alterations in the shoot apical meristem and developing leaf primordia. Neither roots nor hypocotyls are affected in *dip* mutants, but the shape of the cotyledons is sometimes altered. The *dip* shoot apical meristem is misshapen and has no apparent zonation pattern. Further, the cellular patterns within leaf primordia are frequently disorganized, resulting in leaf primordia that appear wrinkled. Other leaf primordia do not expand laterally, resulting in peg-like leaves.

(**D**) *dip* mutant at 17 days showing wrinkled leaves due to disorganization of the internal cells (p, leaf primordium; c, cotyledons).

(**E**) Longitudinal section through a *dip* apex showing the misshapen shoot apical meristem (*arrow*).

Bar = 10 μm in **A**; 50 μm in **E**; 100 μm in **B**.

J.D. Callos, F.J. Behringer, J. Vasinda, D. Stewart, B.M. Link, and J. I. Medford

B reproduced from Medford et al. (1992) with permission from American Society of Plant Physiologists.

Vegetative Development: Apical Meristem Mutants

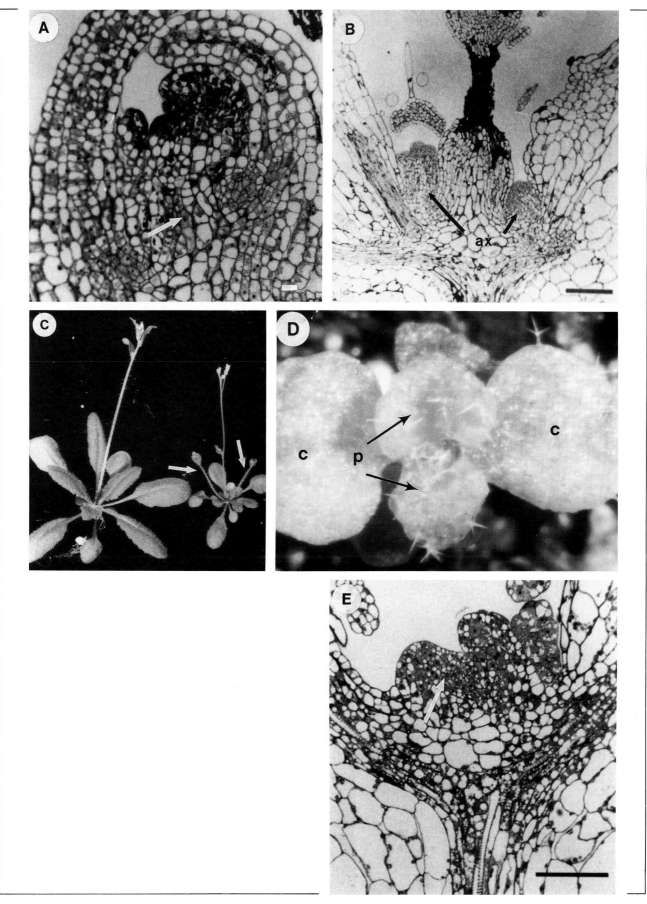

Plate 1.6

25

Plate 1.7
Apical meristem mutants

fully fasciated (fuf)

The recessive mutant *fuf* was generated by ethyl methylsulfonate mutagenesis. The shoot apex is specifically affected in *fuf* mutants. These mutants have an increased number of leaf primordia and a larger shoot apical meristem, compared to wild type. Leaf primordia are initiated in a regular pattern that is different from that of wild type. In addition, the enlarged shoot apical meristem produces a fused and flattened stem upon bolting.

(A) *fuf* mutant at 30 days showing an increase in the number of leaf primordia. In similar growth conditions, wild-type plants produce between 5 and 8 leaves before flowering.

(B) Longitudinal section through a *fuf* apex at 7 days, showing an increase in the size of the shoot apical meristem.

finger leaf (fil)

The recessive mutant *fil* results in a disruption in the shoot apical meristem and developing leaf primordia. The first 2 leaf primordia in *fil* mutants do not expand laterally, resulting in finger-like leaves lacking dorsiventrality. In addition, trichomes do not develop on these leaves. Following the formation of the first 2 leaf primordia, a short, leaf-like stem forms as the plant apparently begins to bolt. Two leaf-like structures form at the distal end of the leaf-like stem, with no obvious meristem remaining. Trichomes do develop on the distal surfaces of these leaf-like structures, but trichomes fail to develop all along their adaxial surfaces, as occurs in normal leaves. After the plant apparently bolts, small leaf-like structures develop from the surface and at the base of the leaf-like stem that vary in phenotype from peg-like outgrowths (with or without a distal trichome) to near-normal leaves.

(C) *fil* apex at 10 days. The first pair of leaf primordia (p) are finger-like and lack dorsiventrality.

(D) *fil* apex at 10 days. Short leaf-like stem bearing 2 leaf-like structures (l) at its distal end (t, trichome).

(E) *fil* leaf-like stem at 21 days. Leaf-like structures develop on the surface and at the base of the leaf-like stem.

Bar = 50 μm in **D**; 100 μm in **B**, **E**; 500 μm in **C**.

J.D. Callos, F.J. Behringer, J. Vasinda, D. Stewart, B.M. Link, and J.I. Medford

A and **B** reproduced from Medford et al. (1992) with permission from American Society of Plant Physiologists.

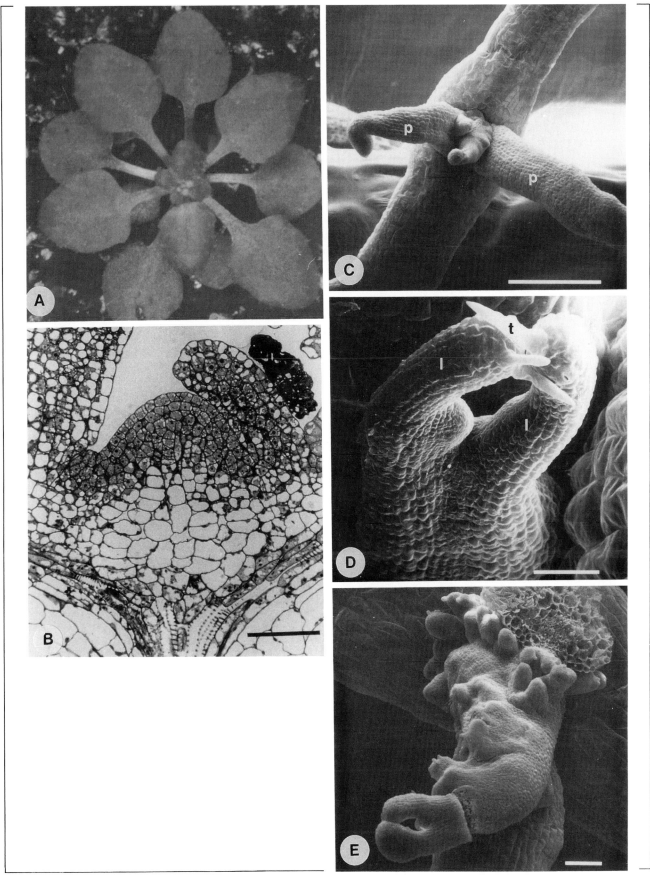

Plate 1.7

27

Plate 1.8
Early development
of the leaf

(**A–C**) Transverse sections of young *Arabidopsis* leaves from the apex of seedlings.

(**A**) 5 days after sowing.
(**B**) 6 days after sowing.
(**C**) 8 days after sowing.

The first pair of leaves, which are surrounded by cotyledonary tissue, expand quickly and by day 6 the differentiation of the mesophyll into palisade and spongy mesophyll is evident. By day 8 a distinct layer of palisade mesophyll cells (*arrows*) is visible beneath the upper epidermis on the inside surface of the expanding leaf. The few vacuolated cells are restricted to the epidermis and the mesophyll tissue between the developing bundle and the abaxial epidermis. The individual large cells (*arrowheads*), which are part of the upper epidermis in all three pictures, are early, differentiating, trichome leaf hairs. Plants were grown at 20°C with a 5°C night depression, a 16 hour day/8 hour night regime, and 70% relative humidity.

Bar = 50 μm.

K.A. Pyke, J.L. Marrison, and R.M. Leech

Reproduced from Pyke et al. (1991) with permission from Oxford University Press.

Vegetative Development: The Leaf

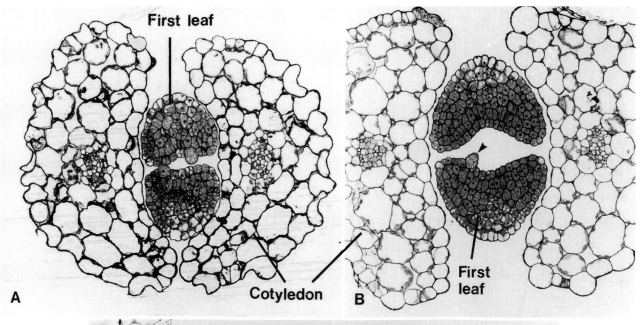

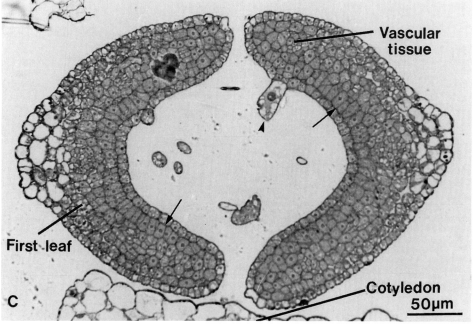

Plate 1.8   29

Plate 1.9
Differentiation of the mesophyll

Paradermal section from the first leaf of *Arabidopsis* harvested 9 days after sowing. The leaf has been sectioned just below the upper epidermis, and most of the mesophyll cells seen in the upper part of the leaf are palisade cells. The enlarged cells at the middle of the top surface are part of a trichome. Cells toward the base of the leaf are still largely mitotic, while cells nearer the tip of the leaf are differentiated and undergoing expansion. The zone of mitotic cells in the basal region extends throughout the thickness of the leaf. No evidence of marginal or apical (near the leaf tip) meristems is seen in leaves of this developmental stage. The transition from the mitotic phase to the expansion phase is marked by the appearance of vacuoles. Vascular strands are also evident. The section covers 75% of the area of the leaf.

Bar = 100 μm.

K.A. Pyke, J.L. Marrison, and R.M. Leech

Reproduced from Pyke et al. (1991) with permission from Oxford University Press.

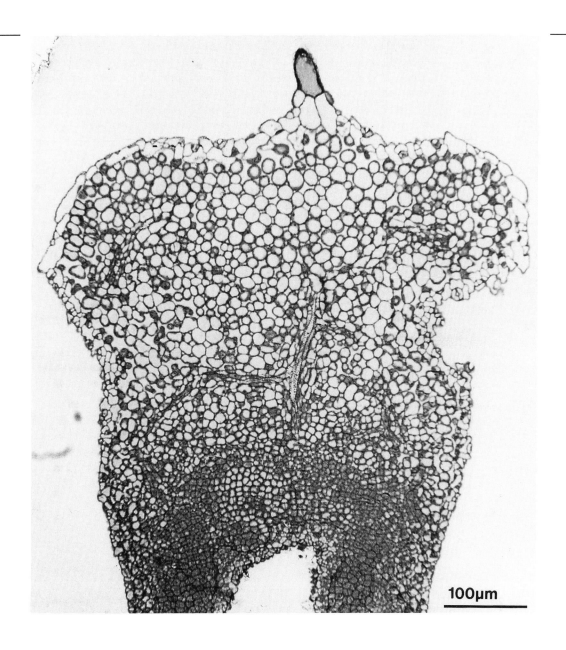

100μm

Plate 1.9

31

Plate 1.10
Differentiation of the mesophyll

Series of transverse sections cut across the entire width of a 9-day-old first leaf of *Arabidopsis*. The sections were cut at varying distances from the leaf base. This distance is expressed as a proportion of total leaf length. The sections show that cells from near the leaf base (**D, E**) are small, largely undifferentiated, and still in division, whereas cells from near the leaf tip (**A, B**) are vacuolated and undergoing expansion. In the mesophyll cells in the section from nearest the leaf tip (**A**), chloroplasts are visible around the margins of the cells.

From to day 9 through day 20, cell expansion becomes the important parameter leading to the increase in leaf size. The onset of mesophyll expansion begins at the leaf tip and progresses towards the leaf base, producing a gradient of cell size and developmental state in the mesophyll. This results in wide variation of cell size throughout the mature leaf. The relative amount of airspace in the mesophyll per unit leaf area increases dramatically during this phase of cell expansion. The lamina reaches its final size (about 30 mm²) by day 18, but the leaf continues to increase in thickness through day 20, with the increase in thickness correlated with the elongation of the palisade cells perpendicular to the lamina. The mature first leaf (about 22 days old) is composed of 60.5% mesophyll, 25.5% airspace, 12.5% epidermis, and 1.5% vascular tissue by volume. The total number of cells of the first leaf is estimated at 132,000, of which 36% are mesophyll, 36% are vascular, and 28% are epidermal.

Bar = 100 μm.

K.A. Pyke, J.L. Marrison, and R.M. Leech

Reproduced from Pyke et al. (1991) with permission from Oxford University Press.

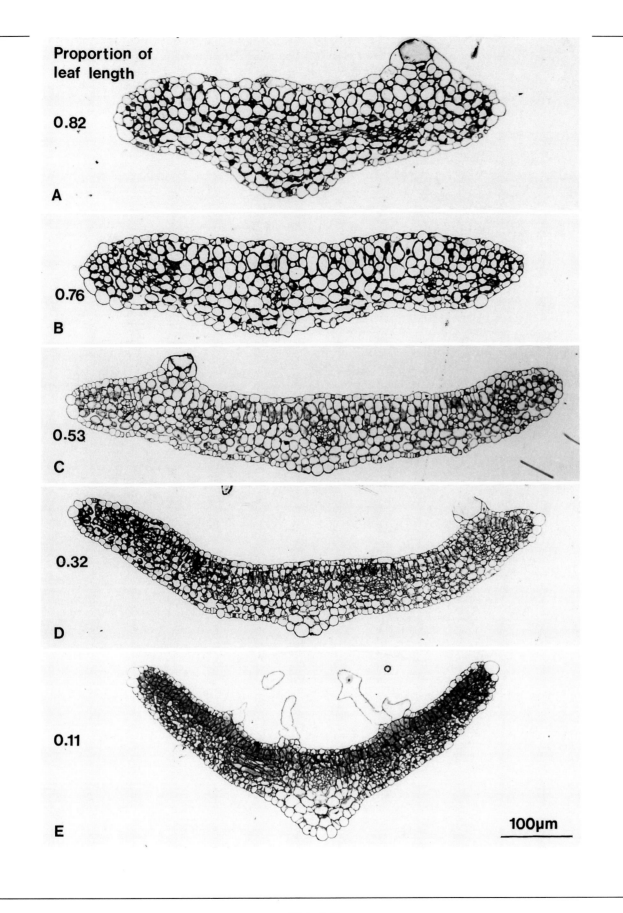

Proportion of
leaf length

0.82

A

0.76

B

0.53

C

0.32

D

0.11

E

100μm

Plate 1.10

33

Plate 1.11
Stipules

Transmission electron micrograph of a thin section through the base of young leaves in the apical region of the plant. The main structure in the center of the micrograph is a stipule. To the right are the differentiating cells of this organ and their elaborate endomembrane system. Each young leaf primordium has a pair of flanking stipules. At an early stage of leaf development, these stipules are quite large in comparison to the leaf primordia.

Bar = 5 μm.

K. Plaskitt

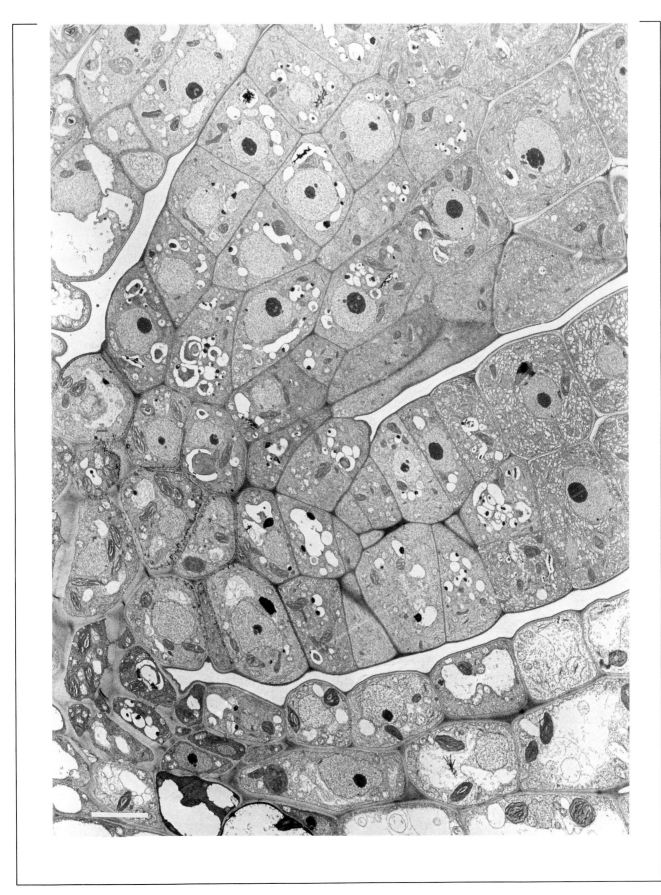

Plate 1.11

Plate 1.12
Stipules

Transmission electron micrograph of mature cells in the stipule. The cells on the right in the micrograph have a highly elaborated endomembrane system. There are large numbers of vesicles and a large fibrous body in each cell. The function of this elaborated membrane system is obscure. On the left are the vacuolate cells of the young leaf. Note the clear, electron-dense cuticle on both the leaf and the stipule.

Bar = 2 μm.

K. Plaskitt

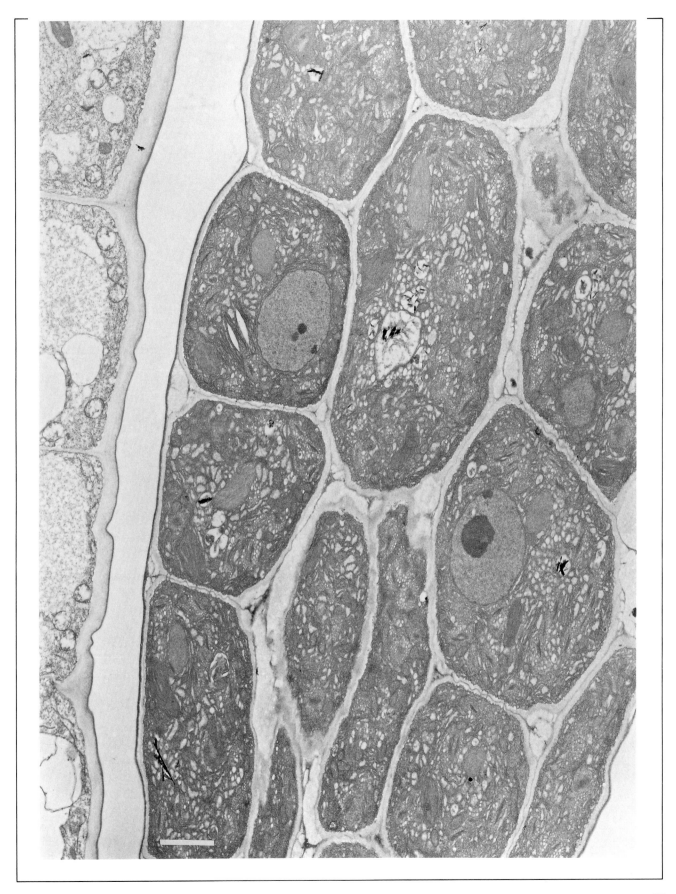

Plate 1.12

Plate 1.13
**Morphology of the
expanded first leaves**

(**A**) Cross-section of a fully expanded first leaf. Clearly differentiated adaxial (upper) palisade mesophyll (p) and abaxial (lower) spongy mesophyll (s) are evident. Stellate trichomes are visible on the adaxial surface of the leaf.

(**B**) Reticulate pattern of vasculature of the first leaf. The vasculature differentiates as a single strand of tracheary elements along the midvein that exhibit acropetal lignification, followed by the formation of annular and spiral secondary wall thickenings (Dharmawardhana et al., 1992). The differentiation and lignification of the peripheral xylem elements may proceed either acropetally or basipetally (Dharmawardhana et al., 1992). The composition of the lignin in the bolting stem of *Arabidopsis* plants is typical of that of herbaceous plants, consisting mainly of guaiacyl and syringyl moeities (Dharmawardhana et al., 1992).

(**C–H**) Epidermal surfaces of fully expanded first leaves.

(**C–E**) Adaxial surfaces.

(**F–H**) Abaxial surfaces.

The epidermis of the leaf consists of irregularly shaped cells with interspersed stomata, with the adaxial and abaxial surfaces easily distinguishable. Stellate trichomes on the adaxial surface are visible in **C**, and elongated epidermal cells of the abaxial surface overlying the midvein can be seen in **F**.

Bar = 20 μm in **E**, **H**; 50 μm in **A**, **D**, **G**, 200 μm in **C**, **F**.

J.L. Bowman

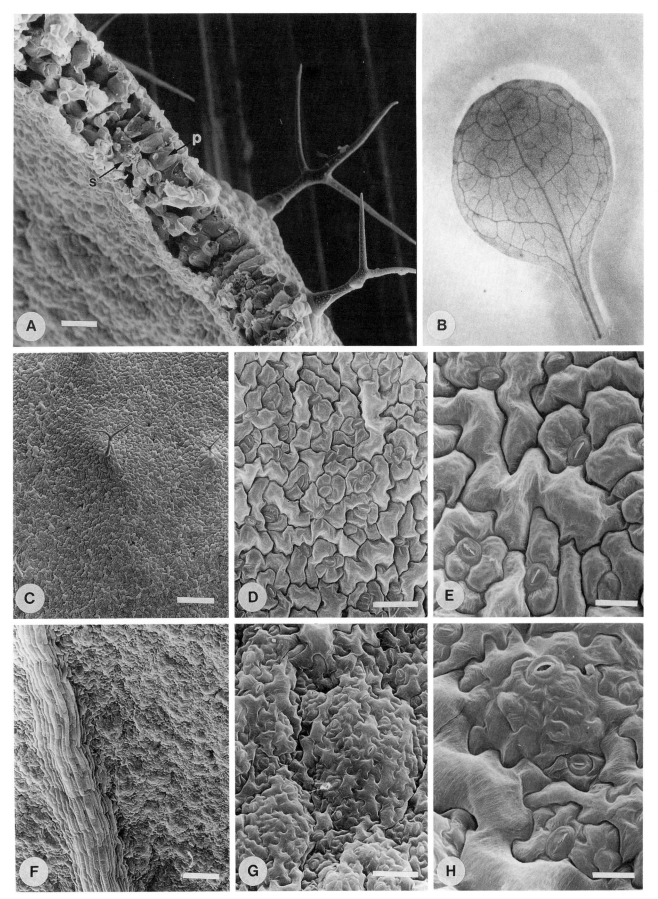

Plate 1.13

Plate 1.14
Epidermal surface

Cryo-scanning electron micrograph of the surface of a rosette leaf. Stomata are clearly visible, as are the epidermal cells which are of variable size and shape. There is very little evidence for epicuticular waxes on the leaf of *Arabidopsis*.

Bar = 30 μm.

P. Linstead, L. Dolan, and K. Roberts

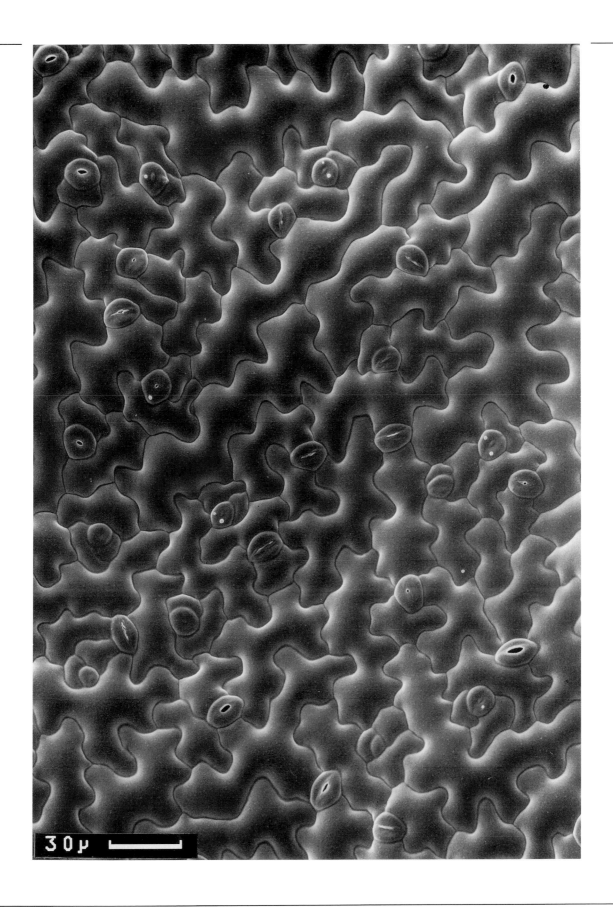

Plate 1.14

Plate 1.15
Epidermal surface

Cryo-scanning electron micrograph of the leaf surface. A 3-branched tri-chome, with its characteristic surface pattern, is prominent.

Bar = 100 μm.

P. Linstead, L. Dolan, and K. Roberts

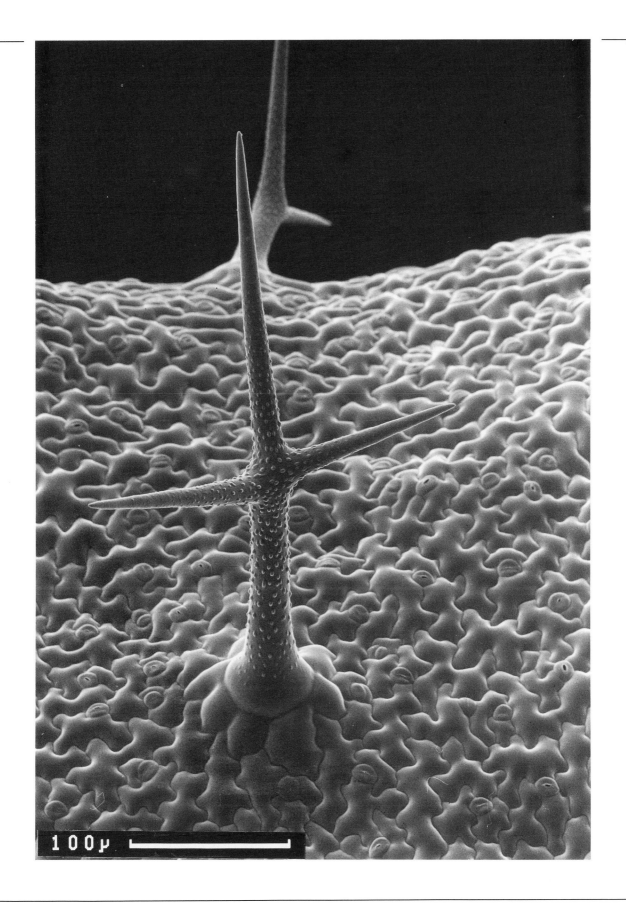

100μ

Plate 1.15

Plate 1.16
Seedling development

A comparison of 7-day-old dark-grown (etiolated) (*left*) and light-grown (*right*) *Arabidopsis* seedlings (Pepper et al., 1993). Hypocotyl (hy), cotyledons (co), and leaf primordia (lp) are indicated. Etiolated seedlings are characterized by the presence of elongated hypocotyls, folded cotyledons, an apical hook, etioplasts, and the absence of leaves. As seen in the electron micrograph in **A**, etioplasts are small, irregularly-shaped plastids containing a large, central, apracrystalline array termed the prolamellar body (Chory et al., 1989b). Light-grown plants have short hypocotyls and expanded cotyledons, and the plants develop leaves. The plastids in the cotyledons and leaves of light-grown plants develop into chloroplasts. As seen in **B**, the chloroplasts are larger than etioplasts, lens-shaped, and lack the prolamellar body. Thylakoid membranes can be seen as granal stacks (Chory et al., 1989b). Plastids which are not in the leaves or cotyledons follow a different developmental course. For example, amyloplasts (**C**) are starch-containing plastids that develop in the roots (Chory and Peto, 1990).

Bar = 1 μm.

D.S. Poole and J. Chory

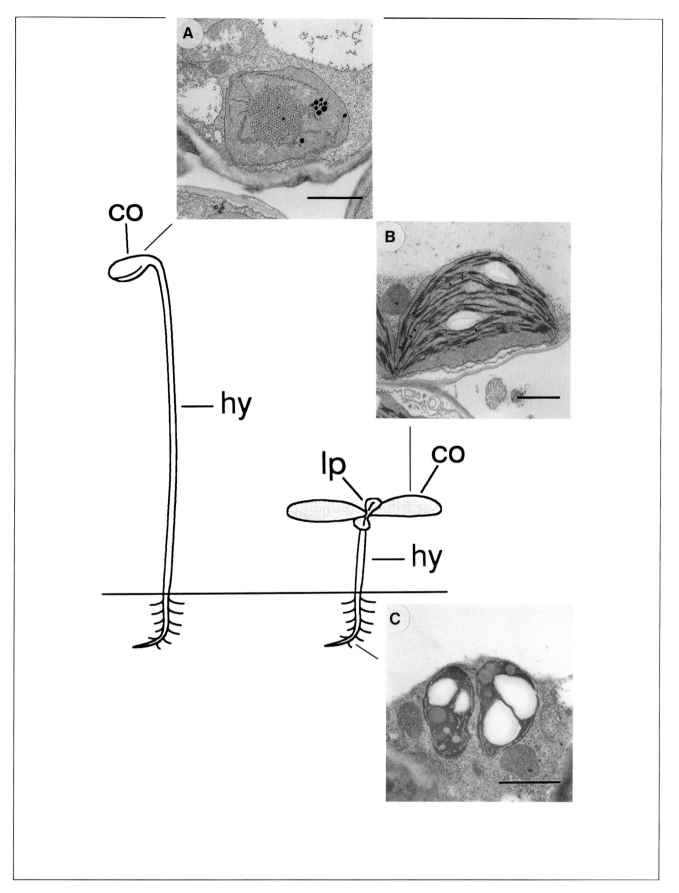

Plate 1.16

45

Plate 1.17
Chloroplast development

Electron micrographs of representative chloroplasts during development of wild-type plants.

(**A–D**) Plastids from plants grown in continuous light (a mixture of fluorescent and incandescent lights; see Chory et al., 1991).

(**A**) Eight days postgermination.

(**B**) Fourteen days postgermination.

(**C**) Nineteen days postgermination.

(**D**) Twenty-two days postgermination. At 22 days, the thylakoid membranes of some of the chloroplasts begin to break down as senescence begins in the leaves. This corresponded to the first flowering of the plants.

(**E–G**) Plastids undergoing the etioplast-to-chloroplast transition. Wild-type plants were grown in the dark for 7 days and then transferred to continuous white light (Susek et al., submitted).

(**E**) A typical etioplast observed when plants were not transferred to the light.

(**F**) After 6 hours in the light, the prolamellar body is gone; thylakoid membranes have formed and display some stacking.

(**G**) After 24 hours, further stacking of the membranes is seen.

Bars = 1 μm.

D.S. Poole and J. Chory

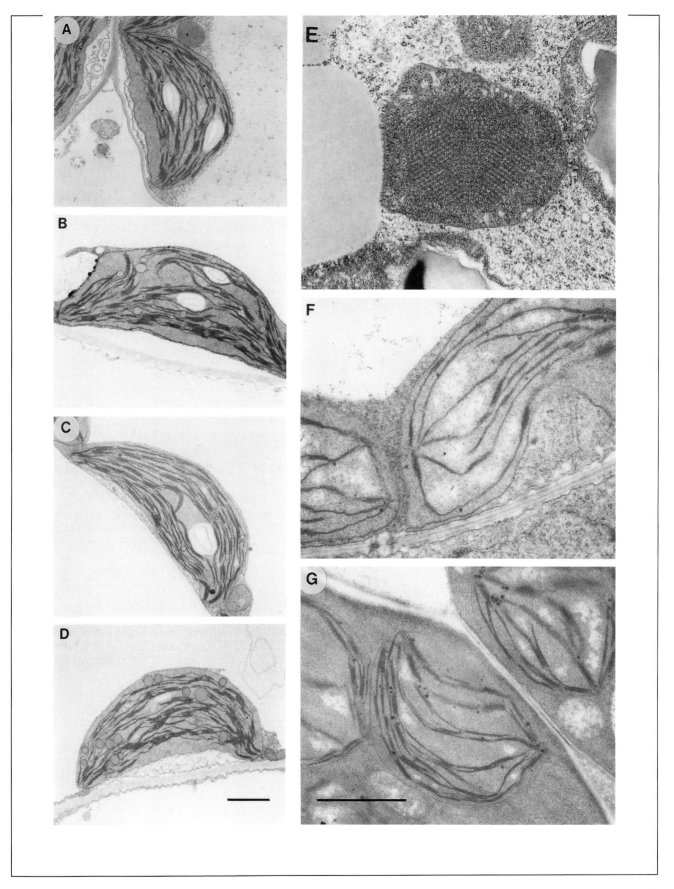

Plate 1.17

47

Plate 1.18
de-etiolated mutants

Dark-grown wild-type and *de-etiolated* mutant plants. Mutant seedlings were selected that, when grown in the dark, showed some characteristics of light-grown wild-type plants (Chory et al., 1989b; Chory et al., 1991; Deng et al., 1991; Wei and Deng, 1992; Hou et al., 1993).

(**A**) and (**B**) From left to right: wild type, *de-etiolated1-1* (*det1-1*) mutant, *de-etiolated2-1* (*det2-1*) mutant, and *constitutive photomorphogenic1* (*cop1*) mutant seedlings. The *cop9* mutant displays a similar phenotype (not shown).

(**A**) After 7 days in the dark.

(**B**) After 21 days in the dark.

All 4 mutants have short hypocotyls and expanded cotyledons when grown in the dark. The *det1* and *cop1* seedlings also produce leaves, while *det2* does not continue beyond production of primary leaf primordia. The *cop9* plants display an adult lethal phenotype, senescing before the transition to flowering. The mutants also accumulate mRNAs for several light-regulated genes (Chory et al., 1989b; Chory et al., 1991; Deng et al., 1991; Wei and Deng, 1992). Mutations at three other loci, *COP2*, *COP3*, and *COP4*, also cause dark-grown seedlings to have expanded cotyledons, a characteristic of wild-type light-grown plants, but do not affect plastid differentiation (Hou et al., 1993). The *cop4* mutants also exhibit agravitropic roots (Hou et al., 1993).

(**C–F**) Representative cotyledon or leaf plastids from dark-grown plants. Note that the plastids in *det1* and *cop1* differentiate into chloroplasts, while the plastids in *det2* differentiate into etioplasts.

(**C**) Wild-type etioplast, 7 days in dark (Chory et al., 1991).

(**D**) *det1-1* plastid, 10 days in dark (Chory et al., 1989b).

(**E**) *det2-1* etioplast, 7 days in dark (Chory et al., 1991).

(**F**) *cop1-1* plastid, 6 days in dark (Deng et al., 1991).

The *COP1* gene encodes a protein with a Zn binding motif and a domain homologous to the β-subunit of trimeric G-proteins (Deng et al., 1992). All of the mutants are recessive and appear to act as repressors of light-responsive development. They, along with the products of additional loci, may serve as negative regulators in the signal transduction pathway which couples light with developmental responses.

Bars = 1 μm.

D.S. Poole and J. Chory

Photo of *cop1* plastid provided by Xing-Wang Deng and Albrecht von Arnim. **E** reproduced by permission from American Society of Plant Physiologists. **F** reproduced by permission from Cold Spring Harbor Laboratory Press © 1991.

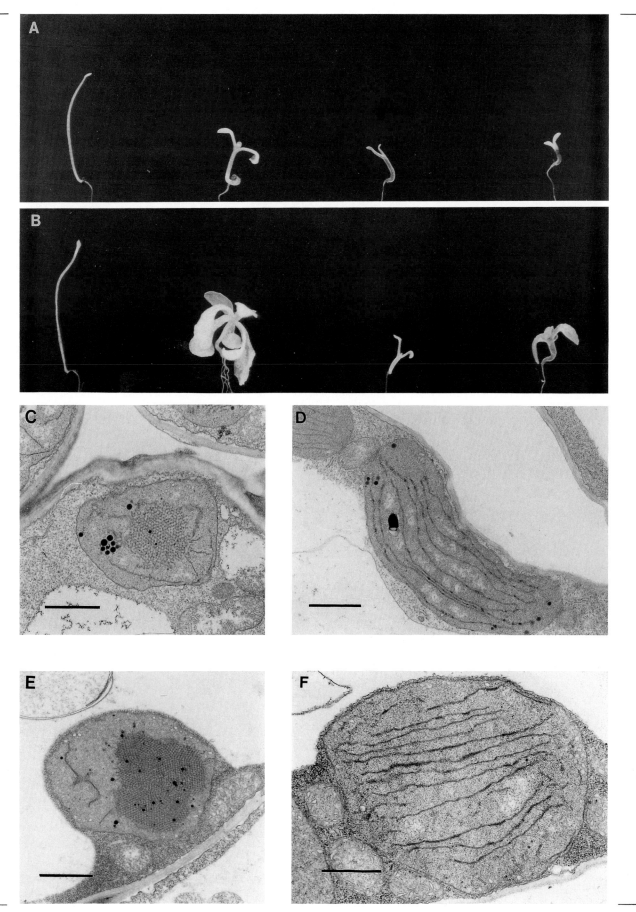

Plate 1.18

Plate 1.19
de-etiolated **mutants**

Adult phenotypes of *de-etiolated* (*det*) mutant plants.

(**A**) From left to right, wild type, *det1-1*, *det1-2*, and a phytochrome-deficient *long hypocotyl* (*hy*) mutant (*hy6*) (Chory, 1991). Both *det1* and *det2* show reduced apical dominance as adult plants. The *det2* plants also have a late-flowering phenotype and delayed leaf senescence (Chory et al., 1991). In contrast, *hy6* plants are pale and have increased apical dominance, fewer leaves, and an early-flowering phenotype.

(**B**) and (**C**) Electron micrographs of representative root plastids from light-grown mutant plants. In addition to the adult phenotypes mentioned above, root plastids in the *det1* and *cop1* mutants develop differently than wild type. They do not have the small size, irregular shape, or large starch granules of wild-type amyloplasts, but instead develop the elongated shape and thylakoid membranes characteristic of chloroplasts.

(**B**) *det1-1* root plastid, 10 days in light (Chory and Peto, 1990).
(**C**) *cop1-1* root plastid, 6 days in light (Deng and Quail, 1992).

Bar = 1 μm.

D.S. Poole and J. Chory

Photo of *cop1* plastid provided by Xing-Wang Deng and Albrecht von Arnim. **C** reproduced by permission from Blackwell Scientific Publications Ltd, © 1992.

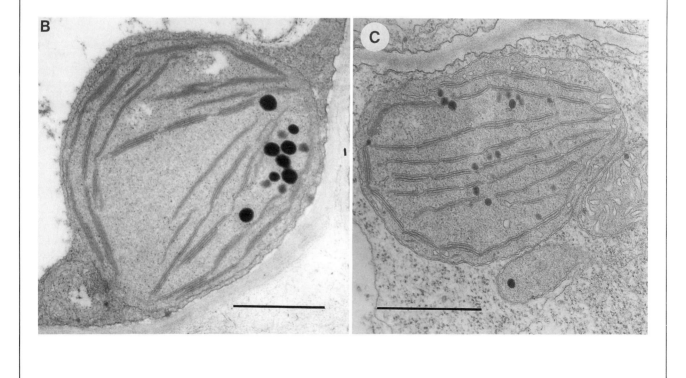

Plate 1.19

51

Plate 1.20

long hypocotyl **mutants**

Light-grown wild type and *long hypocotyl* (*hy*) mutant plants.

(**A**) From left to right, 7-day-old seedlings grown in the light: wild type, *hy1* (21.84N), *hy2* (To76), *hy3* (Bo64), *hy4* (2.23N), and *hy5* (Ci88) (Koornneef et al., 1980; Chory et al., 1989a; Chory, 1992). These mutants were obtained by screening mutagenized seed in white light for seedlings with long hypocotyls, a long hypocotyl being one of the characteristics of a dark-grown seedling. The elongated stems of dark-grown plants are thought to be an adaptive response of underground or shaded plants to reach more favorable photosynthetic environments. The *hy1* and *hy2* mutant plants have significantly lower levels of photoreversible phytochrome activity than wild type, while having near wild-type levels of phytochrome apoprotein (Koornneef et al., 1980; Parks et al., 1989; Chory et al., 1989a). The *hy3* mutants are phytochrome B mutants (Somers et al., 1991; Reed et al., 1993). Mutant *hy4* lacks blue-light inhibition of hypocotyl elongation (Koornneef et al., 1980). Several other mutants that lack blue-light inhibition of hypocotyl elongation have been identified as well (Liscum and Hangarter, 1991). The *hy5* mutants are defective in red-, far-red-, and blue-light mediated inhibition of hypocotyl elongation, while having normal levels of phytochrome (Koornneef et al., 1980). The *hy6* mutants have little detectable photoreversible phytochrome activity, but have near wild-type levels of phytochrome apoprotein (Chory et al., 1989a). The *hy8* mutants, which exhibit the long-hypocotyl phenotype when grown in far-red light, but respond to red light, are deficient in functional phytochrome A (Parks and Quail, 1993).

(**B–D**) Representative chloroplasts from wild type or *hy* mutants grown for three weeks in the light (Chory et al., 1989a).

(**B**) Wild-type chloroplast.

(**C**) *hy1* (21.84N) chloroplast.

(**D**) *hy5* (Ci88) chloroplast. The *hy1* and *hy2* mutant plants have fewer chloroplasts per cell, and the grana in the chloroplast occupy a significantly smaller volume of the chloroplast than for wild type (Chory et al., 1989a). In the *hy5* mutant, the grana occupy a larger volume of the chloroplast than in wild type (C. Peto and J. Chory, unpublished).

The *det* mutations are epistatic to the *hy* mutations (Chory, 1992).

Bar = 1 μm.

D.S. Poole and J. Chory

We are grateful to Charles Peto (Salk) and Abby Ann Sisk (UCSD) for the electron micrographs, and to Marc Lieberman (Salk) for photographs of light- and dark-grown mutants.

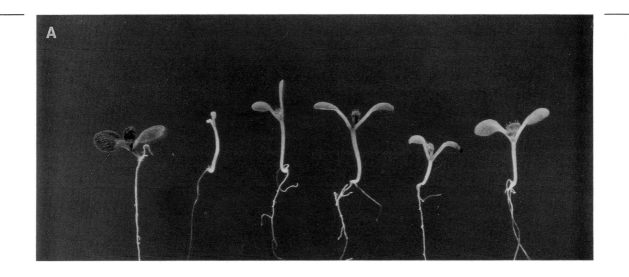

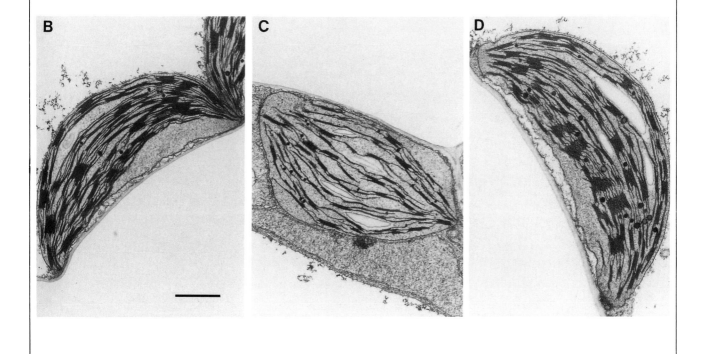

Plate 1.20

53

Plate 1.21
Mutants with altered
chloroplast accumulation

Isolated mesophyll cells from first leaves of wild-type *Arabidopsis* and 3 mutants with altered *accumulation and replication of chloroplasts* (*arc*). Mesophyll cells with conspicuous chloroplasts are viewed with Nomarski differential interference optics.

(**A**) Wild type.

(**B**) *arc1* mutant.

(**C**) *arc2* mutant.

(**D**) *arc3* mutant.

These *arc* mutations are nuclear recessive and cause radical changes in chloroplast number and chloroplast size (Pyke and Leech, 1992). The *arc1* mutant has more and smaller chloroplasts per mesophyll cell than wild type, whereas the *arc3* mutation causes a tenfold reduction in choroplast number, with a large increase in chloroplast size. The *arc2* mutants exhibit an intermediate phenotype.

Bar = 25 μm.

K.A. Pyke, J.L. Marrison, and R.M. Leech

Reproduced from Pyke and Leech (1992) with permission from the American Society of Plant Physiologists.

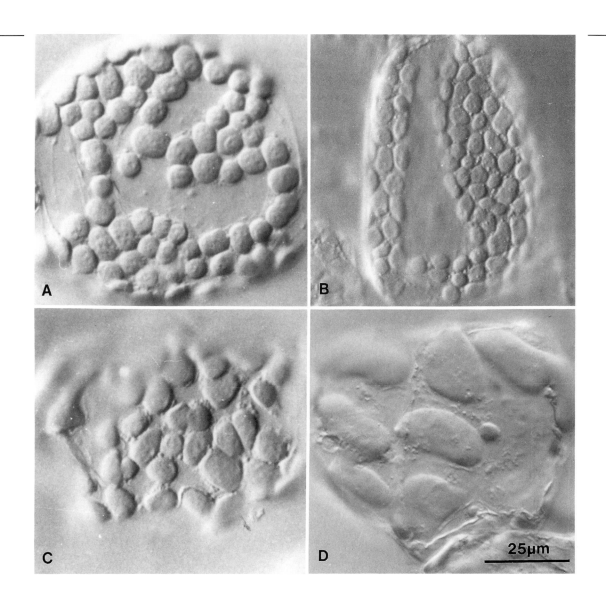

Plate 1.21

Plate 1.22
Wild-type morphology

Arabidopsis thaliana plants are covered with hairs called trichomes. The trichomes are unicellular, non-glandular, persistent, and living. The function of trichomes in *Arabidopsis* is unknown, but trichomes are among the first cells to fully differentiate on developing leaf primordia. In other plant species, this type of trichome development is often correlated with increased insect resistance (Johnson, 1975).

In *Arabidopsis*, trichomes develop on the leaves, stems, cauline or bract-like leaves, and sepals. Leaf trichomes are generally stellate, while trichomes on the stems and sepals are mostly unbranched. In wild-type *Arabidopsis*, trichomes do not form on cotyledons, hypocotyls, abaxial surfaces of the first few leaves, petals, anthers, or pistils. However, *Arabidopsis* can be genetically altered so that trichomes do develop on tissues that are normally glabrous (Lloyd et al., 1992).

The typical leaf trichome extends more than 300 μm out from the surface of the lamina and has a diameter of greater than 50 μm at its base. A group of support cells surround the base of the stalk. The trichome cell wall is thick and is covered with papillae of unknown composition. The cell is highly vacuolated and contains a thin thread of cytoplasm. Cytoplasmic streaming can easily be seen by light microscopy of mature trichomes. The cell has a large nucleus that resides in the stalk of the trichome.

(**A**) Scanning electron micrograph (SEM) of mature leaf trichomes.

(**B**) SEM of mature stem trichomes.

M.D. Marks and J.J. Esch

Reproduced from Marks et al. (1991) with permission from The Company of Biologists Limited and Marks and Feldmann (1989) with permission from the American Society of Plant Physiologists.

Plate 1.22

57

Plate 1.23
Wild-type leaf trichomes
and trichome branching

The trichome branching pattern varies among different geographical races of *Arabidopsis*. For example, most trichomes of Wassilewskija plants have either two or three branches and, rarely, no branches (less than 1%), whereas trichomes with two branches are only rarely observed on plants of ecotype Columbia. Columbia, in contrast to Wassilewskija, also has trichomes with four or more branches.

The number of trichomes on different vegetative leaves varies. The first leaf pair has the fewest trichomes, compared to later leaves. In addition, the first two to three leaves lack trichomes on their abaxial (lower) surface. The number of trichomes on the corresponding leaves of plants of different geographical races also varies. Each leaf of the first pair on plants of the Landsberg *erecta* ecotype has an average of ten trichomes. The corresponding leaves on plants of the Wassilewskija or Columbia ecotypes typically have more than 25 trichomes.

A comparison of three types of trichome branching. Stereoscopic analysis of young third leaves of plants from three *Arabidopsis* ecotypes. Leaves in **A–C** were photographed at a 25× magnification.

(**A**) Landsberg *erecta*.

(**B**) Columbia.

(**C**) Wassilewskija.

(**D**) Field of trichomes containing three branches (*black arrow*) or four branches (*white arrow*) on a Columbia plant.

(**E**) A mature two-branched trichome on a Wassilewskija plant.

M.D. Marks and J.J. Esch

D is reproduced from Marks et al. (1991) with permission from The Company of Biologists Limited.

Plate 1.23

59

Plate 1.24
Development of wild-type leaf trichomes

Trichome maturation proceeds in a basipetal fashion on young leaf primordia. The first apparent step in trichome development is the formation of a bulge in the outer cell wall (**A**). Cross sections through trichomes at this stage in development have revealed an enlarged nucleus. As the trichome expands out from the surface of the leaf, secondary branches emerge from the main branch (**B**, **C**). Trichomes increase not only in length but also in girth, as shown in **D**. When the trichome ceases to expand, 10–20 epidermal cells form a base around the trichome. In the final phase of development, the trichome cell wall thickens and acquires a papillate surface.

Scanning electron micrographs of mature leaf trichomes. All micrographs in the *inset* are shown at the same magnification in order to highlight the diffuse cell expansion of a developing trichome.

(**A**) Expansion of initiating trichome.
(**B**) Initiation of branch (*arrow*).
(**C**) Expansion of branches and stalk.
(**D**) Base of a mature trichome.

M.D. Marks and J.J. Esch

Reproduced from Oppenheimer et al. (1993) with permission from CRC Press.

Vegetative Development: Trichomes

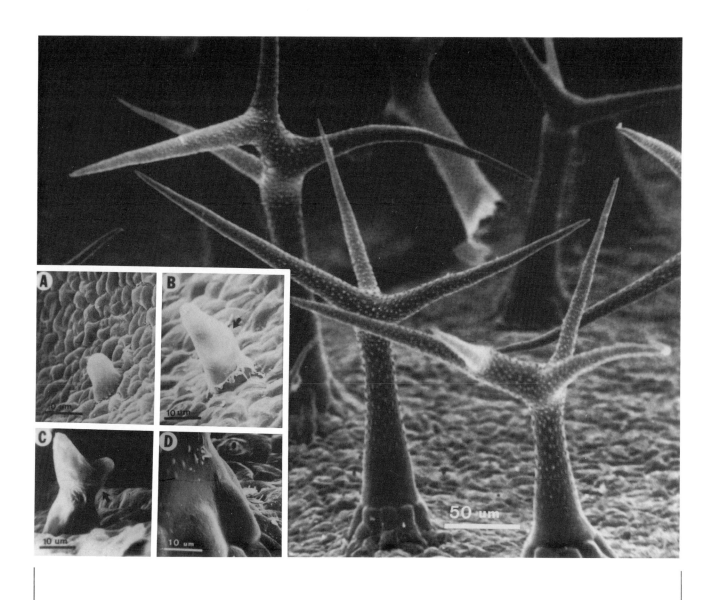

Plate 1.25
Trichome-specific
mutants

Six of the ten mutations that are known to affect trichome development do not appear to affect any other developmental processes. One of these mutants inhibits trichome initiation, while the other five alter trichome expansion.

(**A**) and (**D**) Complete loss of the *GLABRA1* (*GL1*) gene function results in a loss of trichomes on most surfaces (Koornneef et al., 1982a). However, some trichomes form on the margins of the leaves of *gl1* mutants. The *GL1* gene encodes a predicted protein that exhibits sequence similarity to Myb DNA-binding domains (Oppenheimer et al., 1991). Ectopic expression of *GL1* can lead to the production of trichomes on surfaces where they do not normally develop.

(**B**) and (**E**) The *glabra3* (*gl3*) mutation has several effects on trichome formation (Koornneef et al., 1982a). A few aborted trichome-like cells are usually present, along with trichomes with fewer than the normal number of branches. In addition, *gl3* mutants also initiate fewer trichomes than wild-type plants.

(**C**) and (**F**), (**G**) and (**J**) Mutations at two different loci, *DISTORTED1* (*DIS1*) (**C**, **F**) and *DIS2* (**G**, **J**) result in the distorted phenotype (Feenstra, 1978). Trichomes on these plants are swollen, twisted, and have fewer branches than normal.

(**H**) and (**K**) Mutations at the *STALKLESS* (*STL*) locus result in trichomes that essentially lack a stalk (Haughn and Somerville, 1988). The trichomes are more swollen than normal, but they are not twisted like trichomes on *dis* plants.

(**I**) and (**L**) Mutations at the *UNDERDEVELOPED TRICHOME* (*UDT*) locus also affect the branching pattern of the trichomes (Haughn and Somerville, 1988). The trichome phenotype is somewhat subtle, but the trichomes are less branched than normal.

A, **B** and **I** show one leaf of the first leaf pair; **C**, **G**, and **H** show the third leaf. Magnifications in **A**, **B**, **C**, **G**, **H**, **I** are approximately 25×.

M.P. Marks and J.J. Esch

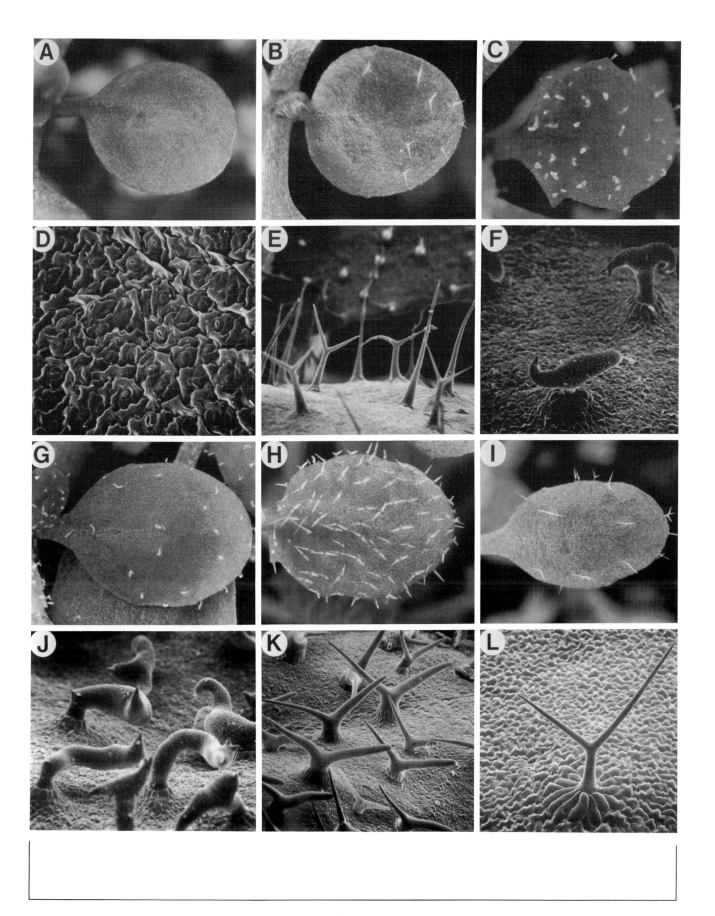

Plate 1.25

63

Plate 1.26
Trichome-specific
mutants

distorted1 (dis1)

Eight recessive mutations alter the shape of the mature trichome. The phenotypes of these altered trichomes range from fewer branches than normal to excessively swollen. These mutations are in genes that are critical for trichome cell expansion. For example, the *distorted1 (dis1)* mutation results in trichomes that are swollen and less branched. The distorted shape of trichomes on this mutant is a result of altered cell expansion.

(**A**) Initiation of a *dis1* trichome.
(**B**) Developing *dis1* trichome showing typical swelling.
(**C**) Mature *dis1* trichome with two branches.
(**D**) An unbranched *dis1* trichome with one aborted branch (*arrow*).

Bar = 5 μm in **A**; 10 μm in **B**; 35 μm in **C**; 40 μm in **D**.

M.D. Marks and J.J. Esch

Reproduced from Marks et al. (1991) with permission from The Company of Biologists Limited.

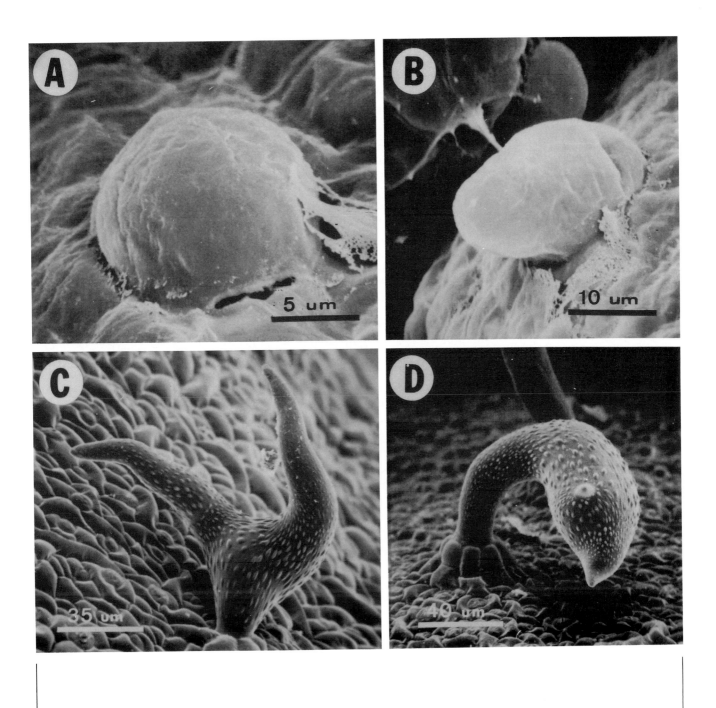

Plate 1.26

65

Plate 1.27
Pleiotropic trichome
mutants

Trichome mutations that are pleiotropic fall into two categories. First, the mutation may be in a gene that is directly required for each of the developmental processes affected. This situation is called direct pleiotropy. The second category is called relational pleiotropy and occurs when the mutation alters a single developmental process that is required for a number of subsequent developmental events. The mutations, *transparent testa glabra* (*ttg*), *glabra2* (*gl2*), and *singed* (*sne*), each of which affects different aspects of plant development, are most likely examples of direct pleiotropy because they affect developmental processes that are not obviously related to one another. However, the trichome phenotype of the *angustifolia* (*an*) mutant may be the result of relational pleiotropy, as the reduced branching in the trichomes may be related to the abnormal leaf development exhibited by *an* mutants.

transparent testa glabra (*ttg*)

Pleiotropic effects of *ttg* mutants: Mutations at the *TTG* locus result in a loss of trichome initiation on most parts of the plant (Koornneef, 1981), with the exception of the leaf margin. The *ttg* mutants also show a loss of seed pigmentation and an absence of the seed mucilage. The *TTG* gene encodes a predicted protein product with a helix-loop-helix DNA binding and dimerization motif (Lloyd et al., 1992). The *TTG* mutant phenotype can be complemented by the R gene *Zea mays* (Lloyd et al., 1992).

(**A**) One leaf of the first leaf pair of a *ttg* plant (25×).
(**B**) Scanning electron micrograph of *ttg* leaf surface.
(**C**) Wild-type (*right*) and *ttg* (*left*) seeds.
(**D**) Wild-type seed mucilage.
(**E**) Lack of seed mucilage on *ttg* seeds.

M.D. Marks and J.J. Esch

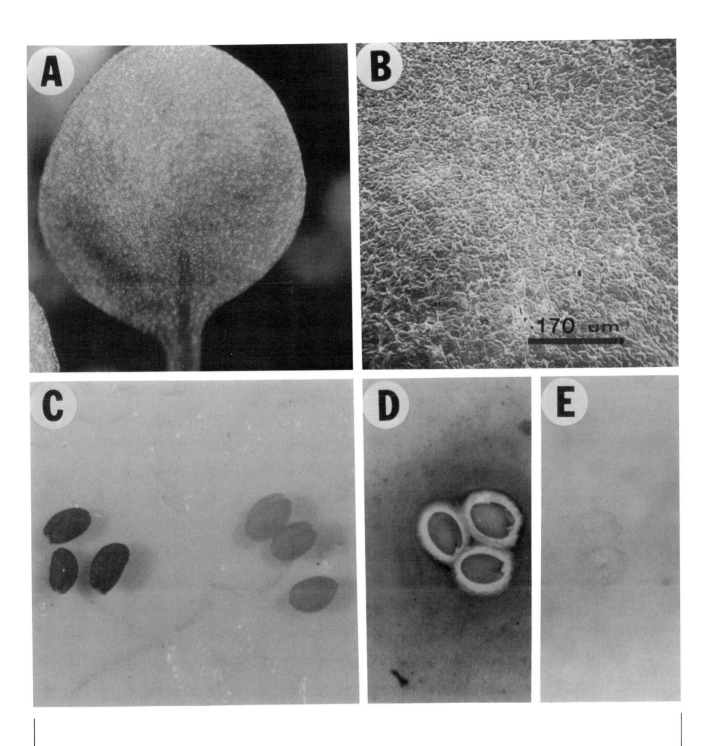

Plate 1.27

67

Plate 1.28
Pleiotropic trichome mutants

glabra2 (gl2)

The *glabra2* (*gl2*) mutants have a complex trichome phenotype and lack seed mucilage (Koornneef et al., 1982). Two types of trichomes are found on *gl2* plants. Many trichomes expand along the surface of the leaves, whereas other trichomes expand in the normal orientation but are less branched than normal.

(**A**) Third leaf of a *gl2* mutant. *Narrow* and *wide arrows* indicate an aborted leaf trichome and an erect trichome, respectively.

(**B**) Aborted *gl2* leaf trichome.

(**C**) Erect *gl2* leaf trichomes showing fewer branches than wild type.

(**D**) Wild-type seed coat mucilage.

(**E**) Lack of seed mucilage on *gl2* mutant.

M.D. Marks and J.J. Esch

B and **C** reproduced with permission from Marks and Esch (1992).

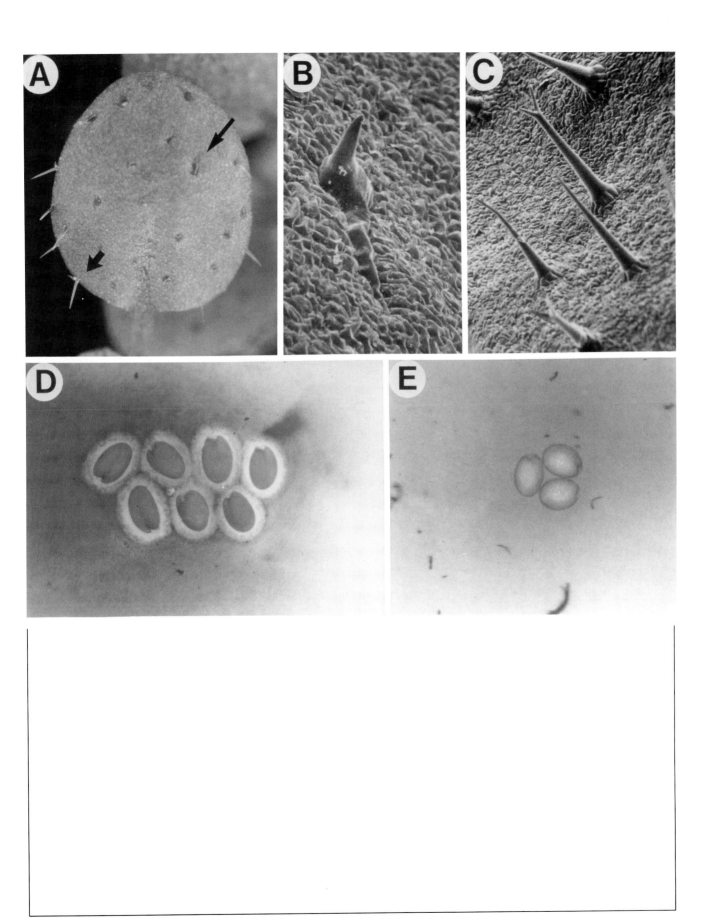

Plate 1.28

69

Plate 1.29
Pleiotropic trichome
mutants

singed (sne)

Mutations at the *SINGED* (*SNE*) locus affect the development of both trichomes and root hairs (previously named *WAVY*, Marks and Esch, 1992). The stalks and branches of trichomes on *sne* mutants are curved instead of straight and the root hairs are much shorter than normal.

(A) Scanning electron micrograph of *sne* leaf trichomes.

(B) Wild-type root hairs.

(C) *sne* mutant root hairs.

(D) Third leaf of *sne* plant.

M.D. Marks and J.J. Esch

A, B, C reproduced with permission from Marks and Esch (1992).

Vegetative Development: Trichomes

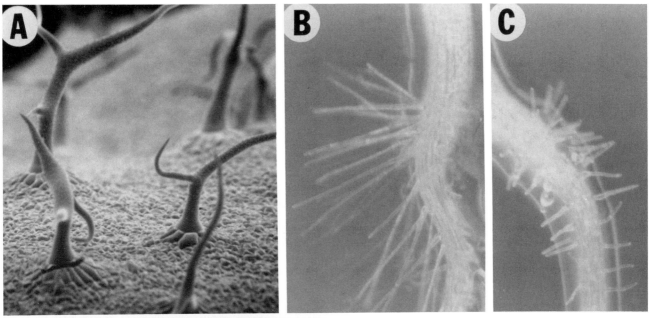

Plate 1.29

71

Plate 1.30
**Pleiotropic trichome
mutants**

angustifolia (*an*)

All of the leaf trichomes of *angustifolia* (*an*) mutants have two branches. However, this mutant also has cotyledons and leaves that are much narrower than wild-type leaves.

(**A**) Third leaf of *an* mutant plant.

(**B**) Scanning electron micrograph of *an* leaf trichome.

(**C**) Wild-type cotyledon.

(**D**) *an* mutant cotyledon.

M.D. Marks and J.J. Esch

Marks and Esch would like to thank David Oppenheimer and Bill Rerie for critically reading their contribution. This work was funded by grants from the National Science Foundation (DCB-9118306), the USDA/CSRS and the University of Nebraska Center for Biotechnology.

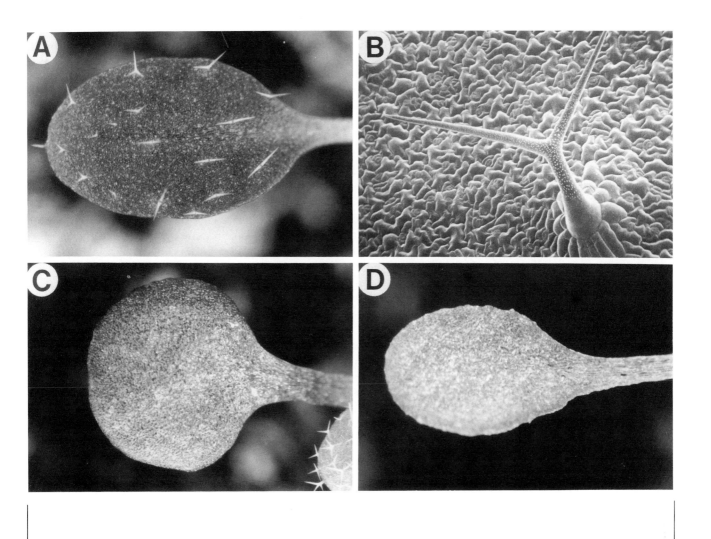

Plate 1.30

73

Plate 1.31
Gibberellin deficient and insensitive mutants

Gibberellins (GAs) have been implicated in many aspects of growth and development such as stem elongation, silique growth (Barendse et al., 1986), germination, and floral induction. Due to the effects of gibberellin on stem elongation and seed germination, gibberellin deficient plants can be selected as nongerminating, GA-sensitive dwarfs. Mutants defective in gibberellin production (*gibberellin deficient*) and sensitivity to gibberellins (*gibberellin insensitive*) have been isolated in *Arabidopsis*.

Recessive mutations at at least five *gibberellin deficient* loci (*GA1, GA2, GA3, GA4*, and *GA5*) result in dwarfs that are sensitive to exogenous gibberellins (Koornneef and van der Veen, 1980). The *ga* mutants are dark green dwarfs with a bushy appearance. Petals and stamens are poorly developed while sepals and carpels are near normal. The plants do not self-fertilize due to a lack of viable pollen. Mutants *ga1, ga2*, and *ga3* are, in general, more extreme in phenotype than *ga4* and *ga5* mutants. Some alleles of *ga1, ga2*, and *ga3* have an absolute GA requirement for germination. Restoration to near wild-type phenotype can be accomplished by repeated application of exogenous GAs. The *ga1-3* mutants also fail to flower when grown under short days, but display only a slight delay in flowering time when grown in long days (Wilson et al., 1992).

In *ga1-3* mutants the biosynthesis of GAs is blocked prior to the formation of *ent*-kaurene, a key intermediate in the GA biosynthetic pathway, and *ga1-3* plants have no detectable endogenous GA activity (Barendse et al., 1986). The cloning of the *GA1* gene will allow identification of the enzyme involved (Sun et al., 1992). In contrast, later steps in the GA biosynthetic pathway are blocked in *ga4* and *ga5* mutants (Talon et al., 1990a).

The partially dominant dwarf mutant *gai* is less sensitive to exogenous gibberellins than wild type (Koornneef et al., 1985; Peng and Harberd, 1993). The phenotype of *gai* plants is similar to that of the gibberellin deficient dwarfs *ga1-1* and *ga1-5* in that *gai* plants exhibit reduced stem growth and reduced seed germination as compared to wild type. However, in contrast to the *ga* mutants, exogenous GA has no effect on the phenotype of *gai* mutants. Endogenous gibberellins of the *gai* mutant differ from wild type in that *gai* mutants accumulate high levels of C_{19}-GAs (Talon et al., 1990b). Since *gai* plants contain endogenous GA activity, they are insensitive to both endogenous and exogenous GAs, suggesting that the *GAI* gene controls a step beyond the synthesis of an active GA, such as a component of the signal transduction pathway from receptor to the stem elongation response.

Shown are wild-type, *ga1-1, ga1-5*, and *gai* plants. Repeated application of exogenous GAs would restore the *ga1-1* and *ga1-5* plants to a nearly wild-type phenotype, but would have little effect on the *gai* plant. (The *ga1-1* plant was germinated in 10 μM GA_{4+7}).

M. Koornneef and J.L. Bowman

Reproduced from Koornneef et al. (1985) with permission from Munksgaard International Publishers.

Vegetative Development: Phytohormones

Plate 1.31

75

Plate 1.32
Abscisic acid deficient
and insensitive mutants

Abscisic acid (ABA) affects several aspects of plant growth, such as water relations (drought stress and stomatal regulation) and seed dormancy. Both *abscisic acid deficient* (*aba*) and *abscisic acid insensitive* (*abi*) mutants have been isolated in *Arabidopsis*.

The *aba* mutants were isolated by selecting germinating lines from mutagenized non-germinating gibberellin responsive dwarf mutants of the *ga1* locus (Koornneef et al., 1982b). Plants homozygous for the recessive *aba-1* mutation (**A**) are characterized by reduced seed dormancy, increased transpiration, symptoms of withering (*arrowhead*), and a lowered ABA content in developing and mature seeds and leaves as compared to wild type (**B**). The onset of seed dormancy correlates with the presence of embryonic ABA and cannot be maternally rescued (Karssen et al., 1983). In addition, *aba* plants lack the ability to cold acclimate, with freezing tolerance not increased by a low temperature pretreatment (Heino et al., 1990).

Mutants at three other loci (*ABI1*, *ABI2*, and *ABI3*) are insensitive to high levels of exogenous ABA (Koornneef et al., 1984; Koornneef et al, 1989a). The *abi1* and *abi2* plants are characterized by a moderate reduction in seed dormancy, loss of a water-stress response (Schnall and Quatrano, 1992), increased transpiration, and symptoms of wilting. In contrast, *abi3* plants display a dramatic reduction of seed dormancy (embryos remain green throughout development and are desiccation intolerant), while exhibiting nearly wild-type water relations (Koornneef et al., 1989a; Nambara et al., 1992). Seeds of *abi3* plants also exhibit a dramatic reduction in both 2S and 12S seed storage proteins (both mRNA and protein) and storage lipids as compared to wild-type seeds (Finkelstein and Somerville, 1990; Nambara et al., 1992). The protein encoded by the *ABI3* gene has characteristics of transcription factors and is similar to the *viviparous-1* protein of maize (Giraudat et al., 1992). Thus, the *abi1* and *abi2* mutants primarily affect the vegetative part of the plant, with moderate alterations in seed development, while the effects of *abi3* mutations are primarily seed specific.

M. Koornneef and J.L. Bowman

Reproduced from Koornneef et al. (1982b) with permission from Springer-Verlag.

Plate 1.32

Plate 1.33
Auxin resistant mutants

auxin resistant (axr1)

Recessive mutations in the *AUXIN RESISTANT1* (*AXR1*) gene confer resistance to the plant hormone auxin (indole acetic acid or 2,4-D) as well as causing a number of morphological defects (Estelle and Somerville, 1987; Lincoln and Estelle, 1990; Lincoln et al., 1990; Lincoln et al., 1992).

(**A**) Seven week-old *Arabidopsis* plants grown with constant illumination: *left*, wild type; *center, axr1-3; right, axr1-12*. The *axr1-3* mutation is a weak allele and *axr1-12* is a strong allele. Mutant plants display several defects compared to wild type. These include a reduction in internode length, a decrease in apical dominance, and smaller, wrinkled leaves. The *axr1* mutants are deficient in a number of other aspects of growth and development. The vascular tissue in the inflorescence of *axr1-12* plants is poorly developed relative to wild type (not shown). Mutant flowers are smaller, and particularly for the stong alleles, seed set is reduced (not shown).

(**B**) *Arabidopsis* seedlings grown in the dark. From *left* to *right*: wild type, *axr1-3, axr1-12*, and *axr1-23*. Both the *axr1-12* and *axr1-23* mutations cause a decrease in hypocotyl elongation relative to wild type, and the loss of the apical hook under these conditions.

(**C**) Light grown *Arabidopsis* seedlings grown on vertical plates to illustrate root branching. The seedlings on the *left* are wild type and those on the *right* are *axr1-3*. Mutant seedlings have significantly fewer lateral roots than wild-type seedlings. The rate of root elongation is greater in the mutants than in wild type, and mutant roots have a slower gravitropic response than wild type (not shown).

M.A. Estelle

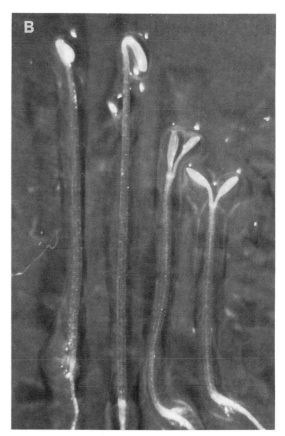

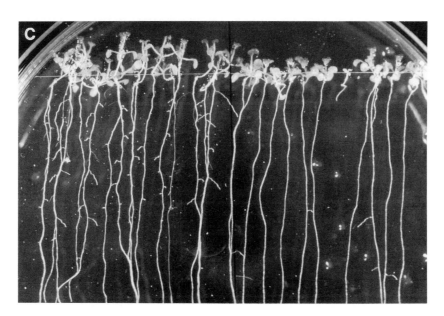

Plate 1.33

Plate 1.34
Auxin resistant mutants

auxin resistant (*axr2*)

The *auxin resistant2-1* (*axr2-1*) mutant is dominant and confers high levels of resistance to indole acetic acid (IAA) and ethylene, as well as resistance to abscisic acid (ABA) (Wilson et al., 1990; Timpte et al., 1992).

(**A**) Seven-week-old *Arabidopsis* plants grown with constant illumination: *left*, wild type; *right*, *axr2-1* mutant. The *axr2-1* mutation has a dramatic effect on morphology. The leaves of mutant plants are smaller, rounder, and have shorter petioles than wild-type leaves. Internode length is much shorter in mutant plants compared to wild type, and *axr2-1* inflorescences do not orient normally. Mutant inflorescences frequently curve downwards and grow back towards the soil, suggesting a defect in gravitropism. The dwarf phenotype is also apparent in dark-grown hypocotyls (not shown). The hypocotyls of *axr2-1* plants are much shorter than wild-type hypocotyls. Root growth is also affected by the mutation. The roots of mutants lack root hairs and are agravitropic (not shown).

(**B–E**) Most aspects of the *axr2-1* morphology can be explained by a defect in cell elongation.

(**B**) and (**C**) Longitudinal sections of mature wild-type (**B**) and *axr2-1* (**C**) inflorescence stems examined with a scanning electron microscope. The pith cells of the mutant stems are much shorter than wild type.

(**D**) and (**E**) Epidermal cells of mature wild-type (**D**) and *axr2-1* (**E**) inflorescence stems. Epidermal cells were peeled from inflorescence stems, stained with toluidine blue, and examined with light microscopy. The epidermal cells of mutant plants are much shorter than those of wild-type stems. The reduction in cell elongation is also apparent in epidermal cells of the dark-grown *axr2-1* hypocotyl (not shown).

Bar = 2.5 μm in **E**; 4 μm in **D**; 160 μm in C; 175 μm in **B**.

M.A. Estelle

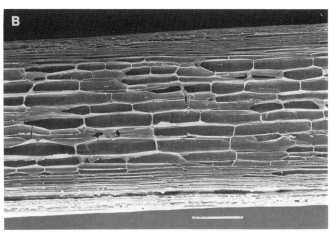

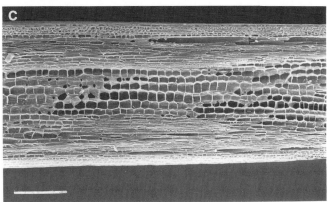

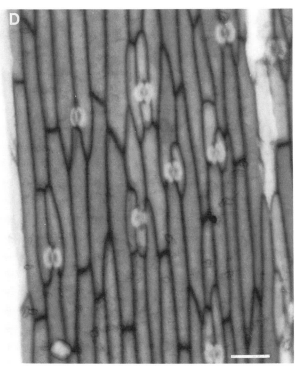

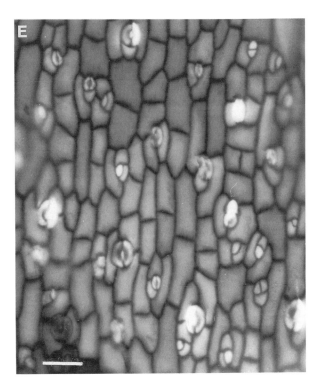

Plate 1.34

Plate 1.35
Primary dwarfs

dwarf (*dwf*) and *twisted dwarf* (*twd*)

A primary dwarf is defined as a plant with a short, robust stem and short, dark green leaves. Primary dwarfs often display a prolonged vegetative phase and delayed senescence, and exhibit a reduction in fertility that can be slight or severe, depending on the affected locus. Examination of the flowers from these mutants suggests that reduced fertility may be due to shortness of the stamen filaments, such that pollen is shed onto the ovary wall rather than onto the stigmatic surface. On the basis of their morphology, dwarf mutants can be classified into one of three types: standard, small, or twisted. Standard dwarfs [Wassilewskija (Ws) ecotype background] usually attain a height of 8–9 cm at 5 weeks, although slightly more growth is possible under certain environmental conditions. (Wild-type Ws plants are approximately 30 cm in height at 5 weeks.) Small dwarfs (Ws background) grow to only 5–6 cm at 5 weeks and their rosette width is smaller than that of other dwarf types. Twisted dwarfs can be similar to standard dwarfs in height, but various organs display twisted growth: stems, petioles, leaves, pedicels, roots, and hypocotyl. A variety of hormone-insensitive primary dwarfs have been isolated, representing 5 distinct loci (standard dwarfs, *dwf1* and *dwf4*; small dwarfs, *dwf2* and *dwf3*; twisted dwarf, *twd1*). Multiple mutant alleles, some of which are due to T-DNA insertion, are known for some of the loci.

(**A**) and (**B**) Seven-day-old seedlings grown either in continuous light (**A**) or darkness (**B**). From left: wild type, *dwf2*, *dwf3*, *dwf4*. When grown in the light, all organs of these mutants are shorter than those of wild-type plants. In dark-grown seedlings the most severe reduction in organ length is seen in the hypocotyl. Interestingly, the lack of hypocotyl elongation in dark-grown mutant seedlings is accompanied by opening of the cotyledons, suggesting that these primary dwarfs may be defective in the regulation of light-dependent developmental processes.

(**C**) Forty-five-day-old plants. From left: *dwf1*, *dwf2*, *dwf3*, *dwf4*, *twd1*. *dwf1* and *dwf4* are standard dwarfs, *dwf2* and *dwf3* are small dwarfs, and *twd1* is a twisted dwarf. Note the reduced rosette diameter in the small dwarfs.

Bar = 5 mm in **A**, **B**; 5 cm in **C**.

K.A. Feldmann and R. Azpiroz

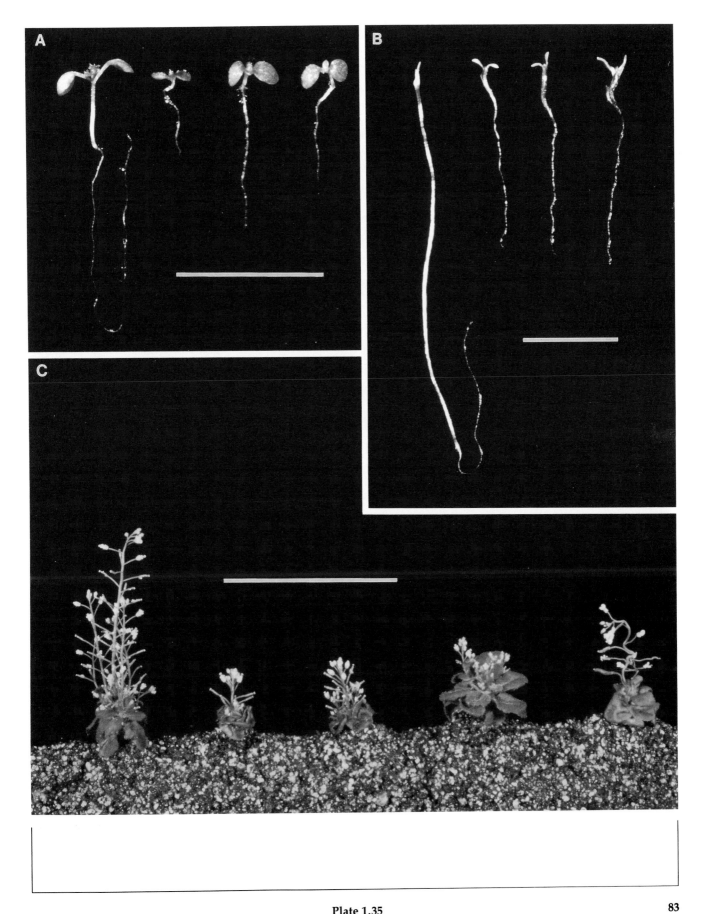

Plate 1.35

Plate 1.36
Primary dwarfs

dwarf (dwf) and *twisted dwarf (twd)*

A dwarf plant can be shorter than wild type because of a reduction in cell size, cell number, or both. Plant hormones, and gibberellins in particular, exert their effect on growth primarily by stimulating cell elongation. These dwarf mutants do not respond to exogenously applied gibberellins. However, they owe their reduced stature, in part, to a defect in cell elongation as shown in these micrographs of safranin-stained tissues.

(A–E) Longitudinal sections of 7-day-old light-grown hypocotyls, all shown at the same magnification.

(F–J) Longitudinal sections of 5-week-old inflorescence stems, all shown at the same magnification.

(A) and (F) Wild type (Wassilewskija ecotype).
(B) and (G) *dwf2*.
(C) and (H) *dwf3*.
(D) and (I) *dwf4*.
(E) and (J) *twd1*.

Mutant hypocotyls and stems have an average cell length between 40% and 60% of wild type, so the small size of these dwarfs may also be due to a reduction in cell number. The diameter of these organs is larger in the mutants, however, and in the case of stems can be as much as 1.5 times that of wild-type.

Bars = 100 μm.

K.A. Feldmann and R. Azpiroz

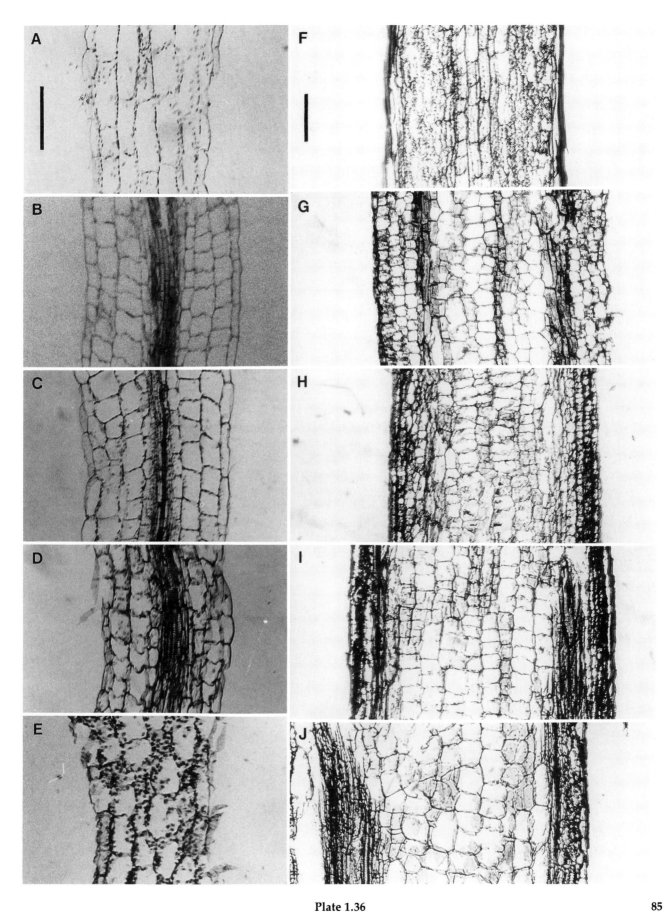

Plate 1.36

Plate 1.37
Epicuticular wax
deposition mutants

eceriferum (*cer*)

Mutants with brighter green stems and siliques than wild type, termed *eceriferum* (*cer*) mutants, are characterized by altered epicuticular wax deposition. Eighty-nine independent recessive mutants, representing 21 loci, fall into four phenotypic classes based on differences in visible amount of wax deposition (degree of glossiness) and sterility when grown under reduced humidity (Koornneef et al., 1989b).

Group I (*cer1*, *cer3*, *cer6*, *cer10*) mutants are characterized by a high degree of glossiness, reduced fertility in low humidity growth conditions, and reduced plant height. Group II (*cer2*, *cer4*, *cer5*, *cer7*, *cer9*) mutants are characterized by a high degree of glossiness with normal fertility.

Group III (*cer8*) mutants are characterized by a moderate degree of glossiness and reduced fertility in low humidity growth conditions.

Group IV (*cer11*, *cer12*, *cer13*, *cer14*, *cer15*, *cer16*, *cer17*, *cer18*, *cer19*, *cer20*) mutants are characterized by a moderate degree of glossiness and normal fertility.

Scanning electron micrographs of wild-type *Arabidopsis* and some *cer* mutants. Epicuticular waxes on the stems of wild-type *Arabidopsis* are characterized by an amorphous ground layer on top of which can be seen irregularly-shaped flat structures with rounded edges, together with tube-shaped structures. Similar structures are observed on wild-type siliques, although the number of structures is lower. No qualitative differences are observed under growth conditions differing in humidity. The *cer* mutants differ in both the amount and/or morphology of wax structures compared to wild type. All structures are absent in most alleles of *cer1*, *cer2*, and *cer6*. Wax structures are greatly reduced in most alleles of *cer3*, *cer7*, *cer10*, and *cer16* mutants, while tube-shaped structures are lacking in *cer5*, *cer8*, *cer9*, and *cer17* mutants. The *cer4* mutants have wax structures of an abnormal morphology. The degree of glossiness is correlated with the severity of alterations in the epicuticular wax structures.

(**A**) Wild-type stem (detail), (**B**) wild-type stem, (**C**) wild-type silique, (**D**) *cer1-2* stem, (**E**) *cer2* stem, (**F**) *cer7* stem, (**G**) *cer16* stem, (**H**) *cer8* stem, (**I**) *cer15* stem, (**J**) *cer15* silique, (**K**) *cer19* stem, and (**L**) *cer4* stem.

Bars = 1 μm.

M. Koornneef and J.L. Bowman

Reproduced from Koornneef et al. (1989b) with permission from Oxford University Press.

Plate 1.37

Plate 1.38
Leaf shape mutants

(A) *brevipedicellus* (*bp*) pedicels are reduced in length and siliques point downward (Koornneef et al., 1983).

(B) Mutant with a phenotype similar to that reported for *fiddlehead* (Lolle et al., 1992). It is not known if this mutation is allelic to *fiddlehead*. Margins of sepals are often postcongenitally fused (*arrowhead*), resulting in mechanical sterility and deformation of floral organs. Occasionally leaf margins may also fuse. In the *fiddlehead* mutant, fusion occurs between epidermal cells and does not involve cytoplasmic union (Lolle et al., 1992). In addition, promiscuous germination of pollen, an occurrence normally restricted to the surface of stigmatic papillae, may occur on vegetative surfaces in *fiddlehead* mutants (Lolle and Cheung, 1993).

(C) Wild-type Landsberg *erecta* plant with both cotyledons and five rosette leaves visible. The first leaves produced are round and entire while the later-produced leaves are spatulate and serrate (Röbbelen, 1957a; Medford et al., 1992). The *erecta* mutation affects growth habit in several ways. The primary effect is that there is a reduction in internode length both on the main stem and in the pedicel.

(D) *angustifolia* (*an*). Leaves and cotyledons are narrow and twisted (Redei, 1962; Lee-Chen and Steinitz-Sears, 1967). Sepals and petals may be similarly affected. This particular plant is also homozygous for *distorted trichomes1* (*dis1*) and *erecta*.

(E) *compacta2* (*cp2*). A compact semi-dwarf with round leaves, this particular plant is also homozygous for *asymmetric leaves* and *erecta*. Mutations at two other loci, *cp1* and *cp3*, have similar phenotypes (Koornneef et al., 1983).

(F) *asymmetric leaves* (*as*). Leaves and cotyledons are asymmetric and lobed (Redei and Hirono, 1964). This particular plant is also homozygous for *long hypocotyl1* (*hy1*) and *erecta*.

Bar = 100 μm (B).

J.L. Bowman and M. Koornneef

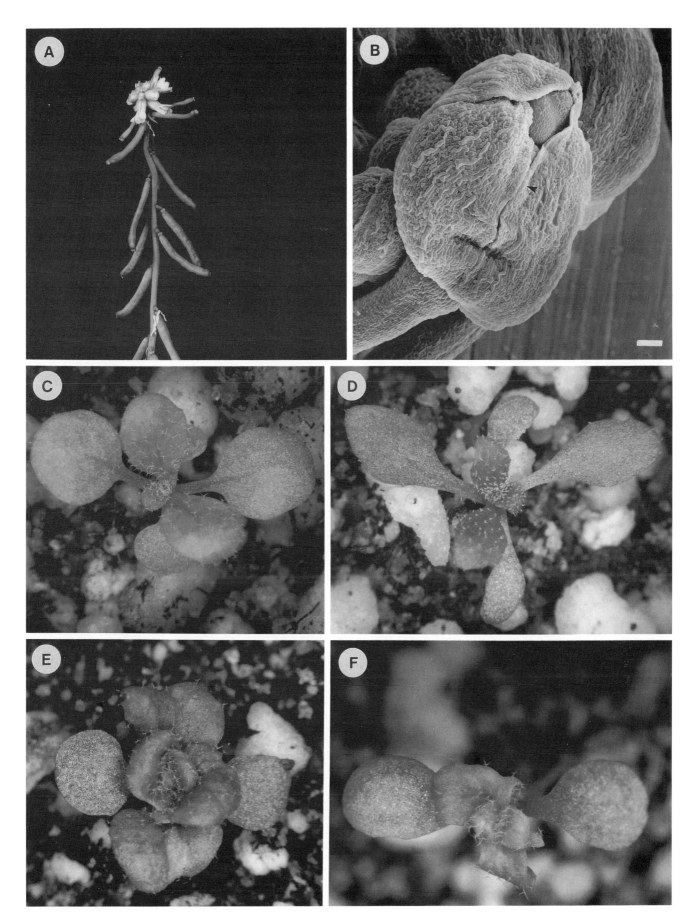

Plate 1.38

2
Roots

Introduction

By the end of embryogenesis, dicot plants have made two meristems, one which will form the aboveground shoot system, and one which will form the root system. Roots of a great variety of forms are found among the flowering plants. For example, prop roots provide support for developing shoot systems in many monocot families such as grasses and palms (Bell, 1991). The stilt roots of mangroves are covered in air-transporting lenticels which facilitate the aeration of the root system growing in anaerobic conditions. More familiar examples of specialized roots are to be found in the storage roots of root vegetables such as carrots and turnips in which the primary root becomes a swollen storage organ. While such a diversity of root form is to be observed in nature, a simpler system has a number of advantages for the study of development and other root-related processes.

Arabidopsis Roots

The small size and cell number of the *Arabidopsis* root make it an ideal system for the study of morphogenesis and the examination of root function. Unlike shoot meristems, the root meristem is not surrounded by developing primordia and axillary meristems but is open and exposed, without the need for surgical removal of tissue. The small size of the root renders it almost transparent and amenable to a variety of experimental manipulations. Individual cells can be observed in real time and their development recorded. Such an approach has proven fruitful in the investigation of the infection mechanism of nematodes in *Arabidopsis* roots (Sijmons et al., 1991).

The radicle (embryonic root meristem) in the Brassicaceae is derived from cells of the proximal suspensor cell and the neighboring cells of the lower part of the embryo proper. Upon germination this meristem initiates growth of the primary root. Lateral root primordia form endogenously (from an internal tissue layer) some distance from the primary root meristem. The new, lateral meristem forms in the pericycle and emerges on the surface by bursting its way through the superficial layers of the endodermis, cortex, and epidermis. The great proliferation of roots on a plant are lateral in origin since there is only one primary root meristem. In this introduction, I briefly describe the structure of the *Arabidopsis* root to assist in the interpretation of the plates that follow.

Origin of the Embryonic Root

The primary root meristem is formed from the derivatives of the lower tier of the octant embryo and the proximal (uppermost) cell of the suspensor (Mansfield and Briarty, 1991). The derivatives of the hypophysis (proximal suspensor cell) form the central root cap (columella), its initials, and the central cells above the root cap that are contiguous with the cortex and endodermis. Cells adjacent to these hypophysis derivatives form the primary root meristem, while the remaining derivatives of the lower octant embryo tier form the majority of the hypocotyl. (See also Embryogenesis chapter.)

Organization of Cells in the Primary Root

Examination of the root with cryo-scanning electron microscopy provides an external view of the developing root, since cell boundaries are well preserved when using this technique. The distal portion of the root is covered by root cap. The root cap is composed of two cell populations which are clonally unrelated. Those cells of the central cap are formed from initials that are derived from the hypophysis. The cells of the lateral root cap (root cap on the sides of the root) are derived from cells of the embryo proper, which also form the epidermal tissues. The outer layer of cells in the root cap is the location of profuse secretion of a range of molecules including complex polysaccharides and glycoproteins. These exudates may act as a lubricant for the growing root as it penetrates the interstitial air spaces of the soil, and have also been shown to modify the microflora in the vicinity of the root (Lynch and Staehelin, 1992). Cells of the lateral root cap are torn apart as a result of the expansion of the underlying epidermal layer which appears at the root surface only after the destruction of the root cap tissue. The emerging epidermal cells are then in the process of elongation.

Root hairs emerge at the basal end of cells at about the stage when cell elongation in the epidermis stops. The epidermal cells are arranged with separate files of hair cells and nonhair cells. The hair-forming cell files are generally interspersed with one or two (and sometimes three) nonhair cells. Once initiated, the hair projection grows by a process known as tip growth. This type of growth is characterized by the polar deposition of membrane and wall material at the apex of the growing hair. Soon after the formation of the hair, the nucleus migrates from the epidermal portion of the cell into the hair, where it becomes elongated and spindle-like in shape. In the early stages of growth the hair is densely cytoplasmic. This cytoplasm is highly differentiated along the length of the hair. The apical region is full of golgi-derived vesicles destined for fusion with the plasma membrane. Next to this is a region rich in golgi, endoplasmic reticulum, and mitochondria. An extensive vacuole develops behind this region of the cytoplasm. After the cessation of hair growth the hair becomes completely vacuolated.

Cellular Organization of the Root Meristem

The *Arabidopsis* root meristem may be considered to be composed of three tiers of cells, each of which gives rise to a discrete set of tissues. The lower tier of cells forms the root cap and epidermis. The middle tier of cells (which includes the central cells as well as the cortical and endodermal initials) is contiguous with the cortical and endodermal files, and the upper tier appears to give rise to the stele tissue (pericycle and vasculature). This organization of cells is considered to be "closed," since cells are restricted to certain tissue blocks, and is characteristic of many (if not all) of the Brassicaceae and many other angiosperm families (von Guttenberg, 1947; Peterson, 1967; Clowes, 1981).

Two remarkable features of the organization of the *Arabidopsis* root are its small size and the relatively invariant number of cell files in the primary root. Our observations, made on transversely sectioned roots of the Columbia ecotype, reveal that the epidermis is composed of between 16 and 23 cell files (average 19), with a cortex and endodermis composed of 8 cell files. The pericycle is composed of 11–13 files (average 12). The epidermis consists of two cell types, root hair cells and nonhair cells. The hair cells are usually, though not always, located over the anticlinal wall between two underlying cortical cells. The location of the hair cells relative to the underlying cortical cells is to be seen in other members of the Brassicaceae (Bünning, 1951).

Lateral Roots Arise in the Pericycle and Vary in Cell Numbers

In many species the tissue organization of lateral roots is more variable than that of primary roots. *Arabidopsis* is no exception. There is more variability to be seen in the numbers of cell files that comprise the lateral root than is observed in the primary root. Whereas there are invariably 8 cell files in the cortex and endodermis of primary roots, cortical cell file numbers in lateral roots varied from 7 to 11, and from 7 to 12 in the endodermis. This variability may be due to the fact that the patterns of cell division in the embryo are tightly regulated and therefore result in the formation of invariant cell numbers, while the pattern of cell division is less tightly regulated in the developing lateral root primordia.

The following plates describe the structure of the wild-type *Arabidopsis* root, followed by illustrations of mutants whose genetic characterization contributes to our understanding of the processes of cell differentiation, expansion, tip growth, and tropism in roots.

L. Dolan

Plate 2.1
Root wild-type structure

(**A**) Cryo-scanning electron micrograph of a young primary root showing the pattern of cells at the tip. The root-cap slime is apparent and the parallel files of cells of the epidermis are also clearly visible.

(**B**) The emergence of a lateral root from the primary root. Lateral roots arise from complex cell division patterns within the pericycle, and eventually the lateral root pushes through the endodermis cortex and epidermal cell layers to burst through the side of the root. The fully elongated cells of the epidermis of the primary root are visible on the right. The primary root is approximately 80 μm in diameter.

Bar = 30 μm

P. Linstead, L. Dolan, and K. Roberts

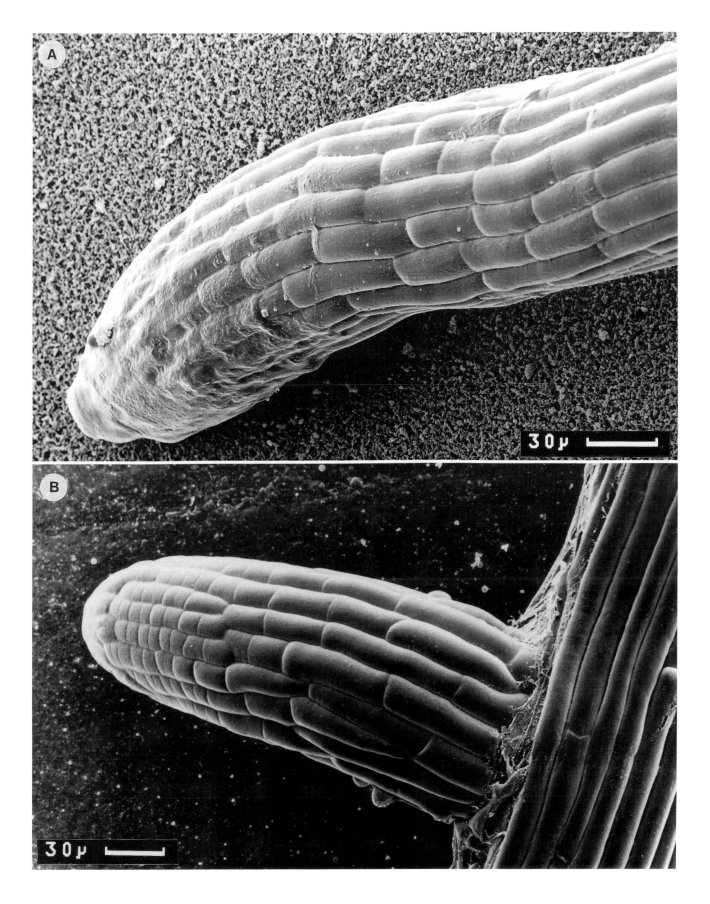

Plate 2.1

Plate 2.2
Root wild-type structure

A transverse section of the primary root approximately 80 μm from the tip. The section was cut on an ultra microtome from resin-embedded material. It was then treated with JIM5, a monoclonal antibody that recognizes pectin in the plant cell wall. The resulting sample was viewed with epifluorescence optics. This method clearly revealed the outlines of all the cells within the root. Notable is the 8-fold symmetry; 8 cortical cells and 8 endodermal cells are clearly visible. The outer walls of the epidermis are particularly rich in pectin. It is also noteworthy that some of the cells in the stele are only about 2 μm in diameter and are among the smallest cells known in higher plants.

Bar = 10 μm

P. Linstead, L. Dolan, and K. Roberts

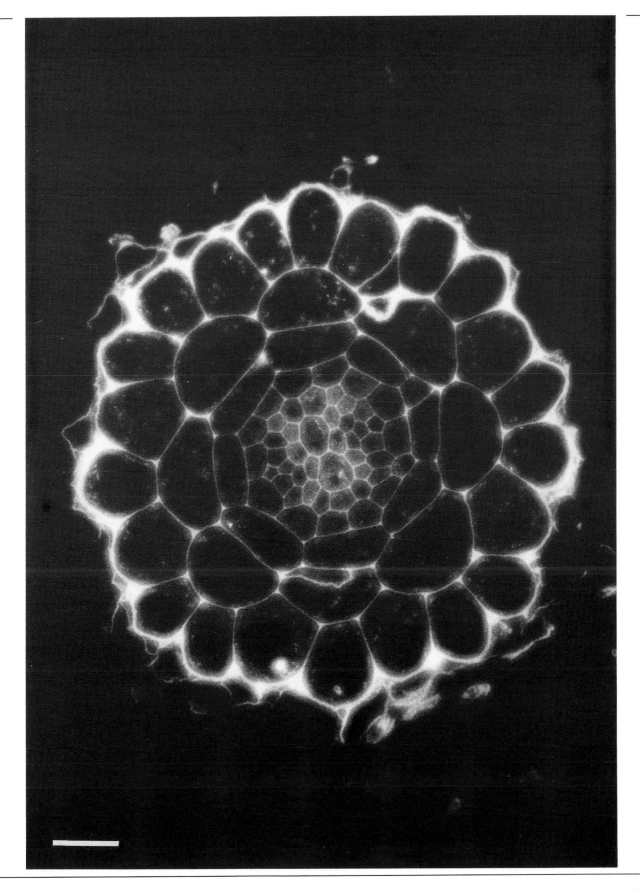

Plate 2.2

99

Plate 2.3
Root wild-type structure

A cross-section and a median longitudinal section of the primary root of *Arabidopsis*. Resin sections were stained with the anti-pectin monoclonal antibody JIM5. The resulting immunofluorescence images were then reverse-printed and photocopied. This gives the appearance of a line drawing but the image is in fact a micrograph of the cell boundaries of all the cells in the root. The anatomy and the origin of the various cell files within the root tip are clearly visible. The curious hanging cells on the longitudinal section are root cap cells which are being sloughed off.

P. Linstead, L. Dolan, and K. Roberts

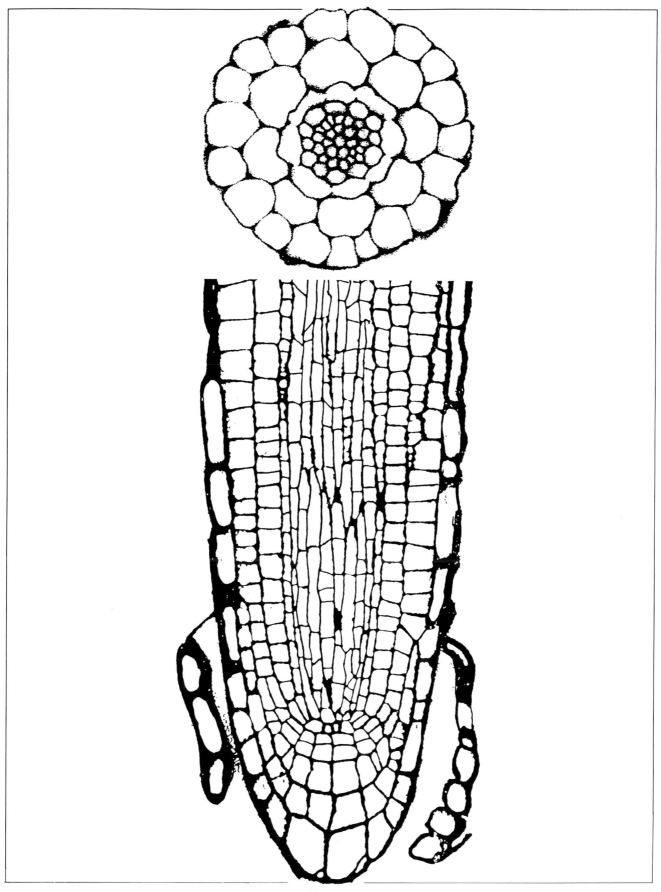

Plate 2.3

Plate 2.4
Root wild-type structure

Transmission electron micrograph of a thin section through the primary root approximately 80 μm from the root tip. Clearly visible within the stele are two "empty" cells that are the protophloem elements. Between these, and at right angles to the axis joining them, is a file of five cells. The two cells at either end of this file will eventually differentiate into the protoxylem elements; the other three cells will later develop into metaxylem. Outside the phloem is a ring of cells, the pericycle, and outside the pericycle are eight endodermal cells (only partly visible in this micrograph).

Bar = 2 μm.

P. Linstead, L. Dolan, and K. Roberts

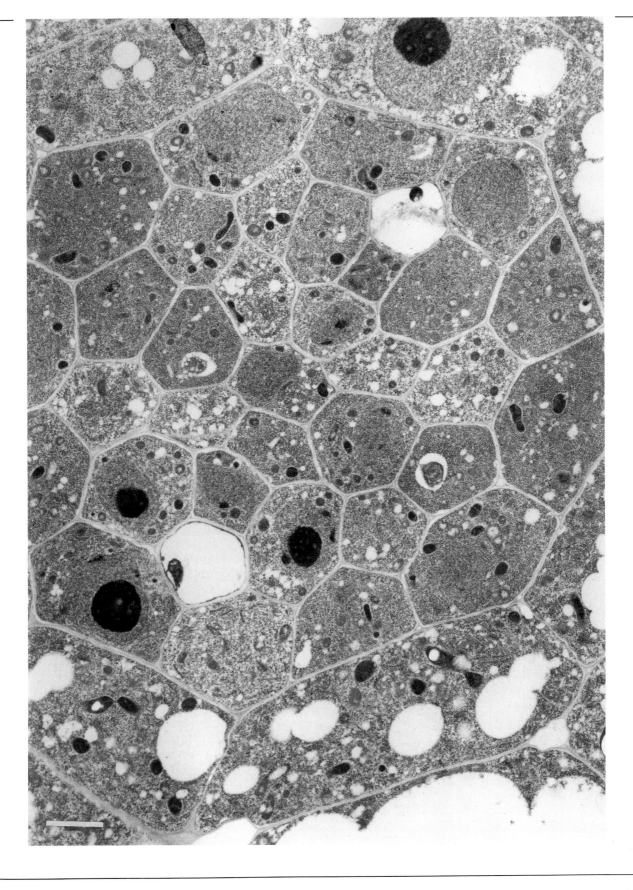

Plate 2.4

Plate 2.5
Microtubule deployment in roots

(**A–C**) Median longitudinal sections of primary roots showing microtubules, as revealed by immunofluorescence microscopy, in the meristem (**A**), the zone of maximum elongation (**B**), and the nongrowing differentiating region (**C**) just basal to the zone of elongation. Note the numerous cells in prophase in **A**, the well-aligned, mostly transverse microtubules in **B**, and the oblique microtubules in **C**.

(**D**) Transverse section of a root approximately 0.5 mm from the tip. Note the single tier of small epidermal cells, a double tier of large cortex cells, a single tier of endodermal cells, and a large number of narrow stellar cells. *Arrow* points to an interphase cell with conspicuous microtubules throughout its cytoplasm. *Arrowhead* points to a transverse section through the mitotic spindle, and *double-arrowhead* points to a transverse section through a phragmoplast.

(**E**) Longitudinal section through a root meristem showing a large number of interphase cells with microtubules prominent not only in the cell cortex, but all through the cytoplasm.

(**F–I**) Longitudinal sections through cells in different stages of mitosis. Note the appearance of the mitotic apparatus in prophase (**F**), metaphase (**G**), anaphase (**H**), and telophase (**I**).

Bar = 5.2 μm in **F**, **G**; 6.3 μm in **H**; 7.6 μm in **I**; 10 μm in **A**, **B**, **C**, **D**; 11.8 μm in **E**. (*bar* located in **D**, *lower right*)

T.I. Baskin

Reproduced from Baskin et al. (1992b) with permission from *Planta*.

Roots: Wild-type Structure

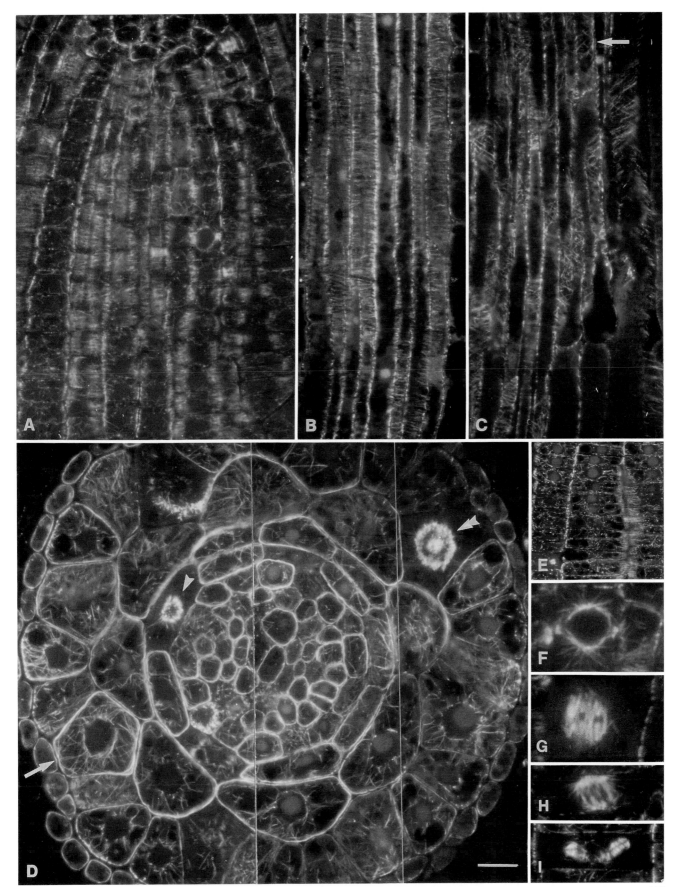

Plate 2.5

105

Plate 2.6
Radially swollen **mutants**

The *radially swollen* phenotype is characterized by extensive swelling of the root apex and a decrease in elongation rate, but has little effect on the root cap or root hairs. These *rsw* lines are temperature sensitive; the plants shown have been photographed following five days growth at the permissive temperature (18°C) followed by two days growth at the restrictive temperature (31°C), unless otherwise indicated. The three lines shown differ in morphological detail and in the kinetics of growth in response to the temperature shift, and they result from single gene recessive mutations at different loci.

(A) Wild type.

(B) *radially swollen1* (*rsw1*).

(C) *rsw2*.

(D) *rsw3*.

(E–H) A comparison of the appearance of *rsw* seedlings grown with (*center panel*) and without (*left panel*) treatment at the restrictive temperature. The seedlings without treatment are indistinguishable from one another, whereas those treated clearly show abnormal root morphology. Hypocotyls are shown (*right panel*) for comparison.

(E) Wild type.

(F) *rsw1*.

(G) *rsw2*.

(H) *rsw3*.

Bar = 150 μm in **A, B, C, D** (*bar* in **D**); 0.85 mm in *right panels* of **E, F, G, H** (*bar* in **H**); 5 mm in *center panels* of **E, F, G, H** (*bar* in **H**).

T.I. Baskin and R. Williamson

Reproduced from Baskin et al. (1992a) with permission from CSIRO Editorial Services.

Roots: Root Mutants

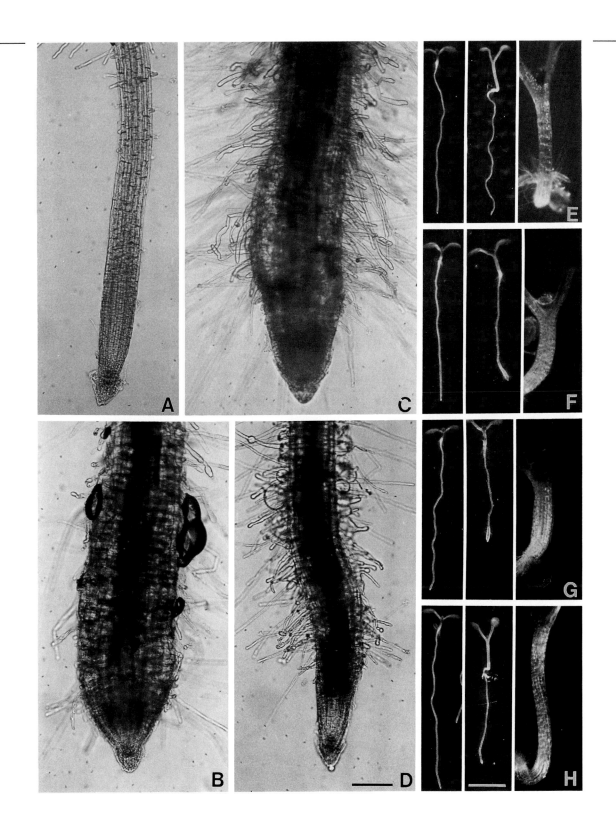

Plate 2.6

107

Plate 2.7

Stunted plant and *root epidermal bulger* mutants

(**A**) and (**E**) Wild-type roots.

(**B**) Dwarfism is a common morphological abnormality known to be caused by mutations in a large number of genes. Roots of plants mutant for *stunted plant* (*stp1*) (**B**) have a meristem and elongation zone about one-half the length of wild type (**A**). Analysis of this phenotype shows that the rates of cell elongation but not of cell division are strongly inhibited. However, for many days after germination, the number of dividing cells increases in the wild-type root meristem but stays constant in *stp1*.

(**C**), (**D**), (**F**), and (**G**) The morphological abnormality in *root epidermal bulger* (*reb1*) mutants is bulging of root epidermal cells. This phenotype is characterized by protuberances in many but not all epidermal cells as seen in *reb1-1* roots (**C**, **F**). Compare with the relatively smooth roots of wild type. Bulging cells first appear in the zone of elongation but the overall morphology of the root is not disturbed, and root hair development can occur (**F**). Expression of the temperature-sensitive allele *reb1-2* is shown in **D** and **G**. In addition to bulging epidermal cells, roots of *reb1-2* plants are characterized by a twisted growth habit. The *reb1-2* plant in **D** was rotated by 90° at the time of transfer to the restrictive temperature (33°C).

Bar = 0.2 mm in **E**, **F**, **G**; 0.4 mm in **A**, **B**, **C**; 0.8 mm in **D**.

T.I. Baskin and R. Williamson

Reproduced from Baskin et al. (1992a) with permission from CSIRO Editorial Services.

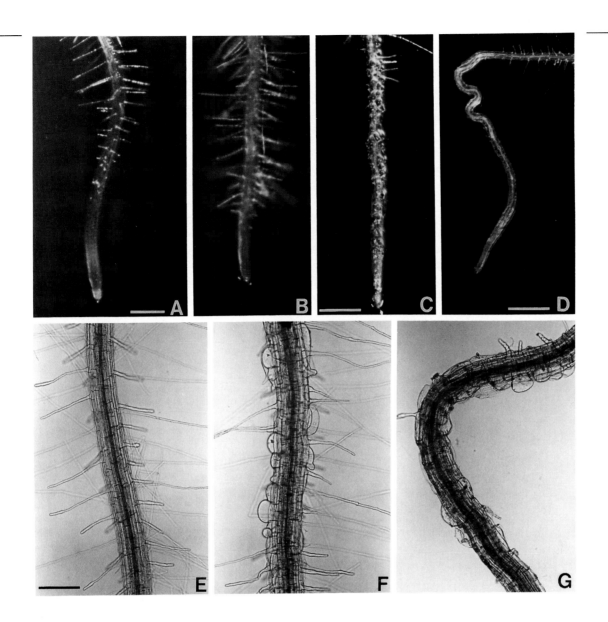

Plate 2.7

109

Plate 2.8
Wild type and root-cell expansion mutants

Whole mounts of wild type and root-expansion mutants.

(**A**) Wild-type root tip. Different regions or zones of development can be distinguished. These are the "meristematic zone" (MZ) in which the initials and quiescent center are located, the "elongation zone" (EZ) in which cells divide and expand, and the "specialization zone" (SZ) in which cells attain their final differentiated state.

(**B**) Root tip of *sabre* mutant. There is relatively uniform expansion along the entire length of the root. Expansion appears to occur primarily in the specialization zone (SZ).

(**C**) Root of *cobra* mutant. The degree of expansion varies along the length of the root. The upper root (UR) appears relatively unexpanded, while the root tip (RT) is grossly expanded. The phenotype is conditional and appears to depend on the rate of root growth. When plants are grown under suboptimal conditions (e.g., low sucrose or low temperature) the expanded phenotype is not expressed.

(**D**) Root of *lion's tail* mutant. As with *cobra*, the degree of expansion differs along the length of the root. Regions of the upper root (UR) are relatively unexpanded, while the root tips (RT) are more expanded. The phenotype is also dependent on the same conditions as *cobra*.

P.N. Benfey

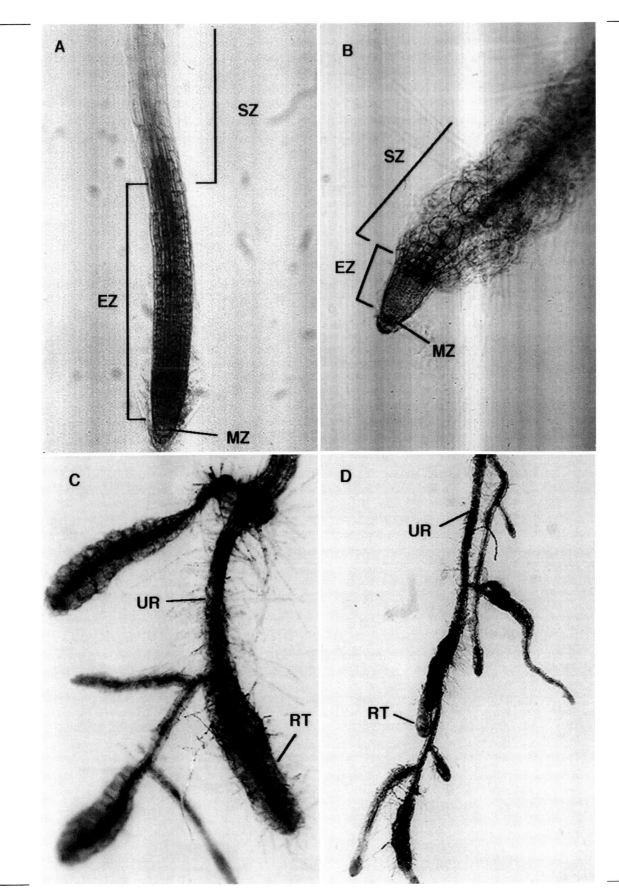

Plate 2.8

111

Plate 2.9
Root-cell expansion
mutants

(A) Transverse fresh section through the specialization zone of a wild-type root (Columbia ecotype). The primary root has a fairly simple cellular organization. Each of the three outer layers, epidermis (E), cortex (C), and endodermis (En), consists of a single cell layer and exhibits radial symmetry. The internal stele layer (S) shows bilateral symmetry. In the primary root, the cortex and endodermal layers have a nearly invariant eight cells.

(B) Transverse fresh section of the *sabre* mutant. The roots have a cross-sectional area approximately five times that of wild type. Expansion is most pronounced in the cortex (C) cell layer.

(C) Transverse fresh section of the *cobra* mutant. The average root diameter is approximately seven times that of wild type. The cell layer with the greatest expansion is the epidermis (E).

(D) Transverse fresh section of the *lion's tail* mutant. The average root diameter is approximately five times that of wild type. The tissue with the greatest expansion is the stele.

(E) Whole mount of the root tip of a *short-root* mutant. The roots of *short-root* are much shorter than those of wild type. The roots grow to approximately 6 mm and then cease growth. The roots that have stopped growing appear differentiated to the tip, as shown in this close-up (compare with wild-type whole mount in Plate 2.8).

Bar = 50 μm

P.N. Benfey

Roots: Root Mutants

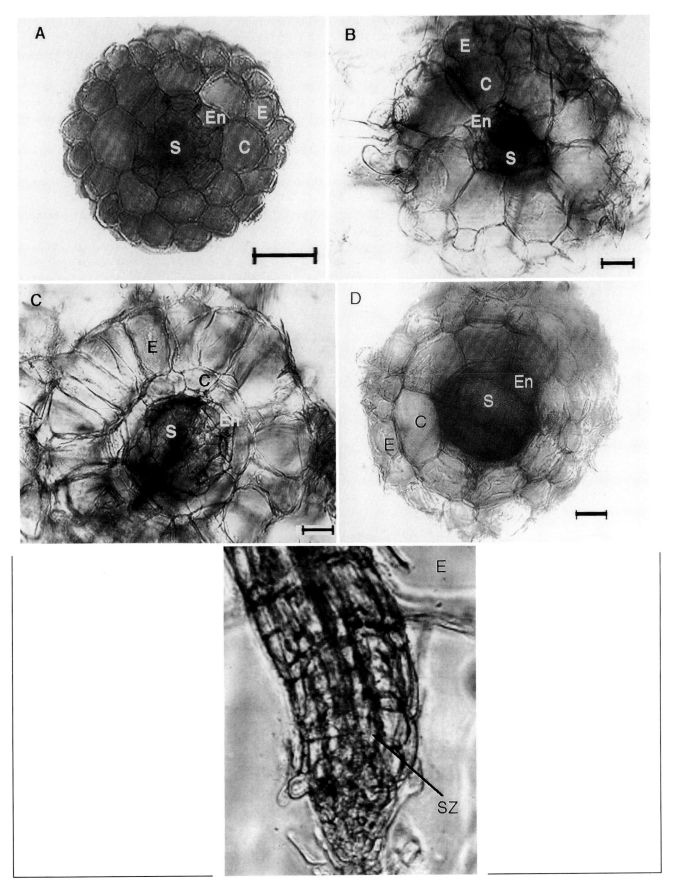

Plate 2.9

Plate 2.10
Wild-type root hairs

Root hairs are thought to play a role in the absorption of water and nutrients by increasing the surface area of the root (Clarkson, 1985). In addition, root hairs serve as attachment sites for soil-borne microbes, such as *Rhizobium* (Bauer, 1981).

(**A**) Root apex of a 4-day-old wild-type *Arabidopsis* (Columbia ecotype) seedling. Root hairs are visible at various stages of development. Root hairs are tubular-shaped extensions of single epidermal cells that emerge and elongate in a region called the root hair zone. In wild-type *Arabidopsis*, root hairs first emerge approximately 1 mm behind the root tip (Schiefelbein and Somerville, 1990). The mature length of root hairs depends largely on the growth conditions, with the Ca^{2+} concentration an important factor (Schiefelbein et al., 1992). In *Arabidopsis* seedlings grown on the surface of agarose-solidified media, root hairs elongate at an approximate rate of 100 μm/h and can reach a length of 1.5 mm (Schiefelbein et al., 1992).

(**B**) Transverse section of a wild-type *Arabidopsis* (Columbia ecotype) root. This section was taken from an immature region of the root prior to the formation of root hairs. The epidermis (e) is the outer ring of cells; the number of epidermal cells varies. The cell layer inside the epidermis is the cortex (c), which invariably consists of eight cells in primary roots. The root epidermis will differentiate to form two distinct cells types: root hair-bearing cells and hair-less cells. In *Arabidopsis* there are eight vertical files of epidermal cells that form root hairs (the eight cells in this figure destined to form root hairs are indicated by *arrows*). Epidermal cells that give rise to root hairs differ from their neighboring hair-less cells in three ways: (1) they are located over the anticlinal walls in the cortex; (2) they are slightly larger; and (3) they are less vacuolated.

(**C**) Scanning electron micrograph of a single, elongating root hair. In *Arabidopsis*, root hairs emerge at the apical end of the epidermal cell (Schiefelbein and Somerville, 1990). The extension of root hairs occurs by tip growth, a highly polarized type of cell expansion characterized by vesicle secretion and cell wall synthesis localized to the cell apex (Heath, 1990).

Bar = 10 μm in **C**; 20 μm in **B**; 200 μm in **A**.

J.W. Schiefelbein

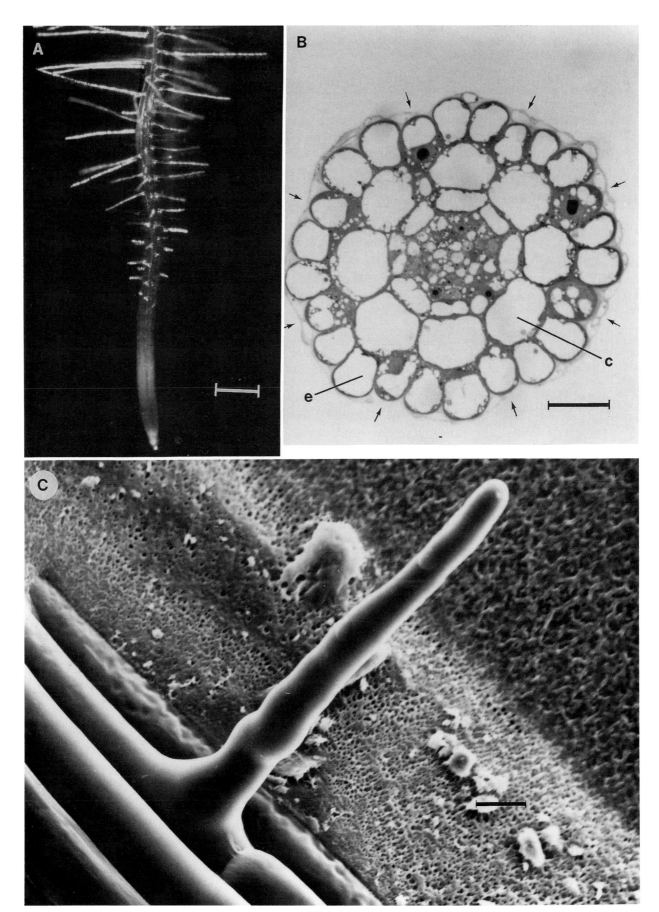

Plate 2.10

Plate 2.11
Root hair mutants

(A–E) Scanning electron micrographs of root hairs from wild-type and mutant *Arabidopsis* seedlings (Schiefelbein and Somerville, 1990). All mutations are recessive.

(F–J) Light micrographs of mature root hairs from wild-type and mutant *Arabidopsis* seedlings.

(A) and (F) Wild-type (Columbia ecotype) root hairs.

(B) and (G) Homozygous *root hair development1* (*rhd1*) mutant root hairs. These hairs are similar in length to the wild type, but they possess a wider shape at their base ("bulbous" hair mutant). The *RHD1* gene product appears to influence the degree of epidermal cell expansion during root hair initiation.

(C) and (H) Homozygous *rhd2* mutant root hairs. These hairs are much shorter than normal ("stubby" hair mutant), and the *RHD2* gene product appears to be required for root hair elongation.

(D) and (I) Homozygous *rhd3* mutant root hairs. These hairs are shorter than the wild type and exhibit a crooked appearance ("wavy" hair mutant). The *RHD3* gene product appears to be required for appropriate control of expansion at the root hair tip.

(E) and (J) Homozygous *rhd4* mutant root hairs. These hairs are shorter than the wild type and display variations in cell diameter along their length ("bulging" hair mutant). Like *RHD3*, the *RHD4* gene seems to encode a product involved in controlling root-hair tip growth.

The double mutant combinations *rhd1 rhd2*, *rhd1 rhd3*, *rh1 rhd4*, and *rhd3 rhd4* display essentially additive effects (Schiefelbein and Somerville, 1990). For example, *rhd1 rhd2* root hairs have a bulbous base (due to *rhd1*) and are reduced in length (due to *rhd2*), resulting in spherical-shaped root hairs (Schiefelbein and Somerville, 1990). In contrast, *rhd2* is epistatic to both *rhd3* and *rhd4* such that *rhd2 rhd3* and *rhd2 rhd4* doubly-mutant root hairs resemble *rhd2* root hairs (Schiefelbein and Somerville, 1990).

Bar = 50 μm.

J.W. Schiefelbein

Reproduced from Schiefelbein and Somerville (1990) with permission from American Society of Plant Physiologists.

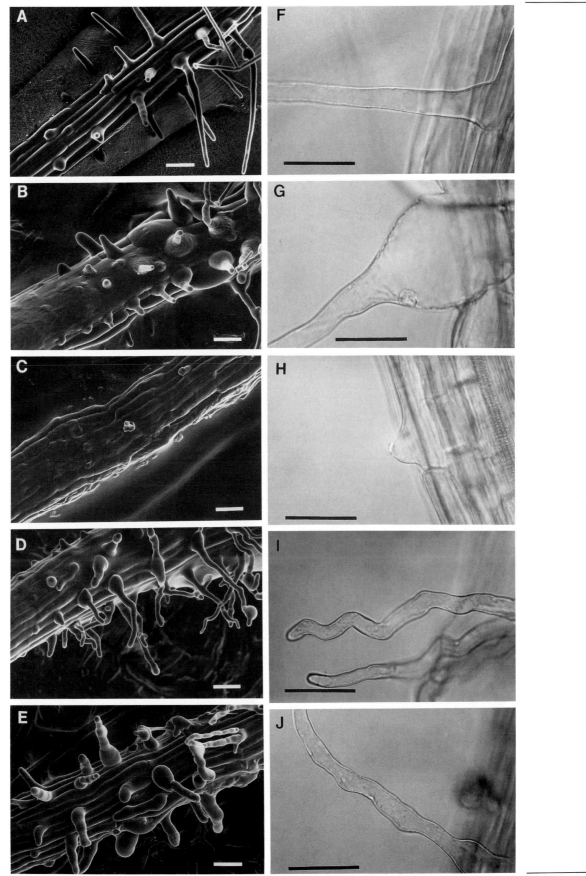

Plate 2.11

Plate 2.12
Wild type and *wavy growth1* **mutants**

Roots of wild-type plants elongate straight down when grown on the surface of hard agar held in a vertical position. If the plates are changed such that they are inclined 45° with respect to the direction of gravity (Figure 1), the path of root growth changes to a sinuous pattern. Shifting the agar plate to an angled position provides a continuous touching stimulus to the roots because the roots begin to bend downward following the direction of gravity and thereby encounter the agar surface, which they cannot penetrate. The touching stimulus induces rotation of the root tips. The rotating force is likely to be dependent on the twisted growth of cells in the elongation zone. Right-handed rotation results in a counter-clockwise curvature, while left-handed rotation results in a clockwise curvature (Figure 2). The periodic reversion of the rotation causes the wavy pattern of growth (Okada and Shimura, 1990; Okada and Shimura, 1992b).

Six mutants were isolated that had abnormal wavy patterns when grown on angled agar plates. Some of the isolated mutants also showed abnormal response to gravity or light. All of the mutants have a single, recessive nuclear mutation (Okada and Shimura, 1990).

Root growth of wild-type and *wavy growth1-1* (*wav1-1*) mutant plants [*wav1* is allelic with *root phototropism1* (*rpt1*); Okada and Shimura, 1992a]. *Arrowheads* indicate the positions of the root tips when the plates were shifted from the vertical to the inclined position. In Plates 2.12–2.16, seedling pictures are all at the same magnification. Close-ups of root growth are also at the same magnification.

(**A**) Seedlings of wild-type plants.
(**B**) Enlarged view of a wild-type root. The twisted pattern of epidermal cells is indicative of the direction of rotation.
(**C**) Seedlings of *wav1-1* mutant.
(**D**) Enlarged view of *wav1-1* root shown in **C**. The wavy growth pattern did not occur and the epidermal cells show little twisted growth.

Bar = 0.5 mm in **B**, 5 mm in **A**.

(*Text continued on p. 121*)

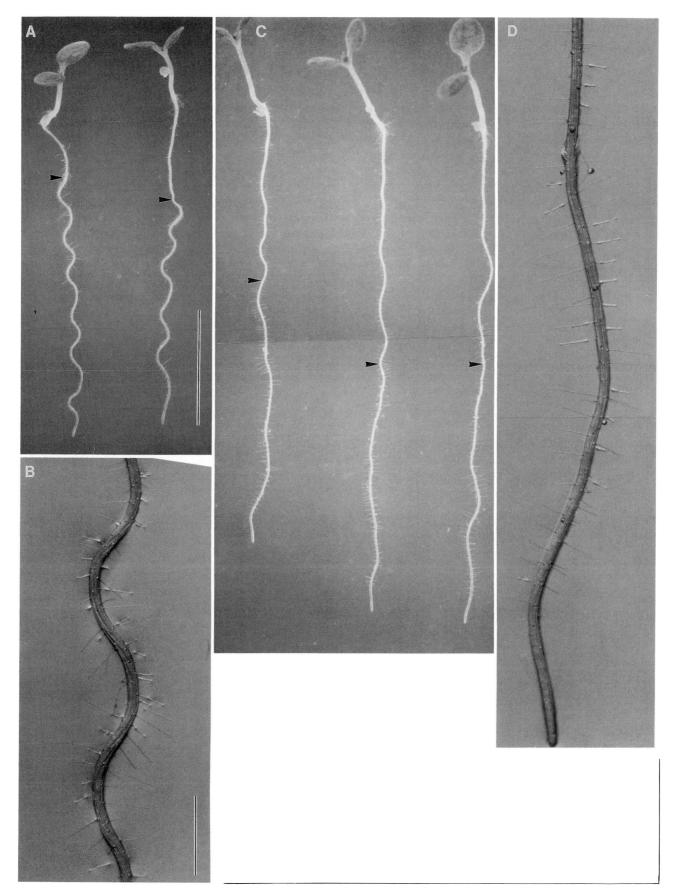

Plate 2.12

119

Plate 2.12
Wild type and *wavy*
growth1 **mutants**
(*continued*)

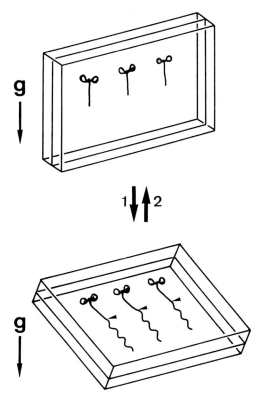

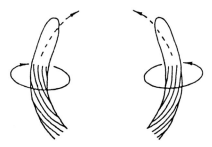

Figure 1. An illustration showing the position of agar plates for assaying touch-induced wavy growth of roots. Seeds are sterilized and placed on 1.5% agar plates containing half strength *Arabidopsis* mineral solution. After cold treating for 2 days in the dark, the plates were incubated in a vertical position for 5–7 days, then for 2 days in an inclined position.

Figure 2. Models showing the relation between the direction of root tip rotation and the direction of curvature. Right-handed rotation results in counter-clockwise curvature, and left-handed rotation results in clockwise curvature.

K. Okada and Y. Shimura

Plate 2.12   121

Plate 2.13
Wavy growth mutants

Root growth of *wavy growth2-1* (*wav1-1*) and *wav3-1* mutant plants. *Arrowheads* indicate the positions of the root tips when the plates were shifted from the vertical to the inclined position. For size bars see Plate 2.12.

(**A**) Seedlings of *wav2-1* mutants. The *wav2-1* mutant develops waves of shorter pitch than those of wild type. Continuous observation using a video recorder revealed that the rate of root tip rotation in *wav2-1* mutants is higher than that of wild type.

(**B**) Enlarged view of *wav2-1* root shown in **A**.

(**C**) Seedlings of *wav3-1* mutant. The *wav3-1* mutant also shows waves of a shorter pitch than those of wild type.

(**D**) Enlarged view of *wav3-1* root shown in **C**.

Both *wav2* and *wav3* mutants often form wavy growth patterns on the surface of agar medium set in the vertical position. It is suggested, therefore, that the two mutants are hypersensitive to touch stimulation. The above ground part of *wav2-1* mutants is smaller than wild type.

K. Okada and Y. Shimura

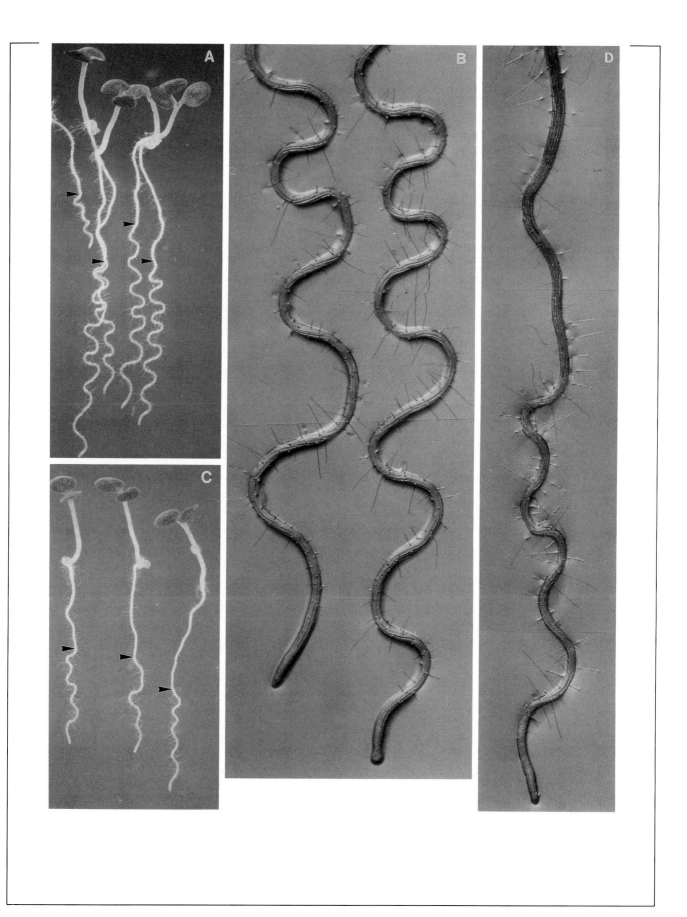

Plate 2.13

Plate 2.14
Wavy growth **mutants**

Root growth of *wavy growth4-1 (wav4-1)* and *wav5* [*wav5* is allelic with *auxin resistant1 (aux1)*; Mirza et al., 1984] mutant plants. *Arrowheads* indicate the positions of the root tips when the plates were shifted from the vertical to the inclined position. For size bars see Plate 2.12.

(**A**) Seedlings of *wav4-1* mutants. The *wav4-1* mutant makes rectangular waves, seemingly due to irregular timing of reversion of root tip rotation (see **C**). In addition, *wav4-1* mutants occasionally have abnormal flowers with 5 or 6 petals.

(**B**) Enlarged view of *wav4-1* root shown in **A**.

(**C**) Seedlings of *wav5-31* mutant. Some *wav5* mutants, such as the *wav5–31* allele, show neither wavy growth nor root-tip rotation. Other alleles exhibit quite different phenotypes (see *wav5*, Plate 2.15). All of the *wav5* mutants isolated display abnormal root gravitropism, but the gravitropic response of the aerial parts of the plants is seemingly normal (Okada and Shimura, 1992a). *wav5-31* exhibits a reduced gravitropic response.

(**D**) Enlarged view of *wav5-31* root shown in **C**.

K. Okada and Y. Shimura

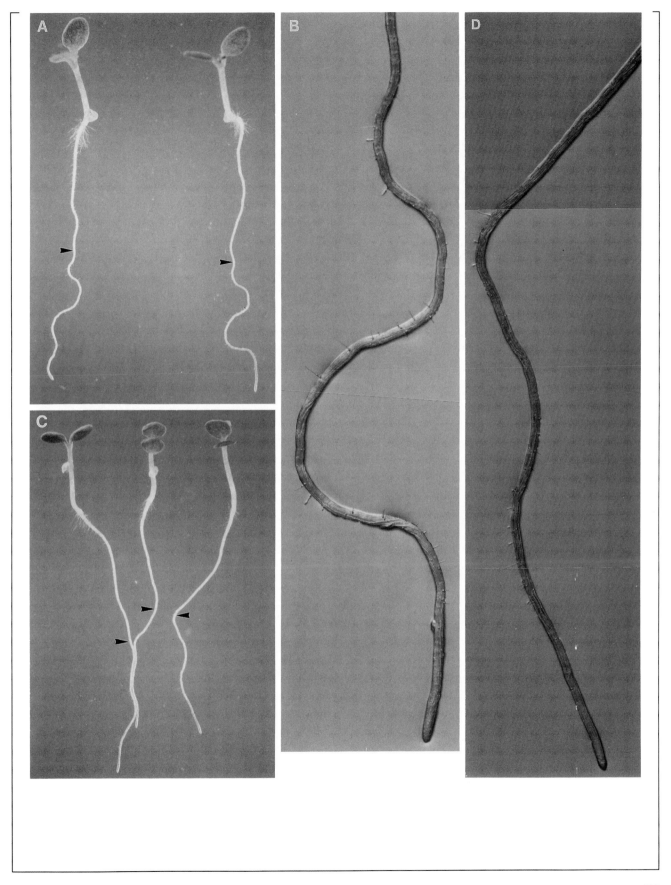

Plate 2.14

125

Plate 2.15

Wavy growth5 (auxin resistant1) **mutants**

Root growth of *wavy growth5* (*wav5*) [*wav5* is allelic with *auxin resistant1* (*aux1*); Mirza et al., 1984] mutant plants. In **A**, *arrowheads 1* and *2* indicate the positions of the root tips when the plates were shifted from the vertical to the inclined position, and their positions when they were shifted back to the vertical position from the inclined position, respectively. For size bars see Plate 2.12.

Some *wav5* mutants, such as the *wav5-31* allele, show neither wavy growth nor root-tip rotation. Other alleles exhibit quite different phenotypes (see below). All of the *wav5* mutants isolated display abnormal root gravitropism, but the gravitropic response of the aerial parts of the plants is seemingly normal (Okada and Shimura, 1992a). *wav5-31* exhibits a reduced gravitropic response.

(**A**) Seedlings of *wav5-33* mutants. The *wav5-33* mutant does not show a root gravitropic response.

(**B**) Enlarged view of *wav5-33* root shown in **A**. Roots of the *wav5-33* mutant exhibit clockwise circles, because the root tips undergo continuous left-handed rotation.

(**C**) Seedlings of *wav5-66* mutants.

(**D**) Enlarged view of *wav5-66* root shown in **C**. Roots of *wav5-66* mutants show three different patterns; clockwise circles, counter-clockwise circles, and no wavy growth.

K. Okada and Y. Shimura

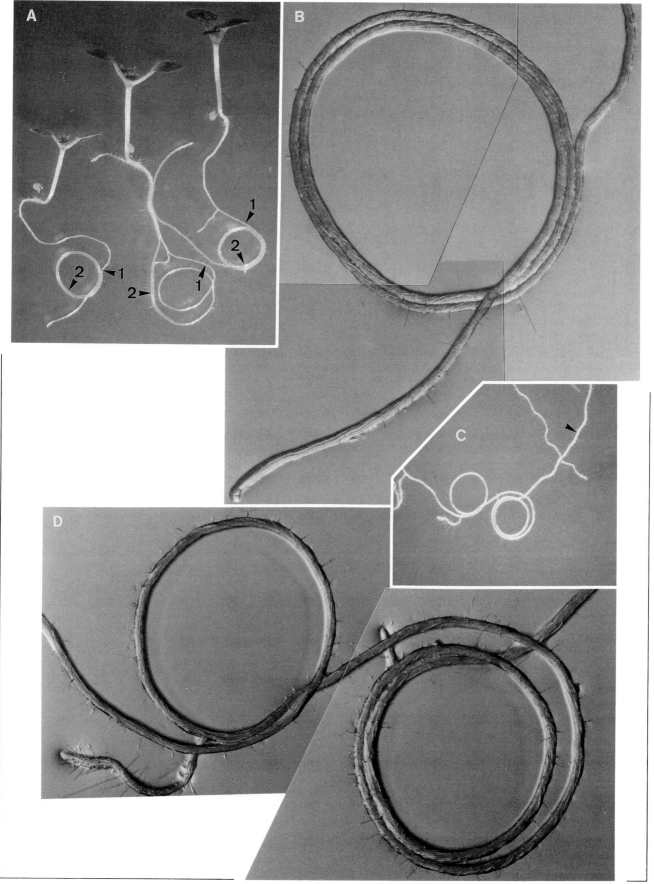

Plate 2.15

Plate 2.16
Wavy growth6
(agravitropic roots1)
mutants

Root growth of *wavy growth6* (*wav6*) [*wav6* is allelic with *agravitropic roots1* (*agr1*); Bell and Maher, 1990] mutant plants. *Arrowheads* indicate the positions of the root tips when the plates were shifted from the vertical to the inclined position. For size bars see Plate 2.12.

The *wav6* mutants have a phenotype similar to that of the *wav5-31* mutants. Roots of *wav6-52* mutants show neither wavy growth nor root tip rotation, and exhibit a reduced gravitropic response (Okada and Shimura, 1992a).

(**A**) Seedlings of *wav6-52* mutants.

(**B**) Enlarged view of *wav6-52* root shown in **A**.

K. Okada and Y. Shimura

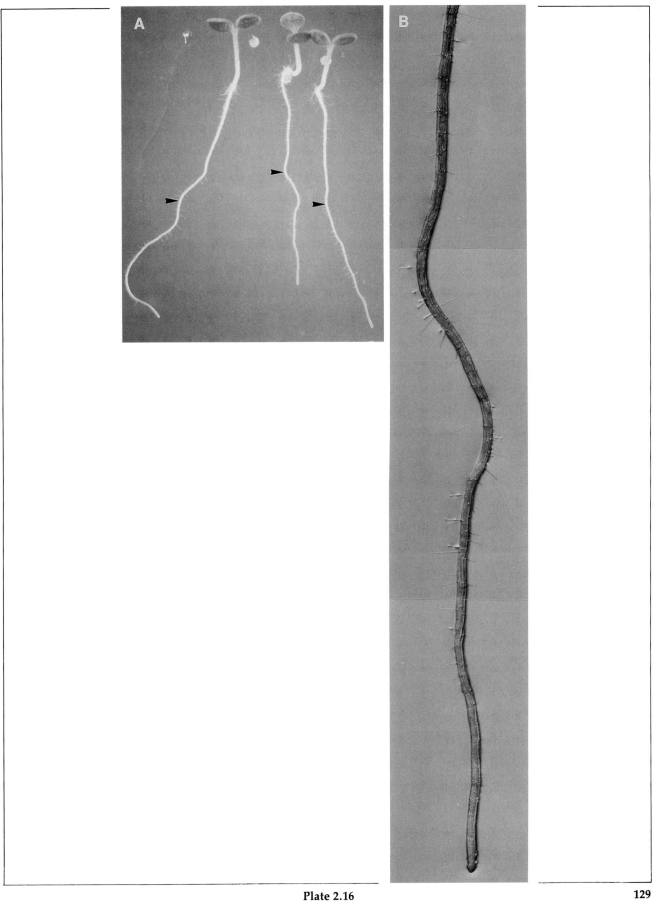

Plate 2.16

Plate 2.17
Root phototropism mutants

Roots of wild-type plants exhibit negative phototropism. The response can be assayed by incubating seedlings on agar plates which are covered by an opaque sheet with an opening at one side. Illumination is provided through the opening (Figure 1). The paths of the roots extend about 30°–45° from the vertical (see **A**; Okada and Shimura, 1992a; Okada and Shimura, 1992b). Blue light was shown to be effective in eliciting the negative phototropic response of the root as well as the positive phototropism of hypocotyls (Okada et al., unpublished).

Using this method, two mutants, *root phototropism1* (*rpt1*) and *rpt2*, were isolated (*rpt1* is allelic with *wavy growth1* (*wav1*); Okada and Shimura, 1992]. The mutants have a single recessive mutation in the nuclear genome.

(**A**) Wild-type seedlings.

(**B**) *rpt1-1* (*upper row*) and *rpt2-1* (*lower row*) mutant seedlings. These mutants' roots show no phototropic response to white or blue light, but their hypocotyls display a normal phototropic response. Both of the mutants display normal gravitropic responses in both the roots and hypocotyls. The roots of *rpt2* mutants exhibit a normal wavy growth pattern on angled plates.

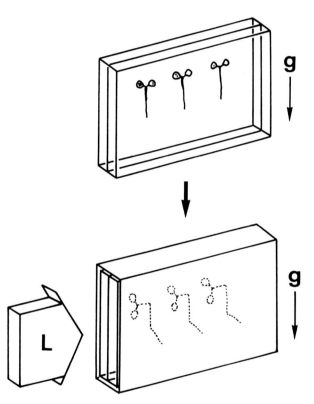

Figure 1. An illustration showing the position of agar plates for assaying the phototropic response of roots. Sterilized seeds are sown on 1.5% agar containing half-strength *Arabidopsis* mineral nutrient solution. After cold treatment for 2 days in the dark, the plates are incubated in a vertical position for 5 days. The plates are then wrapped with an opaque black plastic sheet, and illuminated with white light from one side for 2 days.

K. Okada and Y. Shimura

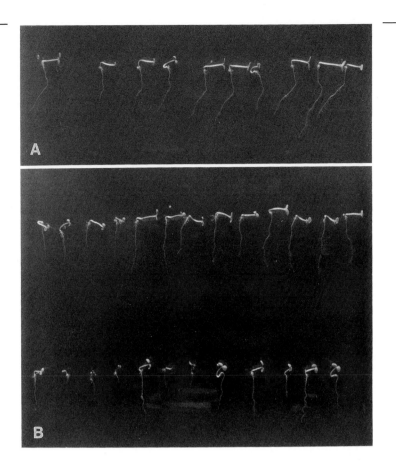

Plate 2.17

131

3
Flowers

Introduction

Arabidopsis thaliana is a rosette annual with separate vegetative and reproductive growth phases. The apical meristem that produced leaves in the vegetative phase switches to producing flowers in the reproductive phase. Thus, the vegetative meristem is apparently converted directly into an inflorescence meristem, which produces the primary inflorescence shoot. There appears to be no fundamental difference in organization between the two types of meristems and the distinction between them may merely be a slight change in shape and the difference in identity of lateral meristems they produce (leaf versus flower). Following germination, the apical meristem produces leaves with little elongation between successive leaves forming a rosette. The number of leaves formed is dependent upon genotype and growth conditions. When the plant becomes florally induced, the apical meristem switches to producing flowers. Subsequent to the production of the first few flower primordia, the plant bolts due to increased internode elongation between the uppermost leaves and between flowers. Thus, the basal positions on the primary inflorescence shoot are occupied by a small number of cauline (stem) leaves and the apical positions by a potentially indeterminate number of flowers. Secondary inflorescence meristems that reiterate the pattern of development of the primary inflorescence shoot develop in the axil of each of the cauline leaves of the primary inflorescence shoot and tertiary inflorescence shoots arise in the axils of the cauline leaves on the secondary inflorescence shoots. In addition, further inflorescence meristems develop in a basipetal manner in the axils of the rosette leaves. The distinction between rosette and cauline leaves may be rather arbitrary as there appears to be a continuum in shape from the first produced rosette leaves through the last produced cauline leaves. The leaves (rosette and cauline) and flowers arise from the flanks of the apical meristem in the same phyllotactic spiral. The inflorescence is an open raceme with the development of individual flowers proceeding acropetally (from basal to apical positions) (Müller, 1961). Flowers are perfect, bractless, and naturally self-fertilize with a low incidence of outcrossing under laboratory conditions (see e.g. Snape and Lawrence, 1971; Relichová, 1978).

The Transition to Flowering

For annual plants, there is no second chance in the decision to stop making leaves and start making flowers. Therefore, *Arabidopsis* plants must choose wisely in the timing of this decision by responding appropriately to conditions that allow maximal seed production. In light of this, it is not surprising that the transition from vegetative to reproductive growth is affected in a complex manner by both environmental (e.g., light and temperature) and internal factors (e.g., age).

The transition from vegetative growth to flowering is promoted by an increase in long-day photoperiods but this is not an absolute requirement for most genotypes, making wild-type *Arabidopsis* a facultative long-day plant (Laibach, 1951; Napp-Zinn, 1969). For example, plants given only a 4- to 5-hour day length will eventually flower (Laibach, 1951). Plants are responsive to photoperiod soon after germination (Laibach, 1951). The spectral quality of the light source is also important. Using either monochromatic light or different ratios of red to far-red light, it has been shown that either far-red light (>700 nm) or blue light generally hastens the transition to flowering, while red light delays the transition (Meijer, 1959; Brown and Klein, 1971; Goto et al., 1991b;

Eskins, 1992; Bagnall, 1992; Bagnall, 1993). Growth in reduced light intensity appears to have little effect on the transition to flowering (Laibach, 1951; Bagnall, 1992). However, growth in complete darkness seems to hasten the transition (Rédei et al., 1974).

Vernalization (transitory exposure of seeds or vegetative plants to cold temperatures) also hastens the transition to flowering. The effect of vernalization varies with plant age (at least in the winter-annual ecotype Stockholm). Response in the imbibed seed and later during vegetative growth is greater than in the plant immediately post-germination (Napp-Zinn, 1969). At least in some cases, the vernalization requirement can be circumvented by application of gibberellins or some base analogs (Bagnall, 1992; Burn et al., 1993).

Additional factors, such as growth temperature and nutrient supply, also affect the timing of the transition to flowering. Lower growth temperatures (e.g., 15°C versus 25°C) can significantly delay the transition to flowering, sometimes more than doubling the number of rosette leaves produced (Westerman and Lawrence, 1971; Araki and Komeda, 1993). Limiting nutrient supplies and high planting densities generally hasten the transition.

Different ecotypes of *Arabidopsis* vary widely in their flowering time as well as in their response to environmental conditions (Laibach, 1951; Napp-Zinn, 1969; Rédei, 1970; Lawrence, 1976; Napp-Zinn, 1985; Lee et al., 1993), presumably reflecting local adaptations. Although all ecotypes tested behave as facultative long-plants, the degree of responsiveness to long days varies between ecotypes (Laibach, 1951; Napp-Zinn, 1985). When grown under continuous light, ecotypes can be divided into four classes based on their responsiveness to vernalization (Napp-Zinn, 1969). Winter annuals exemplified by ecotypes from the northern latitudes, such as Stockholm and Söderland, flower very late and respond greatly to vernalization. Late-summer annuals respond moderately to vernalization, medium-summer annual ecotypes display a slight vernalization response, and early-summer annuals show little vernalization response. However, when grown in short days, even early-summer annuals may respond to vernalization (Martinez-Zapater and Somerville, 1990; Wilson et al., 1992). Conversely, the response to long-day photoperiods is generally greater in winter annual ecotypes.

Crosses between ecotypes showing different requirements for vernalization have indicated that lateness of ecotypes is generally dominant, and the late ecotypes usually differ from the early ecotypes by a small number of both dominant and recessive genes (Rédei, 1970; Karlovská, 1974; Napp-Zinn, 1985 and references therein; Lee et al., 1993). It is not presently known how many different loci confer late flowering phenotypes, since loci identified in crosses between different pairs of ecotypes may correspond to the same locus.

Many of the ecotypes that have been used in molecular biology and genetic experiments, such as Landsberg *erecta*, Columbia, and Wassilewskija, are early flowering genotypes that show little response to vernalization.

Structure of the Inflorescence Meristem

The structure of the dome-shaped inflorescence meristem is similar to that of the vegetative meristem, with three distinct morphological and functional zones evident: the flank meristem, the file meristem, and the central initiation zone (Vaughan, 1952; Vaughan, 1955; Bernier, 1964; Miksche and Brown, 1965; Bernier et al., 1970; Besnard-Wibaut, 1970; Besnard-Wibaut, 1977). Flower primordia are derived from the flank meristem, their initiation brought about by periclinal divisions in the L3 (Vaughan, 1955).

Flowers: Introduction

Compared to that of the vegetative meristem, the file meristem of the inflorescence meristem is very active. It exhibits numerous transverse and longitudinal divisions, reflecting the difference in internode growth generated by the two types of meristems. The file meristem gives rise to the pith of the stem. The first cells produced by the file meristem following the transition to flowering are rectangular with their long axis perpendicular to the major axis of the stem, but later cell elongation reverses this situation (Vaughan, 1955). As in the vegetative meristem, the central initiation zone is responsible for maintaining both the flank and file meristems, as these zones become depleted due to the production of flower primordia and pith, respectively.

The transition from vegetative to inflorescence meristem has been described (Vaughan, 1955; Miksche and Brown, 1965; Besnard-Wibaut, 1977) and is marked by a change in the shape of the apex, from slightly convex to dome-shaped. This change in shape has been attributed to changes in the mitotic activity of the cells within the different zones of the apical meristem (Besnard-Wibaut, 1977).

Some species of the Brassicaceae, such as *Brassica compestris* (Orr, 1978) and *Sinapis alba* (Bernier, 1962), exhibit a clear transitional stage, a loss of the cyto-histochemical zonation pattern, during the vegetative to inflorescence conversion. However, such a transitionary state is not conspicuous in *Arabidopsis thaliana* (Vaughan, 1955), *Capsella bursa-pastoris* (Vaughan, 1955), or *Alyssum maritimum* (Lance-Nougarede, 1961). In these species the zonation patterns are evident throughout the transition. Subsequent to the transition to a dome-shaped structure, flower primordia are initiated by periclinal cell divisions in the L3 of the flank meristem. This contrasts with the earlier-produced leaf primordia, which are initiated by periclinal divisions in the L2 layer of the flank meristem.

Structure of the Flower

The mature flower of *Arabidopsis thaliana* has a simple structure (see Figure 1) typical of the Brassicaceae. The flower has a calyx of four sepals and a corolla of four petals whose positions are alternate and interior to those of the sepals. The androecium consists of four medial, long stamens and two lateral, short stamens. The anthers are two-celled and dehisce longitudinally. Dehiscence of the stamens results in self-fertilization. The superior sessile gynoecium, which occupies the center of the flower, has two carpels. Its two locules are separated by a false septum with ovules arising from parietal placental tissue on each side of the septum.

Despite this seemingly simple floral plan (four sepals, four petals, six stamens, two carpels), there is much controversy surrounding the interpretation of its origin, with several interpretations postulated (e.g., Arber, 1931a; Lawrence, 1951; Endress, 1992 for review).

In one interpretation, the floral organs are in six whorls of basically two organs each (Steinheil, 1839; Alexander, 1952). The sepals comprise two whorls, with the lateral and medial sepals in separate whorls. However, there has been controversy as to whether the lateral or the medial sepals belong to the outermost whorl (Arber, 1931a). The petals comprise a single whorl, but their number has been doubled from two to four. This invokes the concept of *dédoublement*, an increase in the number of organs per whorl by duplication (for review see Endress, 1992). In other words, the presumed ancestor would have two petals, but later subdivision of each petal primordium into two parts, with each part developing into an entire organ, would result in the derived condition of four petals. The stamens occur in two whorls, with the lateral two stamens occupying the outer whorl, and the four medial stamens occupying the inner whorl. Again the medial stamens are doubled in number from two to four. The two carpels occupy the innermost whorl.

In another interpretation, the flower is composed of four whorls of organs, with each organ type simply occupying a separate whorl. In this case, the ancestral condition would have four organs per whorl in the outer three whorls, and two carpels. The six third-whorl stamens are explained by duplication (*dédoublement*) of the medial stamens (De Candolle, 1821). This interpretation has already been employed in recent molecular genetic studies (Bowman et al., 1989), since in wild-type flowers there appear to be no fundamental differences between the two pairs of sepals and the two sets of stamens, respectively.

Likewise there are multiple interpretations concerning the origin of the gynoecium of Brassicaceae flowers (for reviews see Lawrence, 1951; Okada et al., 1989; Meyerowitz et al., 1989). Briefly, one hypothesis is that the gynoecium is composed of two carpels in lateral positions of a single whorl. In another hypothesis, the gynoecium is thought to be composed of four carpels, with two sterile carpels forming the ovary walls, and two fertile carpels forming the placentae. The presumed ancestral form in the four-carpel theory is either two whorls of two carpels, or a single whorl with a four-loculed and -carpelled gynoecium. This latter interpretation (the four-carpel theory) is based mainly on the arrangement of vascular bundles and morphology of the stigma (Lawrence, 1951).

For the purposes of this book, we will use the four-floral-whorl interpretation for the reasons outlined above (Bowman et al., 1989; Smyth et al., 1990), and the two-carpel hypothesis since there seems no compelling reason to adopt a more complex view than necessary for the structure of the gynoecium (Okada et al., 1989).

A note must be made with reference to the term "whorl." Traditionally, in morphological and systematic studies, a whorl is meant to be a cycle of organs that can be delimited from preceding and subsequent cycles (Endress, 1992). However, in recent literature concerning molecular genetics studies of flower development, the term whorl is used to describe a geographic region of the flower where a single organ type develops (Bowman et al., 1989; Meyerowitz et al., 1989; Bowman et al., 1991b; Meyerowitz et al., 1991). These two definitions coincide only in the case where there is merely a single whorl of each organ type. In the descriptions of *Arabidopsis* flowers in this book, the term "whorl" is used to denote a geographic region of the flower in which a single type of floral organ arises, which in the case of the four-whorl interpretation of the *Arabidopsis* flower also corresponds to the traditional sense of the term "whorl."

Phylogenetic analysis reveals that flower architecture is exceptionally consistent throughout the genera of the Brassicaceae (Hedge, 1976; Cronquist, 1981; Endress, 1992). Within the Brassicaceae, the primary differences from the basic floral structure of *Arabidopsis* are (1) a reduction in the number of petals in a few genera; and (2) alterations in the positions and numbers of stamens in many other genera.

In general, stamen numbers may decrease from six to four or two, or the filaments of the medial stamens may be fused. In genera exhibiting a reduction in numbers, either the lateral stamens are lost or the number of medial stamens is reduced from four to two, or both reductions occur. In only one genus is there an increase in the number of stamens.

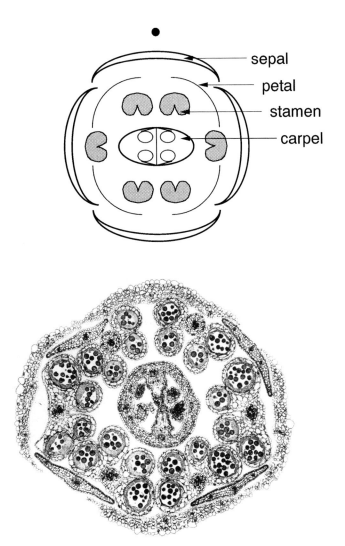

Figure 1. Section and floral diagram of a wild-type *Arabidopsis thaliana* flower. The terms medial, lateral, adaxial, and abaxial in text refer to the positions of the floral organs with respect to the inflorescence meristem whose positions are represented by the dot at the top of the floral diagram.

Development of *Arabidopsis* Flowers

The development of individual *Arabidopsis thaliana* flowers, which is similar to that of other species of the Brassicaceae such as *Cheiranthus cheiri* (Payer, 1857) and *Brassica napus* (Polowick and Sawhney, 1986), has been described from the emergence of the flower primordium through dehiscence of the silique (Vaughan, 1955; Müller, 1961; Bowman et al., 1989; Hill and Lord, 1989; Smyth et al., 1990). Based on morphological criteria, flower development has been divided into several stages (Table 1).

Flower meristems arise from the flanks of the inflorescence meristem in a phyllotactic spiral (stage 1), and soon become demarcated from the inflorescence meristem by a slight indentation (stage 2). The fifteen floral organ primordia develop in a well-defined pattern, sequentially from the floral meristem. The four sepal primordia are the first to arise (stage 3). They appear in a

Table 1. Summary of stages of flower development in *Arabidopsis thaliana* listing the landmark events which define the beginning of each stage and its approximate duration.

Stage	Landmark event at beginning of stage	Duration[a]	Age of flower at end of stage[a]
1	Flower buttress arises	24 h	1 d
2	Flower primordium forms	30 h	2.25 d
3	Sepal primordia arise	18 h	3 d
4	Sepals overlie flower meristem	18 h	3.75 d
5	Petal and stamen primordia arise	6 h	4 d
6	Sepals enclose bud	30 h	5.25 d
7	Long stamen primordia stalked at base	24 h	6.25 d
8	Locules appear in long stamens	24 h	7.25 d
9	Petal primordia stalked at base	60 h	9.75 d
10	Petals level with short stamens	12 h	10.25 d
11	Stigmatic papillae appear	30 h	11.5 d
12	Petals level with long stamens	42 h	13.25 d
13[b]	Bud opens, petals visible, anthesis	6 h	0.5 d
14	Long anthers extend above stigma, defined as zero hours after flowering (HAF)	18 h	1 d
15	Stigma extends above long anthers	24 h	2 d
16	Petals and sepals withering	12 h	2.5 d
17	All organs fall from green siliques	192 h	10.5 d
18	Siliques turn yellow	36 h	12 d
19	Valves separate from dry siliques	up to 24 h	13 d
20	Seeds fall		

[a] Estimated to nearest 6 hours.

[b] Results for stages 13 to 20 (after the flower opens) are summarized from Müller (1961) where they were named B3 to B10. Their timings are given separately because a different strain was grown under different conditions from those used in determining stages 1 to 12. For stage 1 to 12 plants of Landsberg *erecta* ecotype were grown under continuous illumination at 25°C and 70% relative humidity (Smyth et al., 1990). For stages 13 to 20 plants of Dijon ecotype were grown under 16 hours light at 24°C (54% relative humidity) and 8 hours dark at 18°C (58% relative humidity) (Müller, 1961).

stereotypic cruciform pattern, with the abaxial primordium forming slightly before the others. Subsequently, four petal and six stamen primordia are initiated almost simultaneously (stage 5). The remaining floral meristem interior to the outer three whorls comprises the gynoecial primordium. Stage 5–6 flowers consist of four first-whorl primordia that have begun to differentiate into sepals, surrounding eleven undifferentiated organ primordia. At this time, the identities of at least the inner three whorls of organ primordia have not yet been irreversibly determined, their fates being partly or completely unspecified (Bowman et al., 1989; Bowman et al., 1991b). The identity of the first-whorl primordia may be determined slightly earlier (perhaps stage 4) since they are already showing characteristics of sepal development by stage 5. The identity of the floral organ primordia is thought to be determined in accordance with their position within the flower. For some organ types (e.g., stamens), the fates adopted at this stage (about stage 6) of development appear irreversible. However, the specification of cells within other organ types (e.g., petals) appears reversible until relatively late in floral development (Bowman et al., 1989; Bowman, 1991). Following the specification of floral organ identity, each primordium follows an organ-type specific developmental program, with differentiation into both organ-specific cell types and common cell types. Visible

Flowers: Introduction

signs of differentiation occur by stage 7 in stamens and carpels, and by stage 9 in petals. Further details of developmental events taking place in stage 1–12 flowers can be found in legends accompanying plates.

The rate of flower production and development is dependent upon growth conditions and age of the inflorescence meristem. When Landsberg *erecta* plants are grown at 25°C under constant illumination, approximately 1.9 flower primordia are generated by the inflorescence meristem per day (for the first 30 or so flowers produced), and each takes about 13–14 days to progress from the time of emergence from the inflorescence meristem to the time of flower opening at anthesis (Smyth et al., 1990).

Mutants Affecting Flowering

Flowering in *Arabidopsis* can be thought of as a series of developmental steps, the first of which is the transition from vegetative to reproductive growth. Following the transition, flower meristems develop from the flanks of the inflorescence meristem, and their identity as flower meristems is specified. Cells of the flower meristems divide to produce four whorls of organ primordia, in a sequential manner, and in a well-defined pattern. Each of the floral-organ primordia then differentiates into one of the four types of floral organs, depending on its position within the flower. Likewise, within each individual floral organ several cell types develop in stereotypical positions. Mutants have been isolated in *Arabidopsis* that affect one or more of these steps (for reviews see Coen and Meyerowitz, 1991; Meyerowitz et al., 1991; Bowman and Meyerowitz, 1991; Weigel and Meyerowitz, 1993).

The Transition from Vegetative to Reproductive Growth

Several mutants have been isolated among early-flowering ecotypes that delay the time to flowering without affecting other aspects of the life cycle (Koornneef et al., 1991). The phenotypes of the late-flowering mutants are differentially affected by those environmental factors that promote the transition to flowering. It is not presently known whether alleles that confer a late-flowering phenotype in natural populations correspond to the same loci as those identified by mutagenesis of the early ecotypes. Additionally, mutants that accelerate the time to flowering in early ecotypes have been identified, many of which affect other developmental processes (Goto et al., 1991b; Shannon and Meeks-Wagner, 1991; Zagotta et al., 1992).

Production of Flower Primordia

Mutations at two loci, *PIN-FORMED* (Goto et al., 1987; Goto et al., 1991a) and *PINOID*, result in a failure to produce normal flower primordia, although the transition from vegetative to inflorescence meristem occurs. A phenocopy of this phenotype may be obtained by culturing wild-type plants in the presence of auxin polar transport inhibitors (Okada et al., 1991). Other mutants, such as *leafy*, *clavata*, *hanaba taranu*, and *spitzen*, can also have an effect on the formation of flower primordia, especially when in multiply-mutant combinations.

Specification of Flower Meristem Identity

Mutations at two loci, *LEAFY* (Schultz and Haughn, 1991; Weigel et al., 1992; Huala and Sussex, 1992) and *APETALA1* (Irish and Sussex, 1990; Mandel et al., 1992b; Bowman et al., 1993), result in partial transformations of flowers into inflorescence-like structures, although the transformations occur in a strikingly different manner. In addition, a recessive allele of the *CAULIFLOWER* locus enhances *apetala1* mutations such that the transformation of flowers into inflorescences is more complete (Bowman et al., 1993). The opposite phenotype, the transformation of inflorescence meristems into flower meristems, is observed in *terminal flower* mutants (Shannon and Meeks-Wagner, 1991; Alvarez et al., 1992).

Floral Organ Patterning and Identity

Two classes of mutation fall into this category. These are homeotic mutations which result in one type of organ developing in the position normally occupied by another type, and meristic mutations which result in a change in the number of parts or organs without exhibiting replacement of one type for another (Bateson, 1894).

Specification of Floral Organ Identity

The floral homeotic mutants thus far identified in *Arabidopsis* fall into three classes with each of the mutations primarily affecting the identity of floral organs in two adjacent whorls. Mutations in the *AGAMOUS* gene result in a double flower phenotype due to alterations in the third and fourth whorls. The third-whorl organs develop as petals, and the fourth-whorl organs develop as four sepals, interior to which are an indeterminate number of whorls of petals and sepals (Bowman et al., 1989; Bowman et al., 1991b). Mutations in the *APETALA2* gene result in alterations in the first and second whorls. In *apetala2* flowers the first-whorl organs are carpels rather than sepals, and the second-whorl organs are stamens rather than petals (Komaki et al., 1988; Bowman et al., 1989; Kunst et al., 1989b; Bowman et al., 1991b). Mutations in *APETALA1* also result in homeotic alterations in the outer two whorls, with the bract-like organs occupying the first whorl, and staminoid organs the second whorl (Irish and Sussex, 1990; Bowman et al., 1993). Mutations in either of two genes, *APETALA3* or *PISTILLATA*, cause second- and third-whorl defects. In *apetala3* and *pistillata* flowers, the second-whorl organs develop as sepals while the third-whorl organs develop as carpels (Bowman et al., 1989; Hill and Lord, 1989; Bowman et al., 1991b; Jack et al., 1992). It must be noted that in some cases the floral homeotic mutations (e.g., *apetala2* and *agamous*) may also have meristic effects since they cause alterations of organ and whorl numbers.

Most commonly, the homeotic mutations affect the identity of each of the members of a particular whorl in the same manner. For example, the six petals that occupy the third-whorl positions in *agamous* flowers are indistinguishable from each other and from the petals of the second whorl. However, there are two exceptions to this generalization. Although each of the third-whorl organs is carpelloid in weak *apetala3* and *pistillata* mutants, the lateral organs are less carpelloid than the medial organs. Likewise, the developmental fates of the lateral and medial first-whorl organs are different in both *apetala2* and *apetala1* flowers, with the medial organs more carpelloid, and the lateral organs arising lower from the pedicel and more often aborted. However, it is not clear whether these data are likely to elucidate the evolutionary interpretation of the number of whorls in *Arabidopsis* flowers (Arber, 1931b) since teratological studies have rarely provided insight into evolutionary relationships (Carlquist, 1969; Meyerowitz et al., 1989).

Specification of Organ and Whorl Number

The pattern defects exhibited by *superman* mutants can be termed meristic, since there is an increase in stamen number without alterations in floral organ identity. Although the increase in stamen number is often accompanied by a decrease in carpel number, the *SUPERMAN* gene product is not required for carpel development (Schultz et al., 1991; Bowman et al., 1992).

Mutations at three loci (*CLAVATA1*, *CLAVATA2*, *CLAVATA3*) result in an increase in floral organ number in each whorl (McKelvie, 1962; Krickhahn and Napp-Zinn, 1975; Koornneef et al., 1983; Bowman et al., 1988; Leyser and Furner, 1992). However, the phenotypic effects of the *clavata* mutants are not limited to the floral meristem, with vegetative, inflorescence, and floral meristems becoming fasciated in appropriate environmental conditions. Likewise, pleiotropic mutations at other loci (e.g., *FASCIATA1* and *FASCIATA2*) result in a loss of floral organs (Leyser and Furner, 1992).

Other mutations that affect pattern formation in the flower, such as *hanaba taranu*, *fl54* (Komaki et al., 1988), *fl82* (Komaki et al., 1988), *spitzen*, *unusual floral organs*, and *antherless* exhibit meristic alterations, usually loss of organs, but display other developmental abnormalities as well. Another class that can be included here is the group of mutants in which the first few flowers are relatively normal, but more apical ones consist solely of carpelloid organs, with the inflorescence meristem often terminating in a mass of carpelloid tissue (Röbbelen, 1965a).

Differentiation of Floral Organs

Several recessive mutants have been isolated in which morphological development of the gynoecium (Komaki et al., 1988; Okada et al., 1989) or the stamens (Chaudhury et al., 1992) is disrupted. Both the stigma and septum are altered in *fl89* flowers, with the septum failing to fuse completely, and the stigma developing in two distinct regions separated by horn-shaped projections (Komaki et al., 1988). The septum also fails to fuse in the thin ovaries of *fl65* flowers (Okada et al., 1989). Failure of the septum to develop normally in *spatula* flowers is associated with a flattened gynoecium, and in *crabs-claw* flowers, the two carpels of the gynoecium do not fuse normally at their apex. In *antherless* flowers, anthers fail to develop, although filament-like structures may be present (Chaudhury et al., 1992).

Models of Flower Development

In this century two basic classes of models concerning how the identities of the organs of the flower are specified have been proposed (see Meyerowitz et al., 1989). One class of model involves communication between adjacent whorls of developing floral organs, leading to sequential specification of organ identity. For example, developing sepals would send a signal, either biochemical or biophysical, to the next interior whorl of organ primordia and instruct them to become petals, and so on. However, it is clear from the phenotypes of mutants characterized in *Arabidopsis*, as well as from teratological evidence accumulated over the past several centuries, that proper differentiation of any particular whorl cannot be dependent on the correct differentiation of neighboring whorls, either inner or outer.

A class of model that is better supported by the data is one in which the flower primordium is divided into concentric embryonic fields which contain positional information (Bowman et al., 1989; Bowman et al., 1991b). Specifically, for *Arabidopsis*, it has been proposed that the positional information is supplied by the products of at least five floral homeotic genes (*APETALA1*, *APETALA2*, *APETALA3*, *PISTILLATA*, and *AGAMOUS*). The products of the floral homeotic genes have been proposed to function, alone and in combination, in three overlapping fields (each comprised of two adjacent whorls) to specify the identities of the four whorls of floral organs in *Arabidopsis* flowers (Bowman et al., 1989; Bowman et al., 1991b; Bowman et al., 1993). How the regional positional information encoded by the homeotic gene products is translated into organ-type specific cellular differentiation is at present a mystery.

The specification of flower meristem identity also seems to involve positional information and the combinatorial action of at least four genes (*LEAFY*, *APETALA1*, *CAULIFLOWER*, and *APETALA2*). Mutations in these genes result in partial transformations of flowers into inflorescence shoots, and double-mutant combinations between them result in more complete transformations (Irish and Sussex, 1990; Schultz and Haughn, 1991; Weigel et al., 1992; Huala and Sussex, 1992; Mandel et al., 1992b; Bowman et al., 1993). Thus, these genes can be considered homeotic with respect to the identity of floral meristems.

Molecular Cloning of Floral Meristem Identity and Floral-Organ Identity Genes

Several of the floral homeotic genes (*APETALA1*, *AGAMOUS*, *APETALA3*, *PISTILLATA*) belong to a single family that has in common a functional motif called the MADS box (Yanofsky et al., 1990; Jack et al., 1992; Mandel et al., 1992b; Goto and Meyerowitz, personal communication; for reviews see Schwarz-Sommer et al., 1990; Coen and Meyerowitz, 1991). The MADS-box amino acid domain is likely to be involved in DNA binding, as it shares a high level of homology to regions of the vertebrate serum response factor genes (Norman et al., 1988) and to the yeast MCM1 gene (Passmore et al., 1988), both of which have been shown to be DNA-binding transcriptional regulators (Norman et al., 1988; Schröter et al., 1990; Christ and Tye, 1991). Each of the identified MADS-box genes of *Arabidopsis* also contains a second conserved domain, called the K box (Ma et al., 1991), with structural similarity to the coiled-coil domain of keratins (Steinert and Roop, 1988). In contrast, the protein encoded by the *LEAFY* gene is characterized by a proline-rich domain and an acidic domain (Weigel et al., 1992).

The patterns of expression of the cloned floral meristem and floral homeotic genes largely correlate with their respective mutant phenotypes. For example, both *LEAFY* and *APETALA1* are expressed throughout flower meristems prior to the formation of first-whorl organ primordia (Weigel et al., 1992; Mandel et al., 1992b), which is consistent with their proposed role in the specification of floral meristem identity. Similarly, *AGAMOUS* and *APETALA3* expression is restricted to those whorls affected by their respective mutants (Yanofsky et al., 1990; Drews et al., 1991; Jack et al., 1992). Further molecular support for the combinatorial field model of specification of floral organ identity has come from the demonstration that *APETALA2* is a negative regulator of *AGAMOUS* mRNA accumulation in the outer two whorls of the flower (Drews et al., 1991). In contrast, *APETALA2* expression is not limited to those floral whorls which are primarily affected in *apetala2* mutants, indicating that post-transcriptional control may be involved in effecting *APETALA2*'s role in flower development (Okamuro et al., 1993). Although it has been shown that these genes are expressed in all three germ layers (L1, L2, and L3) of the floral meristem, it is unknown if their activity is required in all layers, or whether there are inductive interactions between the layers (Meyerowitz et al., 1991).

J.L. Bowman

Plate 3.1
Inflorescence
development

Immature *Arabidopsis thaliana* inflorescences at the commencement of flowering, and mature inflorescences.

(**A**) Vertical view of the primary inflorescence apex of a 14-day-old plant showing that the oldest bud has advanced to stage 3 (3).

(**B**) Vertical view of the primary inflorescence apex of a 24-day-old plant in which the oldest flower (*top*) has reached stage 14. The sepals and petals of the stage 14 flower have spread during preparation.

(**C**) Higher magnification of **B** showing inflorescence meristem and floral buds of earlier stages.

(**D**) Lateral view of an 18-day-old plant just beginning to bolt. The two secondary inflorescence apices [visible in the axils of the two cauline leaves (cl) of the primary inflorescence shoot, which have been removed] have advanced in concert, with buds reaching stage 5 on each. The cauline leaves on the secondary inflorescence shoots show developing trichomes and stipules (*arrowheads*). Another apex is just visible in the axil of one of the dissected rosette leaves (*arrow, lower left*) although flower development is not yet detectable under its cauline leaf primordia.

(**E**) A 16-day-old plant with four buds beyond stage 5 (sepals closed) on the primary inflorescence apex. Two cauline leaves (cl) have been removed from the main stem although their stipules remain (*arrowheads*). Secondary apices within these cauline leaves have also initiated flower development. These apices are flanked in turn by younger cauline leaves, and their oldest flower primordium (*arrow*) is only at stage 2.

Bars = 20 μm in **A** and **C**, 100 μm in **D** and **E**, 500 μm in **B**.

J.L. Bowman and D.R. Smyth

A, **D**, and **E** reproduced from Smyth et al. (1990) with permission from American Society of Plant Physiology.

Flowers: The Inflorescence

Plate 3.1

Plate 3.2
The inflorescence apical meristem

(**A**) Vertical view of an apical inflorescence meristem (im). The oldest flower in the photo is a stage 4 flower (4) in which the medial sepal primordia (ad, adaxial; ab, abaxial) have begun to overlie the remaining floral meristem. The two lateral sepal primordia (l) are much smaller than the medial sepal primordia at this stage. Three stage 2 flowers (2) as well as a late stage 1 flower (1) and an early stage 1 flower (e) are also visible. The size of the inflorescence meristem varies with ecotype and growth conditions. For example, in Landsberg *erecta* plants the diameter of the inflorescence meristem varies from approximately 40 to over 70 μm in diameter, depending on growth conditions.

The structure of the inflorescence meristem of *Arabidopsis* is typical of that of the Brassicaceae, with distinct morphological and functional zonation, as shown in Figure 1 (p. 151); (Vaughan, 1952; Vaughan, 1955; Bernier, 1964; Miksche and Brown, 1965; Bernier et al., 1970; Besnard-Wibaut, 1970; Besnard-Wibaut, 1977; Orr, 1978). In terms of the tunica-corpus concept (Schmidt, 1924; Satina et al., 1940), the inflorescence meristem of *A. thaliana* has a tunica of two layers of cells (L1 and L2), and a corpus represented by L3 cells (Vaughan, 1955). Cell divisions in the two surface layers (L1 and L2) are nearly exclusively anticlinal (Vaughan, 1955). The L1 cells give rise to the epidermal cells, whereas the L2 and L3 cells give rise to the internal portions of the plant, with the gametes presumably forming from the L2, as is the case in other dicotyledonous plants (Satina and Blakeslee, 1941).

Within the inflorescence meristem, three distinct zones characterized by cell size, cell division rates, and density of histological staining are evident (Vaughan, 1955). In the center of the dome-shaped apex is a bowl-shaped region referred to as the central initiation zone which is characterized by a relatively low cell division rate. Divisions in the central initiation zone contribute to the maintenance of the other two meristematic zones. Surrounding the central initiation zone is the doughnut-shaped flank, or peripheral, meristem which is characterized by a relatively high cell division rate. The cells of the flank meristem, like those of the central initiation zone, exhibit no obvious vacuolation. It is in the flank meristem that flower primordia are initiated by a number of periclinal divisions of L3 cells. The overlying L1 and L2 layers also undergo repeated anticlinal divisions at the point of flower primordia initiation. The flank meristem also gives rise to the procambium of the flower buds as well as the cortex. Underlying the central initiation zone lies the file, or rib, meristem whose lightly-staining cells divide almost exclusively transversely. Cells produced by the file meristem contribute to the pith of the inflorescence stem, and soon become vacuolated.

(**B**) Side view of an apical inflorescence meristem showing the inflorescence meristem, three stage 2 (2), and two stage 1 (1) flower primordia. The abaxial (ab), adaxial (ad), and lateral (l) sepal primordia on the stage 3 flower (3) are also indicated. By late stage 1, growth of flower primordia causes them rise above the level of the inflorescence meristem.

(**C**) Vertical view of an apical inflorescence meristem (im) in which the spiral phyllotaxy is counter-clockwise as opposed to the clockwise spiral phyllotaxy exhibited by the inflorescence meristems in **A** and **B**. The direction of spiralling for any particular plant is random. The direction of spiral for the apical inflorescence meristem is the same as for the rosette leaves, but the direction of spiralling for secondary and tertiary inflorescences of the same plant is not necessarily the same as for the apical meristem (Smyth et al., 1990). The stage of each flower primordium is indicated.

Bars = 10 μm

(*Text continued on p. 151*)

Flowers: The Inflorescence Apical Meristem

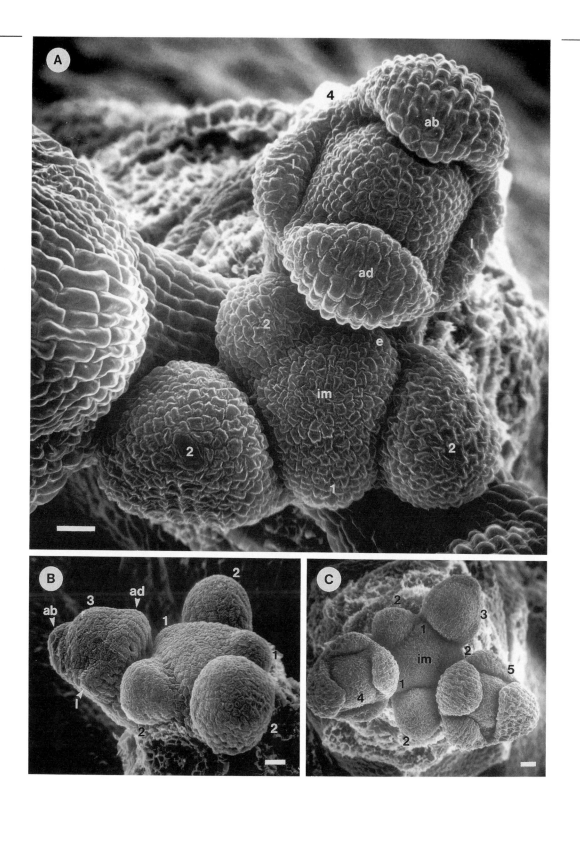

Plate 3.2

Plate 3.2
The inflorescence apical
meristem
(*continued*)

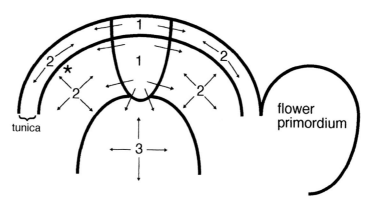

Figure 1. Diagram depicting the structure of the inflorescence meristem. Three distinct zones are indicated: the central initiation zone (1), the flank meristem zone (2), and the file meristem zone (3). The tunica (L1 and L2) contains portions of both the central initiation zone and the flank meristem. The remainders of the central initiation zone and the flank meristem, as well as all of the file meristem, comprise the corpus (L3). *Arrows* denote the main cell division patterns and growth centers. Flower primordia are initiated by periclinal divisions in the L3 of the flank meristem (*). A stage 2 flower primordium is also shown. Adapted from Vaughan (1955).

J.L. Bowman and D.R. Smyth

B and **C** reproduced from Smyth et al. (1990) with permission from American Society of Plant Physiology.

Plate 3.3
Wild-type flower
development: stages 1–5

(A) Young secondary inflorescence (im, inflorescence meristem) with the oldest floral bud at stage two (2). The subtending cauline leaf has been removed but its stipules remain (*arrows*). The tertiary inflorescence meristem (i) is subtended by a developing cauline leaf (cl) and its overall morphology resembles a stage 2 flower.

(B) Side view of inflorescence in **A**. Early-(e), mid-(m), and late-(l) stage 2 flowers are indicated. Stage 2 flowers are defined as flower primordia that are distinct (i.e., an indentation separates the flower meristem from the inflorescence meristem) from the inflorescence meristem (Smyth et al., 1990). A stage 1 flower (1), defined as a flower buttress that has not yet become distinct from the inflorescence meristem, is also indicated.

(C) Stage 1 flower. Close-up of inflorescence in **A** showing a stage 1 (1) and an early stage 2 (2) flower. Flower primordia are initiated by periclinal divisions in the L3 cells of the flanks of the inflorescence meristem (Vaughan, 1955).

(D) Stage 2 flower. Close-up of mid-stage 2 flower (2) in **B**. The flower primordium is separated from the inflorescence meristem (im) by a groove. By late stage 2 the pedicel has become distinct from the floral meristem. A tertiary inflorescence meristem (i) is visible in the back.

(E) Stage 3 flower. The initiation of the sepals, due to periclinal divisions in the L2 layer on the flank of the floral meristem (fm) (Vaughan, 1955; Hill and Lord, 1989), marks the beginning of stage 3. The abaxial sepal (ab) is the first to arise, followed by the nearly simultaneous initiation of the adaxial (ad) and lateral (l) sepals. The position of initiation of the lateral sepals appears to be slightly lower on the pedicel than that of the medial sepals. The cells of the pedicel (pd) become highly vacuolate by stage 3.

(F) Stage 5 flower. Early sepal growth is characterized by periclinal and oblique cell divisions in internal cell layers (Hill and Lord, 1989), and vacuolation in the abaxial cells (Vaughan, 1955). This results in a strong curvature of the sepals over the floral apex such that the sepals entirely enclose the bud before the gynoecial primordium is morphologically initiated. Petal (p) and medial stamen (*arrowheads*) primordia are visible inside the sepals. A late stage 2 flower is also visible to the lower left of the stage 5 flower.

Bar = 10 μm in **C** and **D**, 20 μm in **A**, **B**, **E**, and **F**.

J.L. Bowman

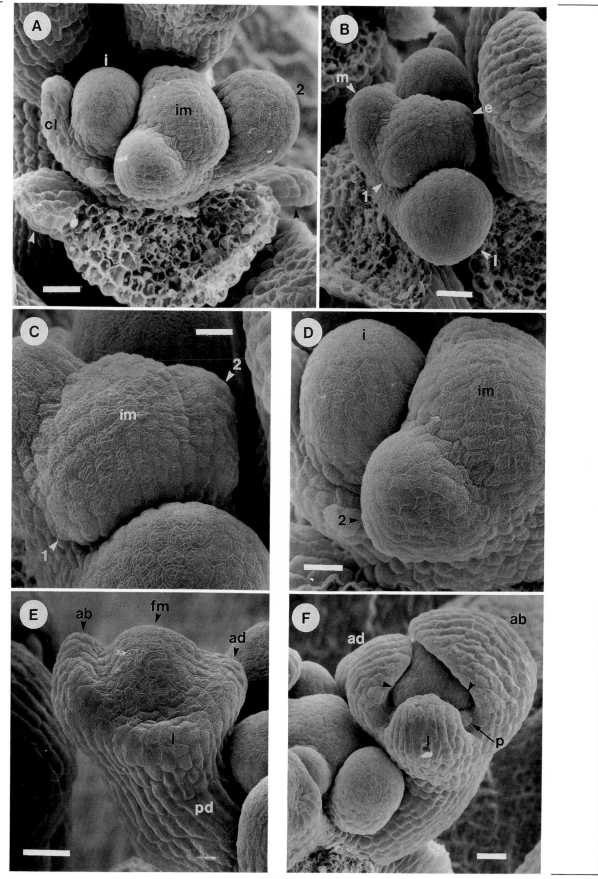

Plate 3.3

153

Plate 3.4
Wild-type flower development: sepal, petal, and stamen initiation

(**A**) Radial section of abaxial sepal initiation by periclinal divisions in L2 (*arrows*) (fa, floral apex; ia, inflorescence apex).

(**B**) Tangential section of wild-type flower showing the early development of lateral sepals (la) (fa, floral apex).

(**C**) Radial section of wild-type flower at an early stage of abaxial (ab) and adaxial (ad) sepal development (fa, floral apex; ia, inflorescence apex).

(**D**) Radial section of wild-type flower just before petal and stamen initiation. The abaxial (ab) and adaxial (ad) sepals nearly cover the floral apex (fa).

(**E**) Petal (p) initiation in a wild-type flower. The *arrow* (*) indicates a recent periclinal division of an elongate L2 cell; the L2 cells on either side of this cell have not yet divided (la, lateral sepal).

(**F**) Wild-type lateral stamen initiation in front of a lateral sepal (la) by periclinal cell divisions in the L2 (*). The section passes obliquely through a medial stamen (ms) already initiated on the other side of the floral meristem (fa, floral apex).

J.P. Hill and E.M. Lord

Reproduced from Hill and Lord (1989) with permission from the National Research Council of Canada.

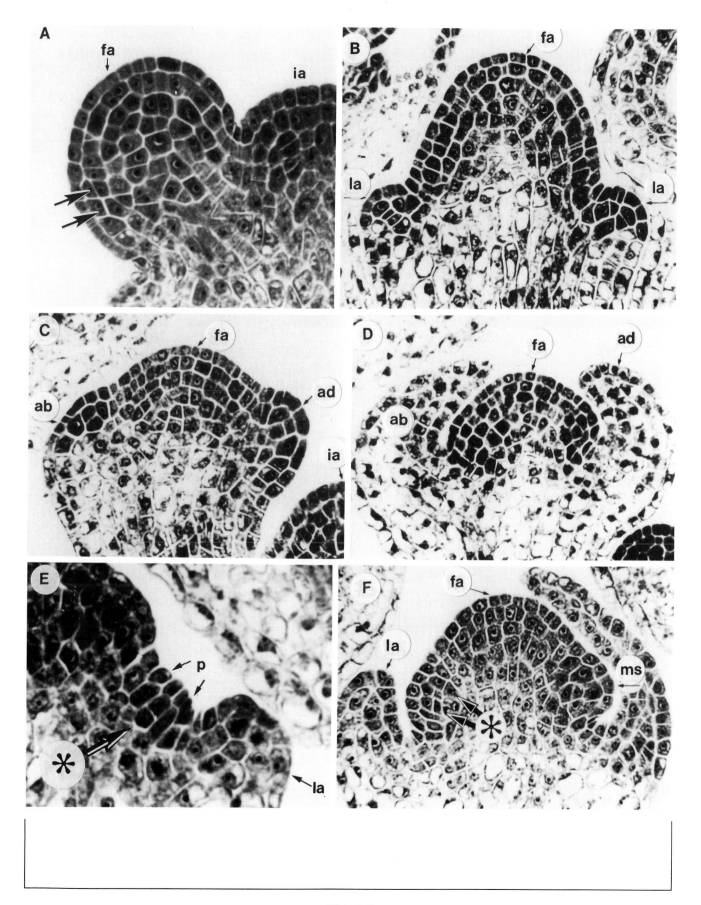

Plate 3.4

155

Plate 3.5
Wild-type flower
development: stages 5–8

(**A**) Stage 5 flower. A bud at stage 5 in which the abaxial and lateral sepals have been removed. The tip of the abaxial sepal usually overlies that of the adaxial sepal (ad), while the lateral sepals meet or overlap underneath the two medial sepals. The medial sepals nearly cover the bud and the four petal and six stamen primordia have differentiated morphologically from the floral meristem. Petals are initiated with a characteristic anticlinal elongation of L2 (and sometimes L3) cells and their subsequent periclinal divisions (Hill and Lord, 1989). Likewise, stamens are initiated by periclinal divisions of L2 (and sometimes L3) cells (Hill and Lord, 1989). One of the small petal primordia (p) is indicated. At this stage, the primordia of the medial stamens (ms) are larger than that of the lateral stamen (ls). The lateral stamen primordia arise slightly later during stage 5 than do the medial stamen primordia, which arise nearly simultaneously with the petal primordia (Vaughan, 1955; Bowman et al., 1989; Hill and Lord, 1989; Smyth et al., 1990). The lateral stamen primordia occupy a position on the receptacle slightly lower than that of the medial stamen primordia but above the level of the petal primordia.

(**B**) Stage 6 flower. Lateral view of a bud at stage 6 in which the sepals had fully enclosed the bud. The stamen primordia (a lateral stamen, ls, is indicated; the other four visible are medial stamen primordia) are now dome-shaped, while the petal primordia (p) are still relatively small. The gynoecium (g) will arise from the remaining floral meristem, a central dome of cells.

(**C**) Stage 7 flower. Medial view of a bud at stage 7 showing that the long stamen primordia are now constricted towards their base (*arrowhead*). The petal primordia (p) have become dome shaped.

(**D**) Stage 7 flower. Vertical view of a stage 7 bud showing that the stamens do not yet show locule ridges on their adaxial surface. Periclinal divisions in the underlying L3 cells initiate the development of the gynoecium (g) which grows vertically as a slotted tube (Hill and Lord, 1989).

(**E**) Stage 8 flower. A bud at stage 8 in which the stamen primordia have increased markedly in size, especially in relation to the petal primordia. The anther (a) and filament (f) regions of the stamens have already differentiated, with the anther portion comprising most of the height of the organ at this stage.

(**F**) Stage 8 flower. Vertical view of a stage 8 bud in which locules (*arrowheads*) are now clearly visible in the stamens. By this stage, pockets of sporogenous tissue are developing in each pollen sac.

Bars = 10 μm.

J.L. Bowman and D.R. Smyth

Reproduced from Smyth et al. (1990) with permission from American Society of Plant Physiology.

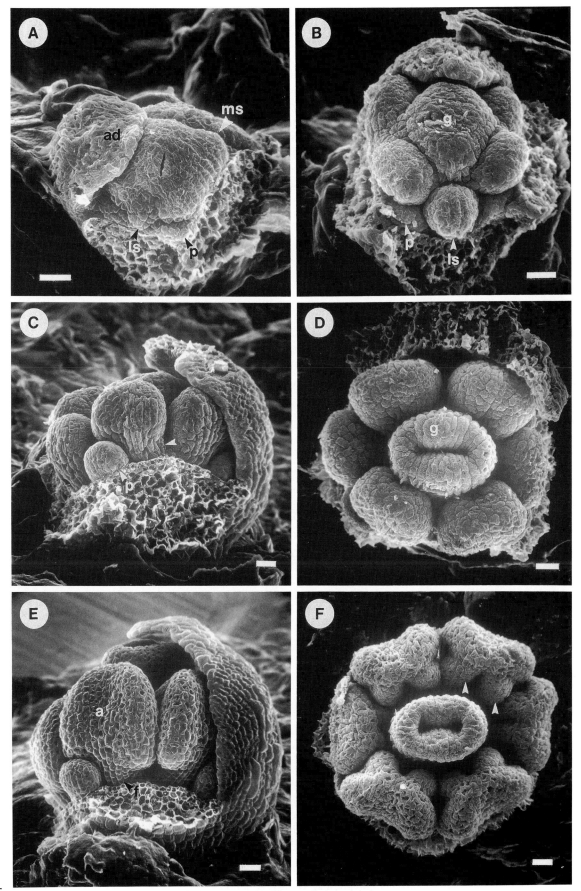

Plate 3.5

Plate 3.6
Wild-type flower development: gynoecial initiation

(**A**) Initiation of a wild-type gynoecium by cell division in the L3 (∗) (ms, medial stamen; ad, adaxial sepal).

(**B**) Early development of the wild-type gynoecium (g). *Arrows* indicate the two principal regions of apical growth visible in the radial plane.

(**C**) Oblique section of a wild-type flower showing medial stamen (ms) primordia, petal (p) primordia, and the gynoecial (g) primordium.

(**D**) Wild-type gynoecium (g) prior to ovule initiation. The pattern of anticlinal cell divisions in L1 and L2 gives the cells lining the central invagination a very uniform appearance (*arrowheads*) (ms, medial stamen).

(**E**) and (**F**) Transverse sections through a wild-type ovary. Periclinal divisions in the subepidermal cells lining the central invagination (*arrowheads*) mark the initiation of ovules (ov) and the development of the central septum (*arrows*).

J.L. Hill and D.R. Lord

Reproduced from Hill and Lord (1989) with permission from the National Research Council of Canada.

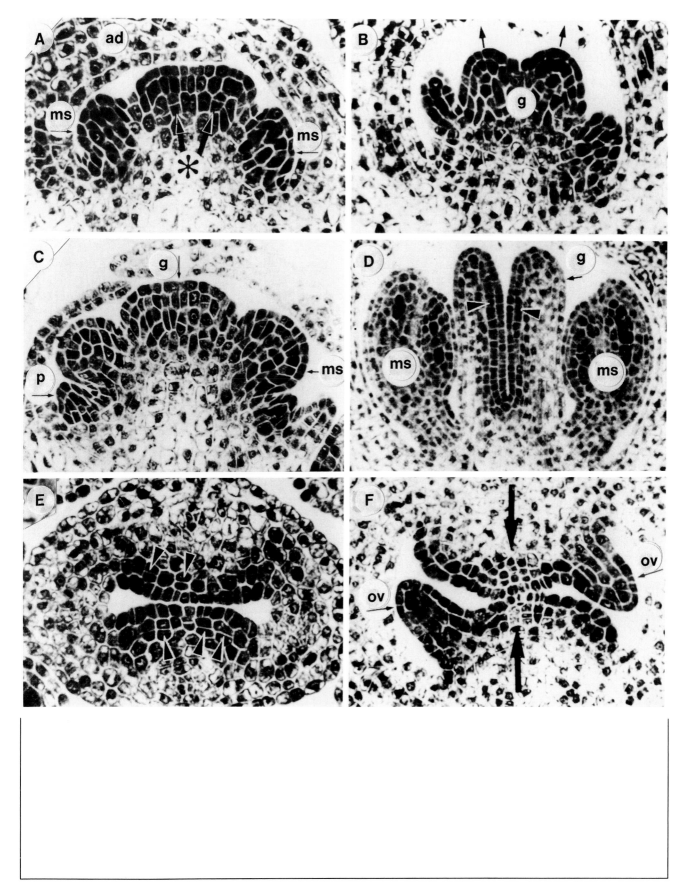

Plate 3.6

Plate 3.7
Wild-type flower
development: stages 9–12

(A) Stage 9 flower. A bud early in stage 9 showing that the petal (p) primordia have become stalked at their bases and form flattened blades as they start growing rapidly. During this stage, the microspore mother cells in the anthers become separated from each other and the tapetum, and undergo meiosis to form tetrads (isobilateral and tetrahedral) of microspores. The gynoecial cylinder becomes constricted at the top and a total of 40–60 ovule primordia arise in four rows from the parietal placental tissue.

(B) Stage 10 flower. A bud in which the petals (p) have just reached the height of the lateral (short) stamens (ls), marking the beginning of stage 10. Buds more than double in size during the long stage 9. In the anthers, the microspores separate from each other to lie freely in the pollen sac. The filaments of the stamens begin to markedly elongate. The gynoecial cylinder becomes closed at the top, and the ovule primordia elongate. By this stage, the characteristic epidermal cells of each of the different floral organs are readily distinguished.

(C) Stage 11 flower. Stigmatic papillae (*arrowhead*) appear on the top of the gynoecium at the start of stage 11. The style of the gynoecium begins to become morphologically distinct from the ovary. Within the ovary, inner and outer integuments are initiated on ovule primordia, and megasporogenesis occurs. Within the anthers, the tapetum of the pollen sac degenerates.

(D) Stage 12 flower. Stage 12 is the final stage before the bud opens. It commences when the petals reach the height of the medial (long) stamens. In the anthers, the endothecium of the pollen sac develops fibrous wall thickenings, and later, the tapetum of the pollen sac is gone and the epidermis of the pollen

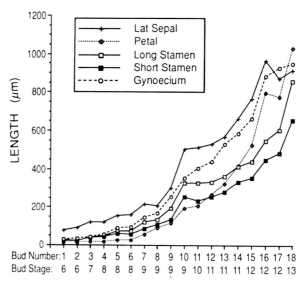

Figure 1. Wild-type flower development: relative sizes and growth rates of floral organs during flower development. The sizes (apical-basal length) of floral organs throughout their development are shown graphically. The data are from all buds of stage six or older on the main inflorescence apex of a single 22-day-old Landsberg *erecta* plant (grown under continuous illumination at 25°C and 70% relative humidity). The youngest bud (bud 1) scored was at stage 6 and the oldest bud (bud 18) had reached stage 13. The stage of each of the other buds is indicated. The lengths of individual floral organs were recorded as the linear distance from the base to the tip of the organ in each case.

Flowers: The Wild-type Flower

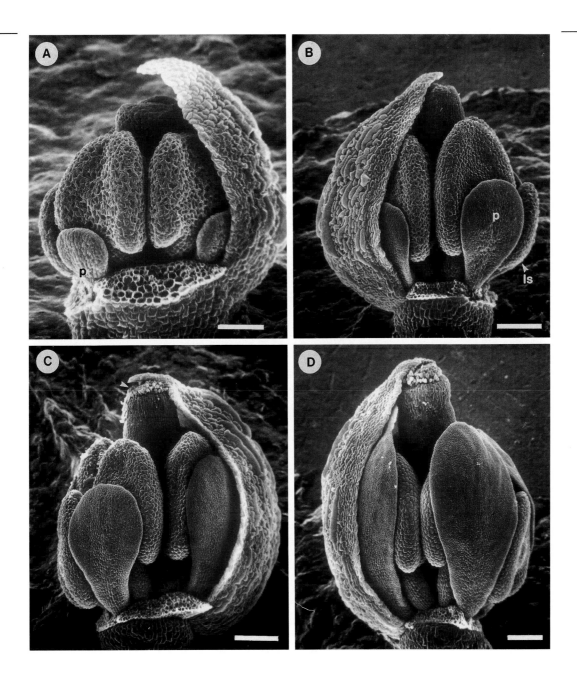

sac shows signs of withering. During stage 12, the integuments of the developing ovule grow to almost cover the nucellus, and megagametogenesis occurs. Ovules begin to show anatropous orientation.

Bar = 50 μm in **A** and 100 μm in **B–D**.

J.L. Bowman and D.R. Smyth

Reproduced from Smyth et al. (1990) with permission from American Society of Plant Physiology.

Plate 3.7

161

Plate 3.8
Wild-type flowers:
epidermal surface
morphology of mature
sepals and petals

Each of the four types of floral organs (sepals, petals, stamens, and carpels) of *Arabidopsis thaliana* has a characteristic epidermal surface morphology in terms of cell size, shape, and texture. This section provides a survey of the various cell types present in the *Arabidopsis* flower.

(A) Adaxial surface of mature petal blade. Petals appear white in daylight.

(B) Higher magnification of the ridged cells which cover the adaxial surface of the blade of mature petals. The cuticular thickenings of cells on the adaxial surface are organized along the conical axis of the cells.

(C) Abaxial surface of mature petal blade. The cells are rounded and cobblestone-like in appearance with prominent cuticular thickenings. Both surfaces of petals lack stomata.

(D) Adaxial surface of a mature petal showing the smooth transition from the elongate cells of the claw (*right*) to the more ovoid and cobblestone-like cells of the blade. The cells at the base of the petal (not visible, to right of photo) resemble those of stamen filaments.

(E) Outer surface of a mature sepal (*left*) showing several long epidermal cells (*arrow*), stomata, and a fringe of smaller cells. The central region appears green while the region covered by the smaller fringe cells (*arrowhead*) appears pale green to white.

(F) Close-up of some elongate cells of the abaxial epidermis of a mature sepal. In some cases these cells may be up to 300 μm in length. Probable progenitors of these cells can be first recognized at stage 8 when the lateral sepals are only about 150 μm tall.

Bar = 5 μm in **B**, 10 μm in **A** and **C**, 20 μm in **D** and **F**, 100 μm in **E**.

J.L. Bowman and D.R. Smyth

E reproduced from Smyth et al. (1990) with permission from American Society of Plant Physiology.

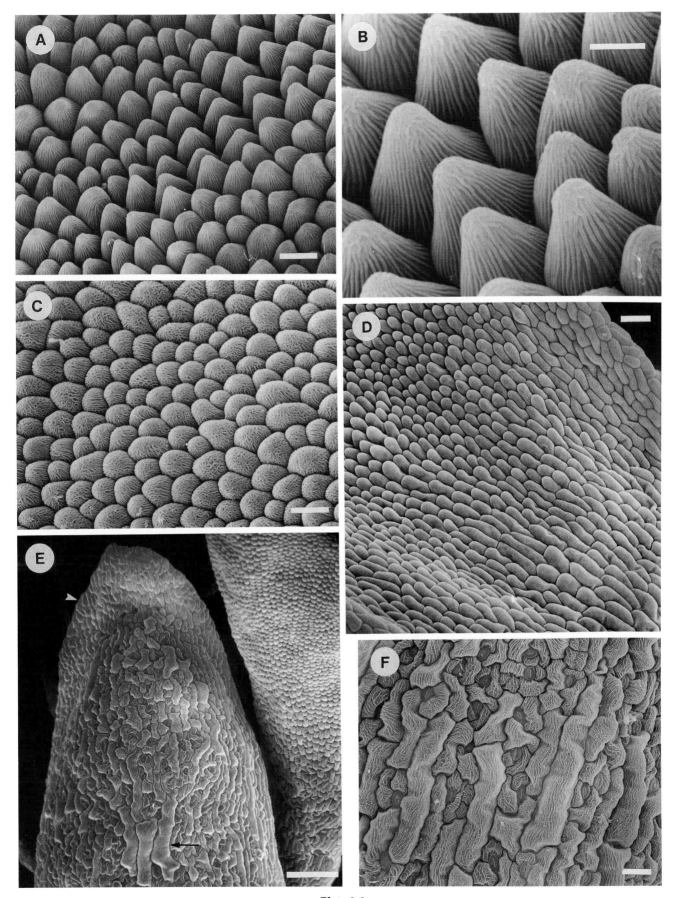

Plate 3.8

Plate 3.9

Wild-type flowers: epidermal surface morphology of stamen and petal cells

(**A**) Anther surface. The interdigitated epidermal cells of the anther have a rugose surface.

(**B**) Petal surface. On the outer surface of the petal, each epidermal cell forms a cone that has epicuticular ridges running from the base to the apex.

S. Craig and A. Chaudhury

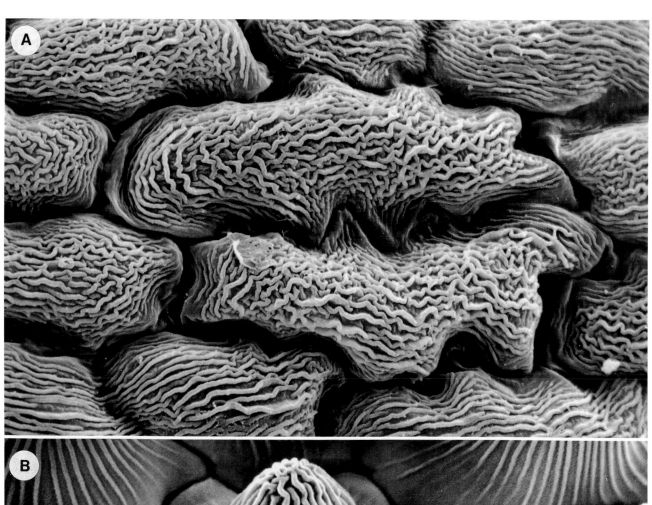

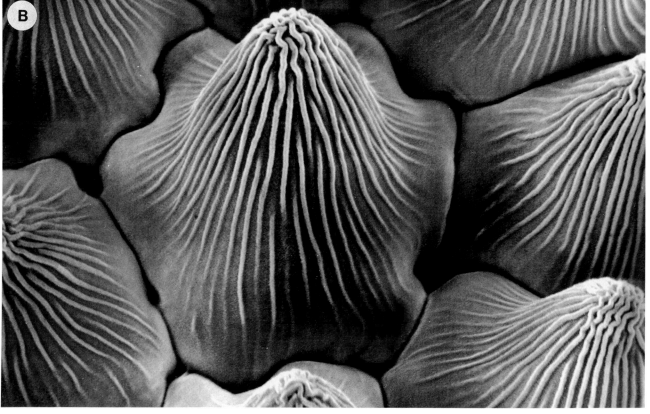

Plate 3.9

Plate 3.10
Wild-type flowers:
epidermal surface
morphology of mature
stamens

(A) Mature stamens after dehiscence, showing epidermal cells of uniform size, and pollen grains.

(B) High magnification of side of anther showing irregularly shaped, uniformly-sized cells. Pollen grains are visible (*upper right*).

(C) Close-up of anther wall after dehiscence, showing pollen grains and degenerated cells lining the pollen sac.

(D) Abaxial view of stamen showing the filament–anther boundary. Stomata (*arrowhead*) are present in the central region of the abaxial surface of the anther.

(E) Adaxial view of anther immediately after dehiscence.

(F) Epidermal cells at the base of the petal (p) are similar in structure to those of a medial stamen filament (f). Nectaries (n) are visible below the stamen filament.

Bar = 20 μm in **B**, **C**, **F**; 50 μm in **A**, **D**, **E**.

J.L. Bowman and D.R. Smyth

A reproduced from Smyth et al. (1990) with permission from American Society of Plant Physiology.

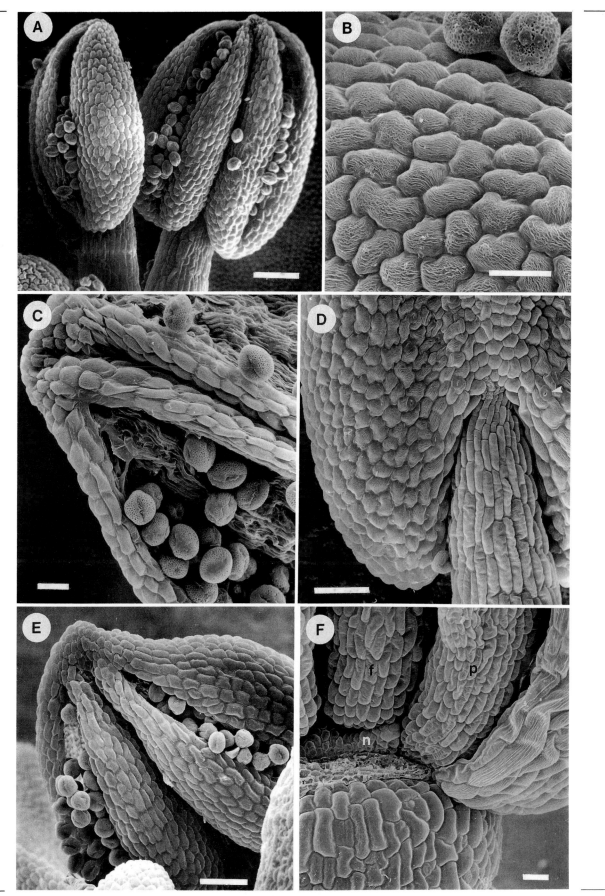

Plate 3.10

Plate 3.11

Wild-type flowers: stigma development and epidermal surfaces of pedicels and inflorescence stems

(**A–D**) Structure of the developing stigma.

(**A**) The surface of the gynoecium is smooth and slotted at the end of stage 9.

(**B**) A cap becomes apparent at stage 10 when the surface is still not closed.

(**C**) The appearance of the densely packed papillae covering the stigmatic surface defines the start of stage 11. The gynoecium closes over and the style becomes morphologically distinct from the ovary.

(**D**) Stigmatic papillae at stage 13. Numerous pollen grains are visible on the stigmatic papillae. The stigmatic papillae are 20–35 μm long when the bud opens at stage 12, and reach a final length of 50–75 μm.

(**E**) and (**F**) Pedicel and inflorescence stem epidermis.

(**E**) The epidermis of the pedicel post-pollination is characterized by long cells in files with interspersed stomata.

(**F**) The epidermis of the inflorescence stem (from a region that has already undergone elongation) is characterized by very long cells in files with interspersed stomata.

Bar = 10 μm in **A**, **B**, **C**; 20 μm in **E**, **F**; 50 μm in **D**.

J.L. Bowman and D.R. Smyth

A, **B**, and **C** reproduced from Smyth et al. (1990) with permission from American Society of Plant Physiology.

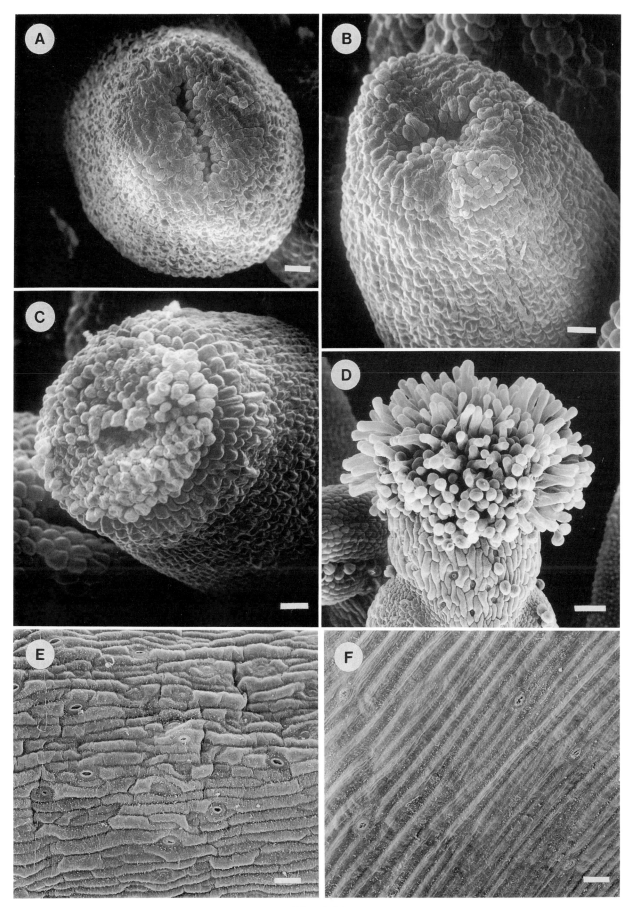

Plate 3.11

Plate 3.12
Wild-type flowers:
epidermal surface
morphology of mature
carpels

(A) The upper parts of a gynoecium in a stage 14 flower showing stigmatic papillae (sg), large cells and stomata on the short style (st), and smaller epidermal cells of the two valves of the ovary (o). Pollen grains are also visible on the stigma.

(B) Close-up of outer epidermal surface of ovary walls of gynoecium in A. At this stage of development the surface is made up of relatively vertical files of epidermal cells. The epidermis of the margin of fusion between the two carpels consists of regular cell files (f).

(C) Higher magnification of B. At this stage of development it appears that stomata are just beginning to develop. There is little visible evidence of stomata at earlier stages prior to fertilization, but stomata are conspicuous on the surface of developing siliques (see F). Epicuticular wax appears as specks on the epidermal cells.

(D) Close-up of the cells of the style of a stage 12 flower. In contrast to the epidermal cells of the ovary walls, the epidermis of the style has prominent cellular cuticular thickenings as well as interspersed stomata.

(E) Surface cells of the septum post-fertilization.

(F) The epidermis of a developing silique from a stage 17 flower consists of elongate cells in files with interspersed stomata. Compare with B (*arrowhead*).

Bar = 10 μm in C, D; 20 μm in B, E; 50 μm in F; 100 μm in A.

J.L. Bowman

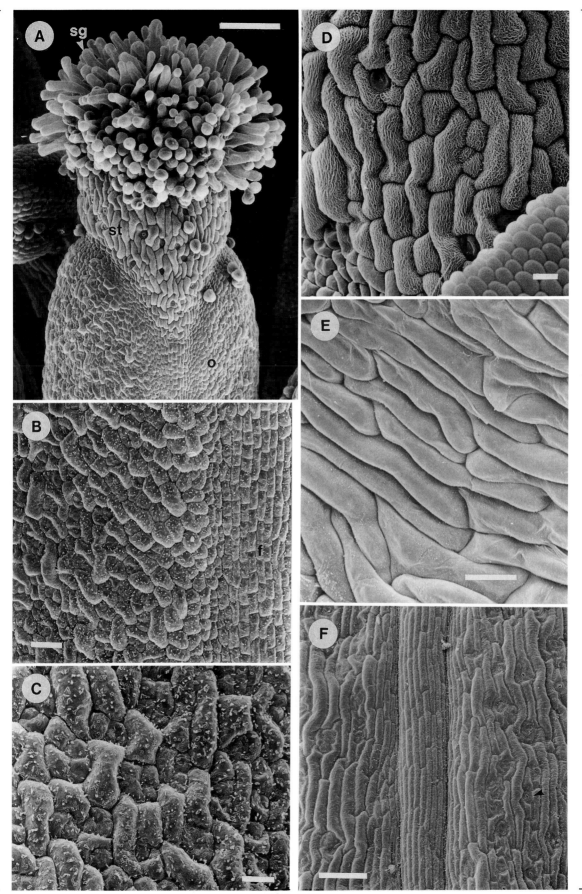

Plate 3.12

Plate 3.13
Wild-type nectary
morphology

Observations of nectarial tissue at the base of mature (open) wild-type flowers of Columbia ecotype, based on Davis (1988; 1992) and Davis and Gunning (1991). All plants were grown under continuous light at constant temperature (22°C).

(**A**) Stylized diagram of the *Arabidopsis* flower in telescoping view, showing one nectary of continuous glandular tissue encircling the androecium. The greatest proliferation of the nectary (ln) occurs external to the lateral stamens and opposite the lateral sepals. Also, up to four multicellular lobes (mn) of the nectary, of various sizes, may arise at positions between the petal bases, abaxial to the filaments of the medial stamens and opposite the medial sepals.

(**B**) Transverse section, stained with toluidine blue, through the receptacle at the level of the medial outgrowths (mn) of the nectary, showing three to four layers of cells (*arrow*) connecting the two medial protruberances which oppose a medial sepal (mse). The medial outgrowths of the nectary are confluent with the upper glandular tissue of the lateral outgrowths (ln). Note the relatively small size of nectarial cells. Two petal claws (p) are apparent. Of the three labeled vascular bundles (vb) of the receptacle, the left-most penetrates the replum of the gynoecium, whereas each of the remaining two bundles supplies a medial stamen.

(**C**) Low-temperature scanning electron micrograph (SEM) of a lateral side of a flower that recently ceased nectar secretion, showing nectarial tissue (*arrows*) that skirts the lateral stamen (ls) and the adaxial surface of the petal claw (p), to connect the lateral (ln) and medial (mn) portions of the nectary. Open modified stomata (*arrowhead*) occur on the gland surface. Three sepals and two petals, whose insertion points are indicated (p), were removed to reveal the gland. In the background, two of the medial stamens (ms) and the base of the gynoecium (g) are evident.

(**D**) SEM of a pre-secretory stage 12–13 flower (Müller, 1961; Smyth et al., 1990) viewed laterally after removal of two petals, a lateral stamen (ls), and a lateral sepal (lse) to reveal a bilobed outgrowth of nectarial tissue (ln) that is confluent around the lateral stamen, including adaxial to this stamen (*arrows*), and connects with a proliferation at the medial position (mn). Note that the cuticular ornamentation of the epidermal cells of the gland is absent around the lateral stamen and between the medial stamens (ms) and the petals, and on the modified stomata (*arrowheads*). At the base of the gynoecium (g), two to three layers of nectarial tissue that succeeds the lateral stamen are evident.

Bar = 20 μm in **B**, **D**; 100 μm in **C**.

A.R. Davis

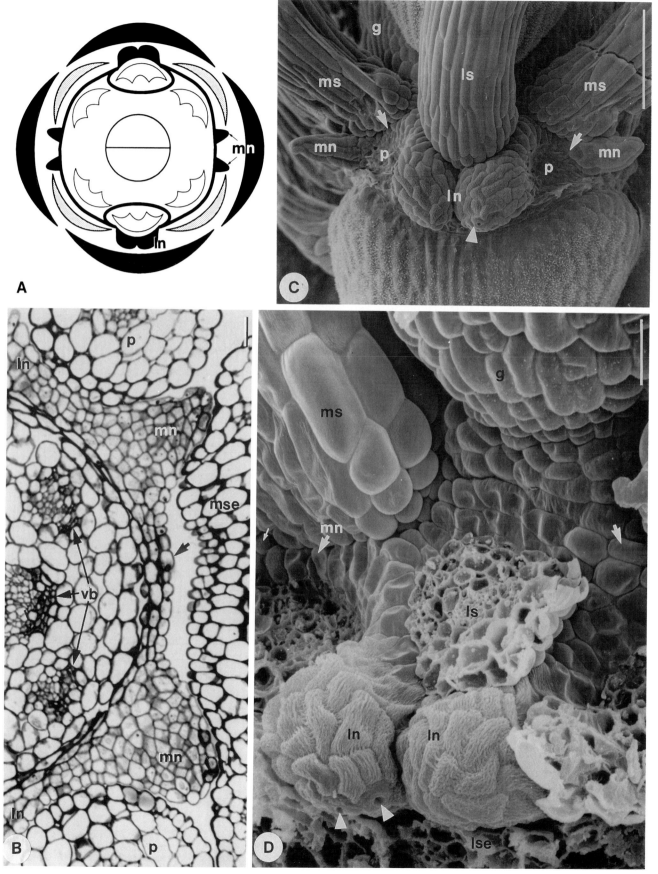

Plate 3.13

Plate 3.14
Wild-type nectary
development

Light and scanning electron micrographs (SEMs) of immature (**A–C, C, G H**), fully open (**F, I**), and spent (**D, E**) flowers of wild-type Columbia ecotype (**A–E, H, I**) and the starchless mutant TC7 (courtesy of Dr. T. Caspar; Caspar et al., 1985) (**F and G**). All plant material was grown at constant temperature (20°C or 22°C) under continuous illumination. Stages of flower development refer to those of Müller (1961), Smith et al. (1990), and Bowman et al. (1991a).

(**A**) Longitudinal section of a bud late in stage 11. Tip of a lateral stamen (ls) exceeds petal height (p). Most nectarial tissue in the lateral position faces the pouch-like base (bottom right) of a lateral sepal, yet also extends (*arrow*) toward the gynoecium (g).

(**B**) Low temperature SEM of half of a bilobed lateral outgrowth (ln) of the nectary of a bud in stage 11. Note the extension of nectarial tissue (*arrows*) between the lateral stamen (ls) and a petal (p). A guard mother cell (*arrowhead*) is evident on the gland but no cuticular patterning of the epidermis of the gland is yet evident.

(**C**) Transverse section through a presecretory bud of late stage 12 flower. Note differences in morphology of lateral nectarial tissue (ln), and inequality of lobe size of the lower outgrowth. Note the relatively small cell diameter in nectarial tissue.

(**D**) Crescent-shaped nectarial tissue (ln) outgrowth at the lateral position amidst growing, bulbous cells of the abscission zones of the lateral stamen (ls), petals, and lateral sepal (lse) in a stage 16 flower. The extension of the nectary (*arrow*) is still visible below the base of the gynoecium (g). Two modified stomata are on the gland surface (*arrowheads*).

(**E**) The other lateral outgrowth (ln) of the same nectary as in **D**, showing gross differences in morphology and disparity in lobe sizes. Modified stomata (*arrowheads*).

(**F**) Longitudinal section through a stage 14 flower that had borne nectar in the pouches of the lateral sepals at fixation, showing sieve elements (*arrows*) penetrating the base of a lateral outgrowth of the nectary (ln). Two modified stomata on the gland surface (*arrowheads*) have been sectioned obliquely, and face the pouch of the lateral sepal (lse). In the receptacle (r), the commonality of source of vasculature on the lateral side of the flower is clear (*bottom center*). Separate bundles derived from this source supply the main vein of the sepal, from which a major source of nectary vasculature originates; the filament of the lateral stamen (ls); and the valve (v) of the gynoecium. An ovule (ov) and the septum base (sp) are evident in the growing silique.

(**G**) Longitudinal section through a lateral outgrowth (ln) of the nectary of a presecretory stage 13 flower, demonstrating vascular innervation by strands of phloem alone (*arrows*). Modified stomata (*arrowheads*) (r, receptacle).

(**H**) Resin replica (for methods see Davis and Gunning, 1992) of the surface of a lateral outgrowth of the nectary of a mature (presecretory) bud showing adjacent and solitary stomata with open pores amid epidermal cells bearing external cuticular ridges.

(**I**) Part of a lateral outgrowth of the nectary of a flower late in stage 14 stained with iodine-potassium iodide after fixation. The guard cells have stained red, indicative of an abundance of amylopectin (starch) in their plastids.

Bar = 10 μm in **B, D, E, H**; 20 μm in **G, I**; 50 μm in **F**; 100 μm in **A, C**.

A.R. Davis

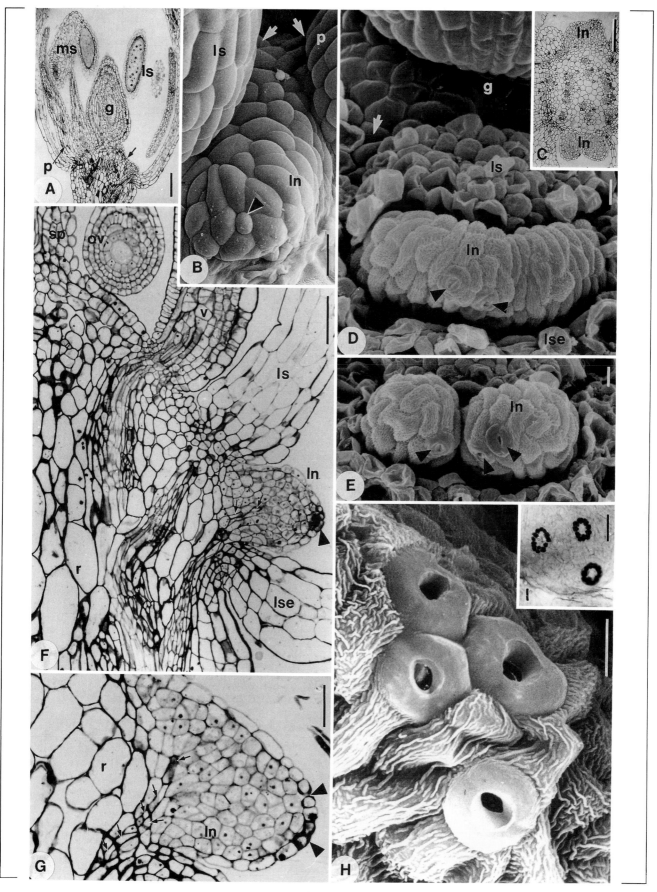

Plate 3.14

Plate 3.15
Wild-type nectary
development

Nectary development (Columbia ecotype), with emphasis on the medial out-growths of the gland. The plants were grown at 20°C or 22°C under constant illumination. Stages of flower development refer to those of Müller (1961), Smyth et al. (1990), and Bowman et al. (1991a).

(**A**) Flower bud (0.8 mm long, when fixed) in late stage 10/early stage 11 as viewed from the medial-lateral perspective after removal of two sepals and pet-als and the lateral stamen. Three stamens, a petal (p), and the gynoecium (g) are visible.

(**B**) Higher magnification of **A** showing remnants of the lateral sepal (lse), petals (p), and a lateral stamen (ls). A deeply-cleft outgrowth (ln) of the nectary is apparent, but does not yet join the small medial protruberance (mn) at the base of a medial stamen (ms) at this stage (g, gynoecium).

(**C**) Only a medial sepal of this stage 12 flower (fresh length, 1.0 mm) has been removed revealing a medial outgrowth of the nectary (*arrow*). Note the pouch-like development at the base of the two lateral sepals, whose receptacu-lar junctions belie that of the medial sepal.

(**D**) Low-temperature scanning electron micrograph (SEM) of a medial out-growth (mn) of the nectary, similar to that of **C**. Note the absence of cuticular striations on the epidermal cells of the gland (ms, medial stamen).

(**E**) SEM of base of a stage 12 bud (2.0 mm long fresh length) with its stigma just visible at the sepal tips. Two medial protruberances (mn) of the nectary are situated abaxially to filaments of the medial stamens (ms) (mse, medial sepal; p, petal).

(**F**) Close-up of the right medial outgrowth (mn) of the nectary of **E**, showing cuticular striations on several epidermal cells which are lacking in the proximal cells of the medial stamens (ms) and petal (p). Compare these epidermal fea-tures to the nectary in **D**.

(**G**) Low-temperature SEM of a longer nectarial protruberance at the medial position (mn), situated near the bases of a medial stamen (ms) and a dissected petal (bottom left) from a flower that had resorbed its nectar. The cuticular pat-terning of the epidermis occurs almost exclusively on the most distal cells.

(**H**) Transverse section of a post-secretory flower in early stage 17 showing a persistent medial outgrowth (mn) of the nectary covered basally by en-croaching, bulbous cells of the abscission zones (az) from a petal (bottom) and medial sepal (*top*). The vacuoles of the nectarial cells are now relatively large and this medial part of the gland would soon have collapsed near the base. Note the lack of a vascular supply entering this medial process, typical of the majority of medial outgrowths. An open stoma (*arrowhead*) on the protruber-ance apex has been sectioned transversely.

(**I**) Low temperature SEM of a medial outgrowth (mn) of the persistent nec-tary of an old, post-secretory flower in stage 17 illustrating the deposition of waxy globules over the epidermal cells of the gland and the abscission zones (az).

Bar = 10 μm in **B**, **F**; 20 μm in **D**, **E**, **G**, **H**, **I**; 50 μm in **A**; 200 μm in **C**.

A.R. Davis

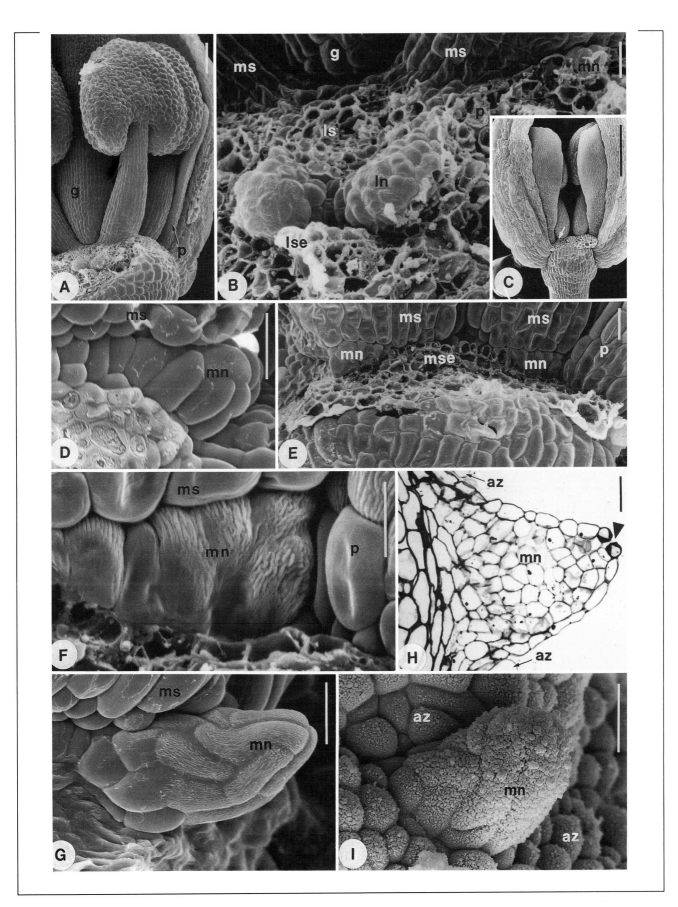

Plate 3.15

177

Plate 3.16
**Timing of the transition
to flowering**

Arabidopsis thaliana is a rosette annual species with separated vegetative and reproductive growth phases. The transition to flowering is promoted by growth in long-day photoperiods and vernalization, but neither treatment is an absolute requirement for most genotypes, making *Arabidopsis* a facultative long-day plant. The transition to flowering is also affected by growth temperature, nutrient availability, and light quality. High temperature, nutrient starvation, and either blue or far-red light promotes the transition, while low temperature and red light delay it.

The number of leaves produced in the rosette is closely correlated with the time to flowering, and is thus used as an indicator of flowering time (Rédei, 1970; Napp-Zinn, 1985; Koornneef et al., 1991). The commonly used laboratory strain Landsberg *erecta*, an early flowering ecotype, produces an average of 6–8 rosette leaves before flowering when grown at 25°C under continuous illumination (Koornneef et al., 1991). In contrast, 15–20 rosette leaves are formed prior to flowering when Landsberg *erecta* plants are grown in short days (8 hr photo periods) at 25°C. In addition, vernalization of Landsberg *erecta* plants causes a slight reduction in the number of rosette leaves to 5–8.

Several monogenic mutants of early ecotypes (e.g. Landsberg *erecta* and Columbia) have been selected on the basis of their late flowering phenotype (Napp-Zinn, 1985; Koornneef et al., 1991). These late flowering mutants exhibit different responses to environmental conditions, such as vernalization and day length, and can be subdivided into epistatic groups (Martinez-Zapater and Somerville, 1990; Koornneef et al., 1991). The response to different environmental conditions is locus specific. Mutant alleles of the *FCA, FPA, FVE,* and *FY* loci are responsive to both vernalization and day length (Koornneef et al., 1991). In contrast, mutant alleles of the *FE, FT, FD, FWA, CONSTANTS (CO),* and *GIGANTEA (GI)* loci show little or no response to day length, and no response to vernalization (Koornneef et al., 1991). Mutants of the second class are less responsive to light quality (red/far-red light ratio) than those of the first class (Martinez-Zapater and Somerville, 1990; Bagnall, 1992; Bagnall, 1993). The mutations are either recessive, semi-dominant (*co*), or almost completely dominant (*fwa*) (Koornneef et al., 1991). Temperature shift experiments with a cold-sensitive mutant allele of *GI* (*gi-2*) suggest that the wild-type *GI* gene product acts during the time that the transition to flowering occurs (Araki and Komeda, 1993). Additionally, *ga1-3* mutants, gibberellin deficient dwarfs, fail to flower when grown under short days, but display only a slight delay in flowering time when grown in long days (Wilson et al., 1992).

Double mutants between various *late flowering* mutants fall into two classes (Koornneef et al., 1991). In the first, double mutants flower at the same time as the later-flowering single mutant. In the second class, the double mutant flowers later than the later-flowering parent. All such double mutants retain the ability to flower. Thus, it appears that the transition to flowering may be promoted by more than one pathway (Koornneef et al., 1991).

Early-flowering mutants have also been selected in the Landsberg *erecta* background (Zagotta et al., 1992). Most of these early-flowering mutants are morphologically normal (*early flowering1-3; elf1, elf2, elf3*), however, one, *terminal flower,* also exhibits an abnormal inflorescence morphology (Shannon and Meeks-Wagner, 1991; Alvarez et al., 1992). In addition, some photomorphogenic mutants (*hy2, hy3*) are also early flowering (Goto et al., 1991b).

A phenotype typical of a late-flowering mutant in an early-flowering ecotype background (*right*) is shown in comparison to wild type (*left*).

M. Koornneef and J.L. Bowman

Flowers: Floral Transition Mutants

Plate 3.16

179

Plate 3.17
The *PIN-FORMED* gene

The *pin-formed* (*pin*) mutant has structural abnormalities in different organs at different stages of plant growth. Two independent alleles have been isolated, one from the Enkeim ecotype (*pin-1*: Goto et al., 1987; Okada et al., 1991; Goto et al., 1991a), and one from the Landsberg ecotype (*pin-2*: Okada et al., 1991). The phenotypes of the two alleles are almost identical. Genetic analysis has shown that the mutations are recessive, and the *PIN* locus is located on chromosome 1 (M. Komaki, K. Okada, Y. Shimura, unpublished).

Morphological phenotypes include aberrant cotyledon, leaf, and flower structures. Cotyledons are often fused, and, in some cases, three cotyledons are formed. In addition, the first leaves form a trumpet-like funnel.

(**A**) Seedling with partially fused cotyledons.
(**B**) Young *pin-1* plant. One of the cotyledons is not fully developed and the first leaves are fused to form a trumpet-like funnel structure.
(**C**) Young plant with three cotyledons.

The rosette leaves and cauline leaves are often deformed, with the *pin* mutant leaves often wider than those of wild type. The major vascular bundles are well-developed, but are abnormally branched. Phyllotaxis of the inflorescence is also abnormal. In wild-type plants, the cauline leaves develop within the spiral phyllotaxy of the inflorescence. In contrast, cauline leaves and axillary shoots often form in "opposite" positions along the inflorescence axis.

(**D**) Wider than normal rosette leaves of *pin* plant. The vascular bundles are well-developed.
(**E**) Cauline leaf of abnormal morphology on *pin* plant.
(**F**) Inflorescence axis in which two cauline leaves with axillary shoots have formed at the same level on the primary inflorescence stem.

K. Okada and Y. Shimura

Plate 3.17

Plate 3.18
The *PIN-FORMED* gene

Abnormal inflorescence morphology is the most conspicuous phenotype of *pin-formed* (*pin*) mutants. The name of the mutant is derived from the major type of inflorescence shoot formed in the mutants, that which bears no flowers or flower-like structures.

(A) *pin* mutants growing in pots. There are no obvious traces of pedicels or floral buds at or near the top of the axes.

(B) Apical region of *pin* inflorescence axes.

(C) Some inflorescences bear abnormal flowers. Some of the flowers have no stamens and have petals about 2–3 times wider than normal. The number of petals is variable (from 0 to 10). In this case, the flower has three wide petals and no stamens.

(D) Pistil-like structures with stigmatic tissue but no accompanying sepals, petals, or stamens, may develop at the top of the inflorescence axis. The formation of floral structures at the apex of the inflorescence axis is also observed in *terminal flower* mutants (Shannon and Meeks-Wagner, 1991; Alvarez et al., 1992).

(E) The terminal pistil-like structures lack ovules and are therefore sterile, as shown in this transverse section.

(F) Both pin-formed axes and deformed flowers are formed on the same plant. In addition, fasciation of the inflorescence axis is often observed.

Although the biochemical and physiological basis for the *pin* mutant phenotype is not fully understood, reduced activity of the polar transport of auxin is likely involved. Direct measurement of the transport activity showed that the inflorescence axis of the mutant has about 10% of that of wild type (Okada et al., 1991). That many of the characteristics of the mutant phenotype can be mimicked in wild-type plants if they are grown in the presence of auxin transport inhibitors (Okada et al., 1991) supports the idea that polar auxin transport activity is required for the early steps of flower primordia formation.

K. Okada and Y. Shimura

Plate 3.18

183

Plate 3.19
The *PINOID* gene

Recessive mutations in the *PINOID* (*PID*) gene affect embryogenesis, vegetative growth, and inflorescence development in *Arabidopsis* plants. About 40% of *pinoid-1* seedlings scored had cotyledon abnormalities, such as fused, additional, or fewer cotyledons. The most dramatic effect of the *pid-1* mutation is on inflorescence structure. The number of cauline leaves and flowers produced on a *pid-1* inflorescence is greatly reduced compared with wild-type plants. Approximately 20% of *pid-1* plants scored did not produce any flowers on the primary inflorescence axis. On some *pid-1* plants neither cauline leaves nor flowers developed. The *pid-1* inflorescence may thus appear as a bare, pin-like structure. Close examination of the primary stem apex reveals the presence of ridges of tissue which are presumably primordia arrested in development. In addition, the growth of the inflorescence apex of a *pid-1* mutant plant is reduced and the stem rarely reaches the height of a comparable wild-type inflorescence. When flowers are present in a *pid-1* mutant, the phyllotaxis associated with their production is aberrant and the flowers often appear to arise as a "bunch" above which the apex shows only limited further growth. The structure of the flowers is also abnormal. The number of sepals is reduced compared with wild type (average 2.2 per flower), as is the number of stamens (average 3.7). The gynoecium often has no ovary and arises as a thin, stem-like structure topped with a large mass of stigmatic papillae. Sometimes a portion or a whole carpel is seen on the side of this stem (average number of carpels was 0.12). In contrast to all the other floral organs, the number of petals produced is greater than in wild type (average 7.6). Because of the reduction in stamen number and poor female fertility, *pid-1* flowers are rarely self fertile.

Three *pinoid* alleles are illustrated. *pid-1* and *pid-2* are in Landsberg *erecta* background, *pid-3* is in the Columbia ecotype. The *pid-2* mutation is less severely affected compared with the *pid-1* and *pid-3* mutations. The *pinoid* gene was named because *pid* mutants resemble *pin-formed-1* mutants (Goto et al., 1991a; Okada et al., 1991). The two loci are not allelic.

(**A**) and (**B**) Scanning electron micrographs of wild-type and *pid-1* seedlings.

(**A**) Vertical view of a wild-type seedling, with two cotyledons and two first leaves (with trichomes) in opposite orientation.

(**B**) Vertical view of a *pid-1* seedling, with three cotyledons and three leaves (with trichomes). About one quarter of *pid-1* plants have three cotyledons.

(**C–E**) Scanning electron micrographs of 10-day-old wild type, *pid-1*, and *pid-2* inflorescence apices.

(**C**) Vertical view of the inflorescence apex of a 10-day-old wild-type plant. Note the primary inflorescence meristem (im), secondary inflorescence apices in the axils of the cauline leaves (s), the first flower (f), and the phyllotaxis of secondary apices and flowers.

(**D**) Inflorescence apex of a 10-day-old *pid-2* plant. Note the primary inflorescence meristem (im), secondary inflorescences (s), and the first flower (f). In wild type, the four sepal primordia are individually distinct in stage 3–4 flowers. However, in this stage 3 *pid-2* flower (f) they are developing as a continuous rim around the flower primordium. The phyllotaxis of the secondary apices and first few flowers produced at this *pid-2* apex appears similar to that of wild type. In stronger *pid-1* and *pid-3* mutants it is usually abnormal.

Bar = 50 μm in **C, D, E, F**; 100 μm in **G, H**; 500 μm in **A, B, I**.

(*Text continued on p. 187*)

Flowers: Inflorescence Mutants

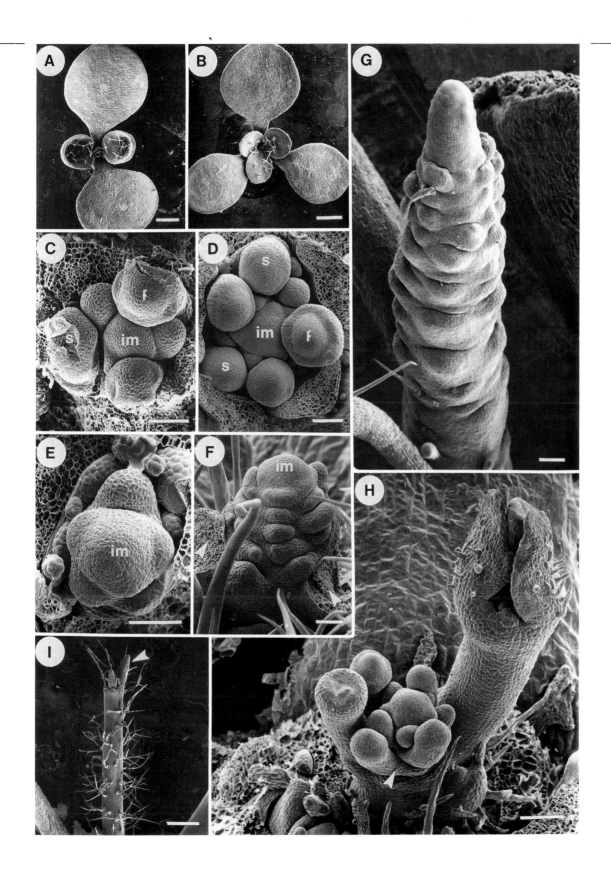

Plate 3.19

Plate 3.19
The *PINOID* gene
(*continued*)

(E) Inflorescence apex of a 10-day-old *pid-1* plant. The main inflorescence apex (im) is apparently growing but the presumptive secondary apices and flower primordia seem to be arrested at an early stage. The phyllotaxis of these putative primordia also appears abnormal.

(F) Side view of the inflorescence apex of a 12-day-old *pid-2* plant. Note the large scars to the left and right of the apex where developing flowers beyond stage 9 have been removed (*arrowheads*). These flowers had grown higher than the apex. Growth of the inflorescence meristem (im) thus appears to have slowed. The development and differentiation of the putative primordia bulging from the sides of this apex are apparently arrested. In addition, the phyllotaxis of these structures is abnormal.

(G) Mature, arrested inflorescence of a *pid-1* mutant plant. Note the "pin" structure, the presumptive undeveloped primordia on the inflorescence flanks, and the abnormal phyllotaxis associated with these primordia.

(H) Inflorescence apex of a 12-day-old *pid-2* plant. An abnormal basal flower with two sepals is developing to maturity, while the differentiation of more apical primordia appears to have slowed in comparison. However, in this weak mutant some of these primordia are stalked (*arrowhead*) and seem to have developed further than the ridge-like primordia of *pid-1* and *pid-3* mutant plants.

(I) Mature inflorescence of a *pid-3* mutant plant. Note that at the apex some branching has taken place and a number of pin-like structures are present (*arrowhead*). In addition, trichomes are more branched than the simple trichomes normally present on stems. Similar abnormal trichomes are seen in other mutants (see *yakka*).

J. Alvarez, G. Bossinger, and D.R. Smyth

Plate 3.20
The *SPITZEN* gene

Recessive mutations in the *SPITZEN* gene (*SPZ*) affect vegetative and inflorescence development of the *Arabidopsis* plant. Phyllotaxis of leaves and flowers may be disrupted in some *spz-1* individuals, and leaf shape is abnormal. The leaf margins curl down and inwards giving the leaves a narrow, tapered aspect. The mutation most drastically affects inflorescence development. A range of structures are produced, from a thin, tapered organ to a part-flower to a normal flower. The part-flowers may consist of only a whorl of sepals, or whorls of sepals and petals, or whorls of sepals, petals, and stamens. The organ whorls produced are mostly normal in terms of organ number and individual organ structure. It seems that the production of floral organ primordia has been terminated at different times during development in different flowers. The first forty "flowers" of 10 *spz-1* plants were scored and the range of flower types was categorized into six classes:

(1) A notch on the stem with no identifiable flower structures (9%).
(2) A thin, tapered, thread-like structure (15%).
(3) A pedicel with 1–4 sepals but no other flower organs (22%).
(4) A pedicel with 2–4 sepals and 1–4 petals but no other flower organs (11%).
(5) A pedicel with 3–4 sepals, 3–4 petals, and 1–6 stamens but with no carpels (6%).
(6) A complete, wild-type flower (37%).

Thus, *spz-1* mutants are able to produce normal flowers, but this ability is often curtailed. This might be due to a reduction in flower meristem size, perhaps resulting from a reduction in the duration of its growth, as there is no evidence of suppression of organ primordia once initiated. The abnormal flowers produced do not have a gynoecium, and consequently, inflorescence stems of a *spz-1* plant may have a large number of naked pedicels that do not have a maturing silique at their apex. This gives the inflorescence a "spikey" appearance, the German word for which, "spitzen," has been used as a name for the mutation. *spz-1* was induced in the Columbia ecotype. Allelism tests have not been performed between *spitzen* and *fl54* or *antherless*.

(**A–F**) Scanning electron micrographs of *spz-1* inflorescence and flowers.
(**A**) Severely affected *spz-1* inflorescence apex. Note the thin, pointed structures on the inflorescence stem along with two abnormal flowers in which only sepals are visible.
(**B**) Vertical view of a *spz-1* inflorescence apex. Note the development of apparently normal flowers (f) and abnormal flowers (*arrowheads*) from the same apex. Only two sepals can be seen developing in the outer whorl of each of the abnormal flowers. In addition there is no meristematic dome internal to these, that in wild type would give rise to the other floral organs.
(**C**) Lateral view of a thin, tapered organ produced by a *spz-1* inflorescence.
(**D**) Vertical view of a *spz-1* flower which has terminated with a stamen at its center (a, anther; fl, filament). Note the nectaries at the base of the filament (*arrowheads*). The claws (c) of four petals originating outside the nectaries are also evident.
(**E**) Lateral view of a *spz-1* flower that is wild type in appearance (one petal has been removed). The central gynoecium is elongating following fertilization.
(**F**) Lateral view of a *spz-1* flower. The outer sepal, petal, and stamen whorls have the wild-type complement of organs. A gynoecium, however, is absent.

Bar = 50 μm in **B**, **C**, **D**; 500 μm in **A**, **E**, **F**.

J. Alvarez

Plate 3.20

189

Plate 3.21
The *FL54* gene

Recessive mutations in the *FL54* gene cause variable phenotypes in the pattern of floral bud formation, as well as in the shapes, numbers, and positions of floral organs (Komaki et al., 1988). The *FL54* locus maps to chromosome 2 (M.K. Komaki, K. Okada, Y. Shimura, unpublished). Allelism tests have not been performed between *fl54* and *spitzen* or *antherless*.

(A) and (B) The structure of the inflorescence is disrupted in *fl54* mutants. Flowers are clustered in relatively short segments along the inflorescence axis. Between the clusters of flowers are segments in which short, green filaments and some sepal-like structures with a short stalk are clustered. Clusters of flowers have developed at the top and at the middle of the inflorescence axis shown in **B**. In between these two clusters are filamentous organs and sepal-like organs.

(C) The shape and orientation of the sepal-like structures are similar to those of the abaxial sepal in wild-type flowers.

(D) and (E) Within individual flowers, the numbers of sepals, petals, and stamens are often reduced. The remaining organs are not necessarily positioned symmetrically. In some cases, short filaments and knobs are visible on the flower receptacles, at positions corresponding to the missing organs. These structures might be floral organs whose development was aborted in the early stages of floral morphogenesis.

Several kinds of homeotic changes are observed in the floral organs of *fl54* mutants. Most of the stamens lack anthers, and consist of merely a filamentous structure (*arrow* in **D**). In some flowers, the top of the filaments is white and somewhat petal-like. In other cases, filaments are green and are capped with stigmatic papillae. Some stamens are extremely thin and are attached to the ovary wall. The number of petals is reduced to only one or two, and pollen sacs are seen occasionally along the margins of the petals. In other flowers, petals are green and sepaloid. Most of the flowers have three sepals, which are slender and have no, or only a few, trichomes. It appears that the pistils are less affected by the mutation, and set seeds well by means of artificial pollination.

K. Okada and Y. Shimura

Plate 3.21

Plate 3.22
The *YAKKA* gene

The *yakka-1* (*yak-1*) mutation is recessive and produces a range of phenotypic alterations in vegetative and inflorescence development. The *yak-1* plants produce many more rosette leaves than do wild-type plants. A sample of ten *yak-1* plants revealed an average of 20 ± 0.87 rosette leaves compared to an average of 7 ± 0.02 in ten wild-type plants grown under the same conditions. The inflorescence development of *yak-1* plants is also substantially altered. All *yak-1* inflorescences are highly fasciated and, consequently, most commonly emerge as a strap-like structure. The fasciation appears to become more severe as the inflorescence apex ages. The number of cauline leaves produced by a *yak-1* plant is also greater than in wild type, an average of 8 ± 1.1 compared with 3 ± 0.4 in wild type. The *yak-1* plants do not always produce flowers on the primary inflorescence stem, so the apex may terminate in a pin-like structure. In the majority of *yak-1* plants, however, a few flowers may be produced before the apex of the inflorescence terminates in a highly fasciated, chaotic clump of abnormal flowers and flower parts.

The phenotype of individual flowers is also different from wild type. The outer whorl often has an abnormal number of sepals, and sepal–petal mosaics occasionally arise. Petals are often reduced in number and size. The number of stamens is often greater than in wild type, especially in flowers where there are less than two carpels. The carpels are most often unfused and their number varies considerably but is frequently greater than two. This indicates that there is a reduction in floral determinacy in many flowers. Some flowers appear phenotypically similar to those of the flower mutation *superman*.

One of the more interesting features of the *yakka-1* mutation is the greatly repressed development of inflorescence shoots in the axils of the rosette and cauline leaves. In wild-type plants a number of these shoots usually arise, especially in vigorously growing plants. The shoots appear in a basipetal sequence, in the axils of the upper leaves first (Alvarez et al., 1992). Close scrutiny of axils of *yak-1* rosette leaves reveals that only rarely are rosette inflorescence primordia in evidence. In addition, if they do arise, they do not necessarily develop in a basipetal sequence.

The *yak-1* mutant was induced in the Columbia ecotype, and is named for its resemblance to the yakka, or Australian grass tree (*Xanthorrhoea*).

(**A–D**) Scanning electron micrographs of *yak-1* inflorescence structures.

(**A**) Lateral view of a *yak-1* inflorescence just after bolting. Note the large number of trichomes on the stem. The apex has become highly fasciated after the production of cauline leaves and a number of abnormal flowers.

(**B**) Lateral view of a *yak-1* inflorescence. The apex has developed into a pin-like structure. Note the abnormal compound trichomes on the stem (*arrowhead*). Similar trichome abnormalities may be observed in other *Arabidopsis* mutants (see *pinoid*).

(**C**) Lateral view of a *yak-1* inflorescence apex. Note the highly fasciated, strap-like stem and the aggregate of flowers at the apex.

(**D**) Enlarged view of a *yak-1* inflorescence apex shown in **A**. The apex has become highly fasciated producing large numbers of cylindrical, finger-like projections.

(**E**) and (**F**) Scanning electron micrographs of *yak-1* flowers.

Bar = 100 μm in **D**, **E**, **F**; 500 μm in **A**, **B**, **C**.

(*Text continued on p. 195*)

Plate 3.22

Plate 3.22
The *YAKKA* gene
(*continued*)

(E) Vertical view of a *yak-1* flower. One of the outer whorl organs is a sepal–petal mosaic (*arrowhead*). There is an increase in the number of stamens, and the carpels are unfused and reduced in size. The phenotype is similar to that observed in *superman* mutants.

(F) Vertical view of a *yak-1* flower. There are four sepaloid organs and four petaloid organs (two are obscured). Note that there more than six stamens and more than two carpels.

J. Alvarez

Plate 3.23
The *TERMINAL*
***FLOWER* gene**

Mutants of the *TERMINAL FLOWER* (*TFL*) gene result in conversion of the inflorescence apical meristem into a flower meristem (Shannon and Meeks-Wagner 1991; Alvarez et al., 1992). Normally, the inflorescence meristem produces flower primordia on its flanks in indefinite numbers. However, in *tfl* mutants a few flanking flower primordia may be produced before flower organ primordia "invade" the apex. The terminal flower produced is usually an aggregation of abnormal numbers of floral organs. There are often two or three separate gynoecia, each surrounded by up to six stamens. A few sepals and petals occur outside the carpels and stamens. Mosaic organs consisting of separate patches of floral organ tissue are common. The phenotype of these flowers may reflect the larger size of the inflorescence apical meristem in comparison with a flower meristem, thus allowing more floral organs to arise, and causing mosaic organs to develop where floral organ primordia overlap the fields of activity of floral homeotic genes (Alvarez et al., 1992).

In *tfl* mutants the main inflorescence apex becomes determinate and shows no further growth. The secondary inflorescence apices arising in the axils of the cauline leaves also become determinate, but they are usually converted directly into a single, near-normal flower.

The *terminal flower* mutant phenotype is markedly affected by environmental conditions. In plants grown under short days (Shannon and Meeks-Wagner, 1991) or at 15°C (Alvarez et al., 1992), termination occurs only after many normal flowers have been produced. The wild-type *TERMINAL FLOWER* gene may normally repress floral induction. Consistent with this, in *tfl* mutants induction occurs earlier than in wild type under conditions favoring flowering (Shannon and Meeks-Wagner, 1991). Also, termination is slowed down in *tfl* plants under environmental conditions unfavorable for floral induction (short days and low temperatures), although it does occur eventually.

(**A–F**) Early development of the inflorescence of *terminal flower*. Scanning electron micrographs of the main inflorescence of wild-type (**A**, **C**, **E**) and *tfl-2* mutant (**B**, **D**, **F**) plants. Individual wild-type Landsberg *erecta* plants were 12, 13, and 14 days old shown in **A**, **C**, and **E**, respectively; mutant plants were 11, 13, and 15 days old, shown in **B**, **D**, **F**, respectively. Individual flowers (f) and secondary branch apices (s) are indicated, the former numbered in apparent order of development. In some cases the older flowers have been removed to reveal underlying structures.

(**A**) and (**B**). No differences are yet visible between the wild-type (**A**) and *tfl-2* mutant (**B**) inflorescence apices. Several apparently normal flower primordia (f1–f4) have arisen from the flanks of the mutant apex (**B**).

(**C**) and (**D**). Differences are now becoming apparent. In **D**, the main inflorescence apex of the mutant plant is developing floral organ primordia around its edges, and the secondary inflorescence apex (s) resembles an early stage 3 floral meristem instead of an inflorescence meristem. (Even so, it is not normal because it has five sepal primordia instead of the normal four.)

(**E**) and (**F**). Three normal flowers (f1–f3) have arisen from the flanks of the mutant apex (**F**), but the apex itself is now being taken over by floral organ primordia. Secondary apices (s) developed into individual flowers.

Bar = 50 μm in **A**, **B**, **C**, **D**, **E**, **F**.

D.R. Smyth and J. Alvarez

Reproduced from Alvarez et al. (1992) with permission from Blackwell Scientific Publications Limited.

Flowers: Inflorescence Mutants

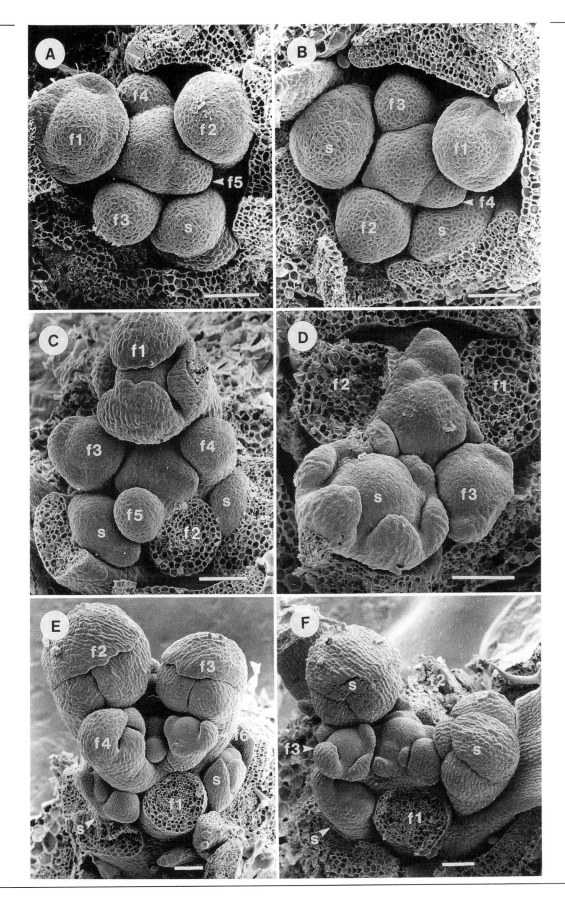

Plate 3.23

197

Plate 3.24
The *TERMINAL*
***FLOWER* gene**

Later development of the inflorescence and floral organs of *terminal flower* mutants.

(**A**) and (**B**) Later developmental stages of the main inflorescence of two *tfl-2* mutant plants. Both plants were 14 days old, although the inflorescence development in **A** was less advanced than that in **B**. Flowers (f) and secondary branch apices (s) are indicated as in preceding figure. Organ primordia developing on the inflorescence apex are tentatively identified as sepals (se), stamens (st), and gynoecia (g).

(**C**) and (**D**) The terminal flower at the main primary apex of a *tfl-2* plant. The plant was 20 days old. The apex is illustrated from above by a scanning electron micrograph (**C**), and a floral diagram interprets the type and location of organs present (**D**). In the floral diagram, sepals are darkly shaded, petals are represented as thin curved lines, stamens have medium shading, and carpels are shown as thick U-shaped organs with ovules attached where they join another carpel. Mosaic organs are represented by combining parts of these symbols. Flowers (s2, s3) arise as secondary "branches" in the axils of cauline leaves, while normal flowers (f1, f2) arise on the inflorescence before the terminal flowers were produced.

(**E**) The determined primary apex of a mature main inflorescence of a *tfl-2* plant after 26 days. Three gynoecia are inserted close together at the apex, and are attached to a common receptacle. The surrounding sepals, petals, and stamens have senesced and fallen.

(**F–H**) Mosaic organs.

(**F**) A sepal–petal mosaic showing a clear lateral division between cell types (se, sepal tissue; pe, petal tissue; tr, trichomes).

(**G**) A petal–stamen mosaic. Two locules of the half stamen sector are visible (pe, petal tissue; st, stamen tissue).

(**H**) A carpel–stamen mosaic. Ovules are visible along the unfused edge of the carpel, and a large number of stigmatic papillae have developed on the apex. The stamen sector has a filament at the base and locules on the facing edge (ca, carpel tissue; st, stamen tissue; sp, stigmatic papillae).

Bar = 50 μm in **A**, **B**; 500 μm in **C**; 100 μm in **E**, **F**, **G**, **H**.

D.R. Smyth and J. Alvarez

A–F, **H** reproduced from Alvarez et al. (1992) with permission from Blackwell Scientific Publications Limited.

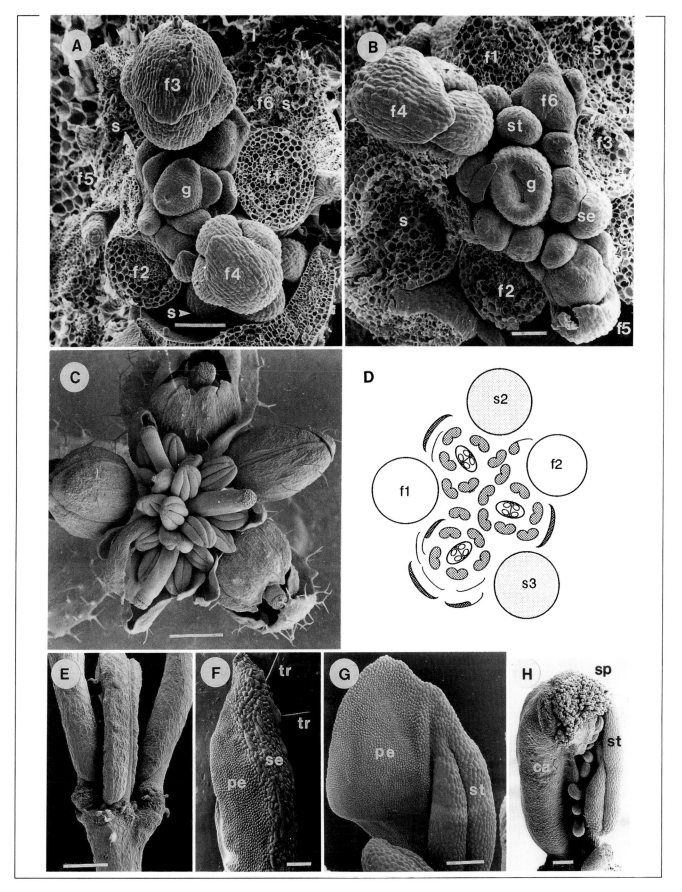

Plate 3.24

Plate 3.25
APETALA1: the *apetala1*
mutants phenotype

Floral development is disrupted in two ways by *apetala1* (*ap1*) mutations: (1) flower meristems are partially or completely converted into inflorescence meristems; and (2) the presence, position, and identity of the outer two whorls of floral organs is altered. Several recessive mutant *ap1* alleles have been analyzed (McKelvie, 1962; Usmanov, 1970; Buggert and Röbbelen, 1970; Irish and Sussex, 1990; Bowman et al., 1993) and can be arranged into a phenotypic series (strong, intermediate, and weak) based both on the extent of inflorescence-like character exhibited by *ap1* flowers and the severity of alterations in floral organ identity in the outer two whorls. The *ap1* mutant phenotype is sensitive to both environmental conditions and acropetal variation, with more complete flower-to-inflorescence meristem conversions occurring at low growth temperatures and in basal stem positions. The brief descriptions below apply for basal flowers on plants grown at 25°C under continuous illumination.

In strong mutant alleles (*ap1-1, ap1-7, ap1-9*), "flower" meristems often give rise to a determinate, branched structure comprised of several individual flowers. These are occasionally transformed completely into inflorescence meristems. In intermediate mutant alleles [*ap1-2, ap1-4, ap1-6, ap1-8* (formerly called *axillaris*; McKelvie, 1962)] the determinate branched structures produced by "flower" meristems consist of fewer individual flowers than do those of strong alleles. In weak mutant alleles (*ap1-3, ap1-5*), flower meristems give rise to single flowers, as in wild type.

The first-whorl organs of strong mutant allele plants are bract-like, those of intermediate mutant allele plants are leaf-like, and those of weak mutant allele plants are mosaic sepaloid organs. The second-whorl organs of strong mutant allele plants are usually absent, those of intermediate mutant allele plants are either leaf-like or staminoid, and those of weak mutant allele plants are staminoid petals.

The *APETALA1* gene has been cloned and is a member of the MADS-box family of genes (Mandel et al., 1992b). The *ap1-1* allele is the result of a splice-site acceptor mutation 3' to the MADS box, the *ap1-2* allele is the result of a missense mutation in the MADS box, and *ap1-3* is the result of an acceptor splice site at a 3' intron (Mandel et al., 1992b).

(A–D) Strong *ap1* mutant flowers

(**A**) Stage 5 *ap1-1* flower. The four first-whorl organ primordia arise in a cruciform pattern although sometimes the pattern is irregular. The lateral first-whorl organ primordia (l) develop from lower on the pedicel than the medial (m). The pattern of the second- and third-whorl organ primordia is also often irregular.

(**B**) Stage 6–7 *ap1-1* flower. The first-whorl organs are bract-like, with flower primordia (f) developing from their axils and stipules at their base (*arrowhead*).

(**C**) Mature *ap1-1* flower. This determinate structure, composed of several individual flowers, is typical of structures that develop from flower meristems in basal positions in strong *ap1* alleles. Secondary flowers develop from the axils of the first-whorl organs of the primary flower (p), as in **B**, with tertiary flowers developing from secondary flowers, and so on. Note the significant internode elongation between the lateral (l) and medial (m) first-whorl organs of the primary flower.

Bar = 20 μm in **A, B, F, G**; 50 μm in **E, J**; 100 μm in **I**; 200 μm in **D, H**; 200 μm in **C**.

(*Text continued on p. 203*)

Plate 3.25

201

Plate 3.25
APETALA1: the *apetala1*
mutants phenotype
(*continued*)

(**D**) Mature *ap1-1* flower typical in apical positions of strong *ap1* mutants. First-whorl organs fail to develop fully, although in many cases organ primordia are present (*arrowheads*). In contrast, no second-whorl organs are present, due to a failure to initiate any second-whorl organ primordia.

(E–J) Weak *ap1* mutant flowers

(**E**) Side view of *ap1-3* inflorescence apex. No axillary flowers are formed in weak mutant alleles, and the pattern of organ primordia formation is similar to that of wild type.

(**F**) Stage 6 *ap1-3* flower. Although the organ primordia are initiated in the proper numbers and positions, the first-whorl organs are narrower than in wild type and fail to enclose the inner-whorl organ primordia. The medial first-whorl organs are often mosaics of central sepaloid tissue and marginal petaloid tissue, as in **J**. The second-whorl organs are petals in basal flowers and petal-stamen mosaics in apical flowers.

(**G**) Stage 6 *ap1-5* flower. The pattern of the second- and third-whorl organ primordia is often irregular in apical flowers of plants with weak alleles, such that the distinction between second- (2) and third-whorl (3) positions may be unclear.

(**H**) Mature apical *ap1-5* flower. The medial first-whorl organs (m) are often mosaics of sepaloid and staminoid, or sepaloid and carpelloid (as in this flower) tissue in apical flowers of plants with weak *ap1* alleles. The second-whorl organs are usually petal–stamen mosaic organs (*arrowhead*).

(**I**) Mosaic organ of petal (p) and stamen (st) tissue found in second-whorl positions of *ap1-5* flowers.

(**J**) Mosiac organ of sepal (se) and petal (p) tissue found in medial first-whorl positions in basal *ap1-3* flowers.

J.L. Bowman

Plate 3.26
CAULIFLOWER: an
enhancer of *APETALA1*

A recessive allele of the *CAULIFLOWER* (*CAL*) locus, *cal-1*, enhances the *ap1* phenotype such that floral meristems are consistently transformed into inflorescence meristems (Bowman et al., 1993). In *ap1-1 cal-1* double mutants, each meristem that in wild type would give rise to a single flower, behaves instead as an inflorescence meristem, termed a "second-order" inflorescence meristem. These second-order inflorescence meristems produce meristems in a phyllotactic spiral. These also behave as inflorescence meristems, i.e., third-order inflorescence meristems. This process may be repeated several times resulting in the production of fourth-, fifth-, and higher-order inflorescence meristems before flowers with an *ap1* mutant phenotype eventually differentiate from some of the meristems. The number of orders of inflorescence meristems produced varies acropetally as well as with environmental conditions. When grown at 25°C, the basal positions of *ap1-1 cal-1* plants may have up to ninth-order inflorescence meristems, while the more apical positions may have only second-order inflorescence meristems. When grown in unfavorable conditions (such as at 30°C), the basal positions may have only second- or third-order inflorescence meristems, while in the apical positions, flowers develop without any supernumerary inflorescence meristems.

ap1-1 plants heterozygous for *cal-1* also display an enhanced phenotype, although the enhancement is less conspicuous. The *cal-1* allele also enhances the phenotype of weak *ap1* alleles. For example, the phenotype of *ap1-5 cal-1* plants is similar to that of *ap1-1* grown at 25°C, such that *ap1-5 cal-1* plants have second-order inflorescence meristems in their basal positions. The phenotype of the floral organs in *ap1-5 cal-1* flowers also resembles that of *ap1-1* flowers.

Plants homozygous for *cal-1* alone have little if any phenotype distinguishable from wild type. Thus, the *CAL* gene appears dispensable in an otherwise wild-type background, but influences flower development in an *ap1* mutant background.

(**A**) Apical view of an *ap1-1 cal-1* plant. The apical inflorescence meristem (i) as well as numerous second-, third-, and higher-order inflorescence meristems are evident.

(**B**) Young *ap1-1 cal-1* apex showing the developmental basis for the *ap1-1 cal-1* phenotype. Meristems that in wild type would develop into individual flowers, are behaving as inflorescence meristems rather than flower meristems. Note these "second-order" inflorescence meristems (2) are morphologically indistinguishable from the apical inflorescence meristem (im).

(**C**) Side view of a much older *ap1-1 cal-1* inflorescence. An apical view of this inflorescence is shown in **A**. At about this stage, when plants are grown at 25°C, flowers with an *ap1* mutant phenotype will begin to differentiate from the higher-order meristems.

(**D**) Close-up of the structure occupying the most basal floral position denoted by *arrowhead* in **C**. This structure developed from a meristem (a second-order inflorescence meristem) that in wild type would have given rise to a single flower.

(**E**) Close-up of the structure produced by a third-order inflorescence denoted by *arrowhead* in **D**. Fourth- (4) and higher-order inflorescence meristems are visible.

(**F**) Perhaps the most striking phenotype occurs when *ap1-1 cal-1* plants are grown at 15°C. In this case no differentiation of flowers is observed even after four months, by which time twelfth-order inflorescence meristems are present. Additionally, cauline-leaf-like organs (cl) with axillary meristems are often produced by the second-, third-, and higher-order inflorescence meristems.

Bar = 50 μm in **B**; 100 μm in **A**, **C**, **F**; 500 μm in **D**, **E**.

J.L. Bowman

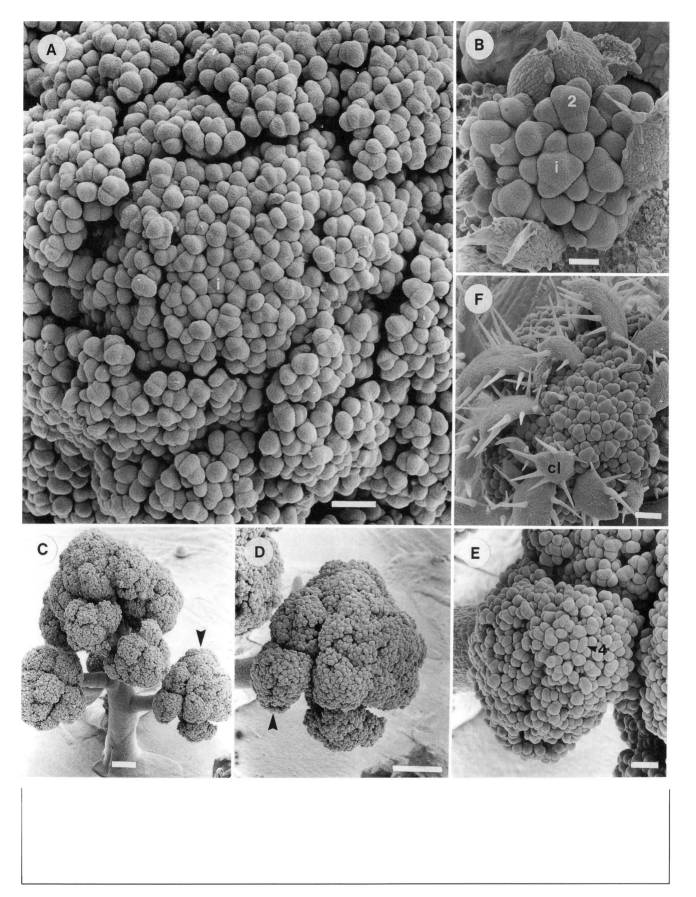

Plate 3.26

205

Plate 3.27
LEAFY: the *leafy* mutant
phenotype

In *leafy* (*lfy*) mutants, the early (basal) arising flowers are completely transformed into supernumerary secondary inflorescence shoots with subtending cauline leaves, whereas the later (more acropetally) arising flowers are partially transformed into inflorescence shoots and have subtending bracts (Schultz and Haughn, 1991; Weigel et al., 1992; Huala and Sussex, 1992). The number of organs in mutant flowers decreases acropetally. In the most apical regions, only bracts or filamentous structures develop, not flowers.

The predicted protein encoded by the *LEAFY* gene is characterized by a proline-rich domain and an acidic domain (Weigel et al., 1992). The *lfy-6* allele is the result of a stop codon near the amino terminus of the predicted protein, and the *lfy-5* allele is the result of a nonconserved missense codon in the acidic domain (Weigel et al., 1992). Shown are scanning electron micrographs of structures arising on the primary inflorescence shoot in plants homozygous for either the strong *lfy-6* or the weak *lfy-5* allele, and homozygous for the *erecta* mutation. All plants were grown under continuous light at 25°C except those shown in **B** and **E**, which were grown at 16°C.

(**A**) Secondary inflorescence shoot dissected from the axil of a cauline leaf in *lfy-6*. It repeats the pattern of the primary inflorescence shoot with supernumerary cauline leaves.

(**B**) Structure intermediate between a secondary inflorescence shoot and a flower in *lfy-5*. These arise in positions along the primary inflorescence shoot between true secondary inflorescence shoots (see **A**) and more normal flowers (see **B–G**). Note secondary flowers (f') as well as sepal-like organs (s). This structure is subtended by a bract (b), which is similar in morphology to a cauline leaf.

(**C**) *lfy-6* flower that is subtended by a bract (b). Most of the organs are sepal- or carpel-like. Stellate trichomes, which are absent from wild-type sepals but occur on leaves, are found on the outer sepal-like organs (*arrowhead*).

(**D**) Partially dissected *lfy-6* flower. Predominantly carpelloid organs, which contain sepaloid sectors as evident from elongated epidermal cells (*arrowhead*), have fused to form a central gynoecium.

(**E**) The phenotype of a *lfy-6* flower at 16°C resembles the phenotype of flowers from plants grown at 25°C (compare to **C**). Most organs are sepal- or carpel-like. Some outer organs have some leaf characteristics (l').

(**F**) Flower from a more acropetal position in *lfy-6*. Note that many of the outer organs in the flower, as well as the subtending bract (b), are carpelloid with ovules and/or stigmatic tissue developing at their margins (*arrowheads*).

(**G**) A *lfy-5* flower that is closer to wild type than the average *lfy-5* flower. Three sepals have been removed. All four types of organs found in wild type are present: sepals (s), petals (p), stamens (st), and carpels, forming the central gynoecium (g). The petal is partially staminoid as evident from the filament-like cells at its base. The extent of the staminoid sector is marked by *arrowheads*. The left stamen (*asterisk*) is petaloid, as evident from the abnormal anther and the shortened filament.

(**H**) Filamentous structure developing on the primary inflorescence shoot in place of a bract in *lfy-6*. It is flanked by stipules (*arrowheads*). No flower has developed in the axil of this structure.

(**I**) Terminal structure of an old inflorescence shoot consisting of several fused carpelloid bracts. Note developing ovules (*arrows*), stigmatic papillae (pa), and stipules (sp).

Bars = 100 μm.

D. Weigel and E.M. Meyerowitz

Parts of this figure were reproduced with permission from Weigel et al. (1992).

Flowers: Floral Meristem Mutants

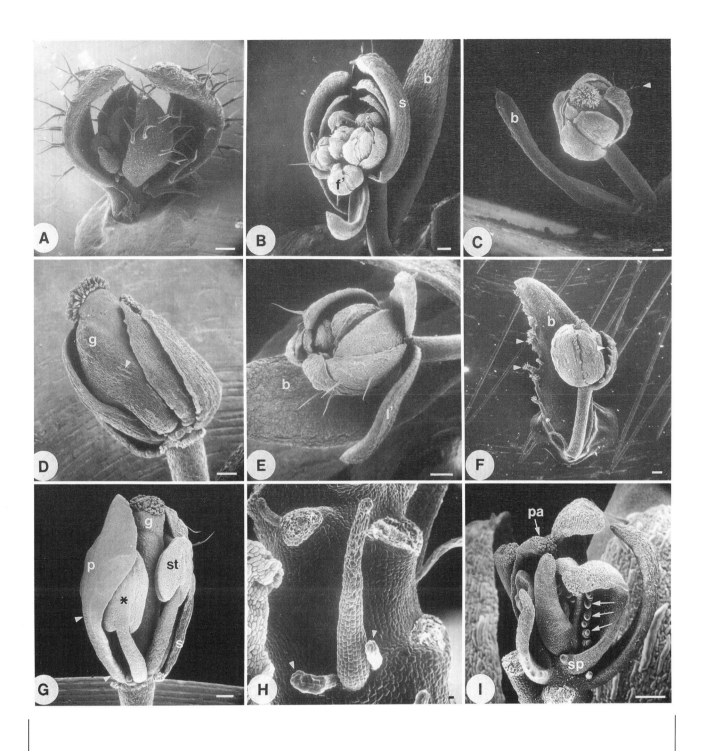

Plate 3.27

207

Plate 3.28
LEAFY: development of
leafy mutants

Scanning electron micrographs of structures arising on the primary inflorescence shoot in plants homozygous for either the strong *lfy-6* or the weak *lfy-5* allele, and homozygous for the *erecta* mutation. All plants were grown under continuous light at 25°C.

(**A**) Primary inflorescence apex of a *lfy-6* plant after it has produced a number of secondary inflorescence shoots. In place of floral buds, bracts (b) arise at the margins of the inflorescence apex. Floral buds arise at the base of the bracts (*arrowhead*) later than in wild type. Most of the floral buds develop four sepal-like outer organs, which arise in a slightly distorted cruciform pattern (*arrows*).

(**B**) Primary inflorescence apex of a *lfy-5* plant. Developing flowers are subtended by rudimentary bracts (*arrowheads*). The pattern in which the first-whorl sepals arise is largely normal.

(**C**) Primary inflorescence apex of an older *lfy-5* plant that has stopped producing flowers. Instead of flowers, this apex has started to produce only bracts (*arrow*), which appear to become progressively carpelloid and fuse. Note that the number and arrangement of the outer organs in the flowers have also become progressively abnormal. In some of the youngest flowers, the outer organ primordia appear to be carpelloid and fuse during early development (*arrowheads*).

(**D**) *lfy-6* flower at a stage equivalent to wild-type stage 5 or 6. Three of the four sepal-like outer organs have been removed. The more interior organs do not arise in whorls as in wild type, but in a more helical pattern.

(**E**) *lfy-6* flower at a later stage, with several sepal-like organs removed. The phyllotaxis of the more interior organs continues to be helical. At least two of the developing organs arise ontogenetically fused (*arrow*), probably reflecting their carpelloid character.

(**F**) *lfy-6* flower in which all organs have arisen. Some sepal-like organs have been removed. Several of the inner organs, which are intermediate between sepals and carpels, arise fused and form an abnormal gynoecium (g).

(**G**) *lfy-5* flower at a stage equivalent to wild-type stage 5. The four outer sepals have been removed. The organs surrounding the prospective gynoecium (g) will probably develop into petal–stamen intermediate organs. Their phyllotaxis appears to be whorled, although it does not resemble the pattern observed for petals and stamens in wild type.

(**H**) *lfy-5* flower at a stage equivalent to wild-type stage 7, with the outer sepals removed. The prospective petal–stamen intermediate organs surrounding the gynoecium (g) continue to develop with the kinetics of wild-type stamens. The gynoecium consists of two carpels, which are slightly distorted.

(**I**) *lfy-5* flower at a stage equivalent to wild-type stage 8. Although all organs surrounding the central gynoecium (g) are of roughly equal size, some are more petaloid (p′), whereas others are more staminoid (s′).

Bars = 10 μm.

D. Weigel and E.M. Meyerowitz

Parts of this figure were reproduced with permission from Weigel et al. (1992).

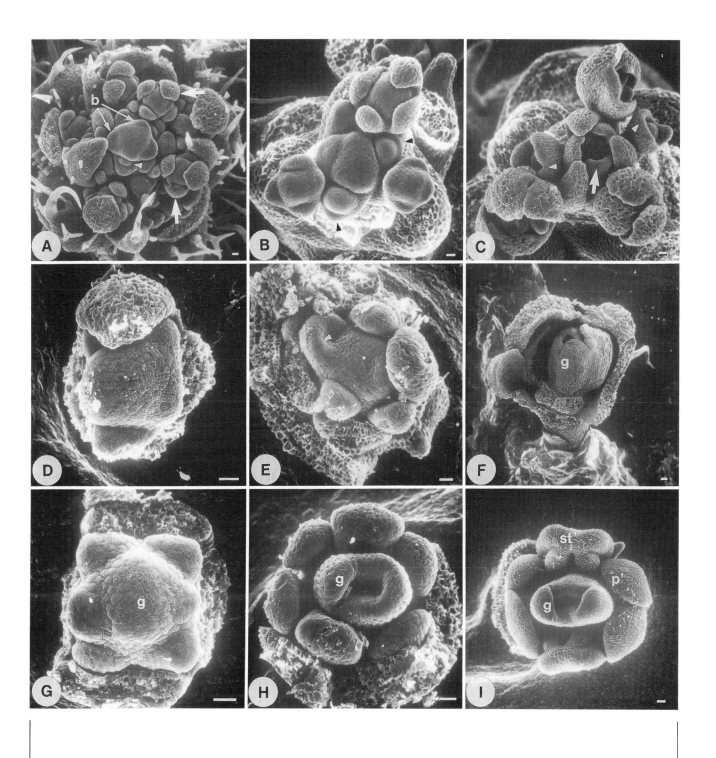

Plate 3.28

209

Plate 3.29

Interactions between
LEAFY and *APETALA1*:
the phenotype of *leafy
apetala1* double mutants

In *lfy ap1* double mutants, later arising (acropetal) flowers are more completely transformed into secondary inflorescence shoots than in either single mutant (Weigel et al., 1992; Huala and Sussex, 1992; Bowman et al., 1993).

Scanning electron micrographs of developing inflorescences and flowers from plants homozygous for the strong *lfy-6* and *ap1-1* alleles, and homozygous for the *erecta* mutation. All plants were grown under continuous light at 25°C.

(**A**) Inflorescence apex. As in *lfy-6* single mutants, bracts develop in place of flowers at the margins of the apex. Buds arise in the axils of these bracts. Note that on the developing flowers, lateral organs (*arrowheads*) are more prominent and more advanced in development than medial organs.

(**B**) *lfy-6 ap1-1* flower at a stage equivalent to wild-type stage 4. The lateral organs (l) arise lower on the floral meristem and are more advanced than the organs that arise in a more medial position (m). The advanced development of the lateral organs is typical for developing secondary inflorescence shoots rather than for developing wild-type flowers, in which this pattern is reversed. A helical phyllotaxis is apparent starting with the left lateral organ, in contrast to the pattern of the four outer organs of *lfy-6* single mutants.

(**C**) More advanced flower with all organs arising in a clear helical phyllotaxis. Some of the outer organs have been partially removed. Note the similarity to the primary inflorescence apex.

(**D**) In some flowers, the organs are filamentous (*arrows*). Note internode elongation between organs, which contrasts with the compressed internodes in wild-type flowers.

(**E**) Mature flower composed of leaf-like organs that are morphologically very similar to the subtending bract (b). Note secondary flowers (*arrowheads*) arising in the axils of the leaf-like organs.

(**F**) Partially dissected flower. Secondary flowers (*arrowheads*) arise in the axils of most outer organs. The more acropetally developing leaf-like organs, as well as organs in the secondary flowers, exhibit increasing carpelloid character with stigmatic papillae (pa) and incipient ovules (*arrows*).

Note: The phenotype of the triple mutant *lfy-6 ap1-1 cauliflower-1* is indistinguishable from that of *lfy-6 ap1-1*.

Bar = 10 μm in **D**, **E**, **F**; 100 μm in **A**, **B**, **C**.

D. Weigel and E.M. Meyerowitz

Parts of this figure were reproduced with permission from Weigel et al. (1992).

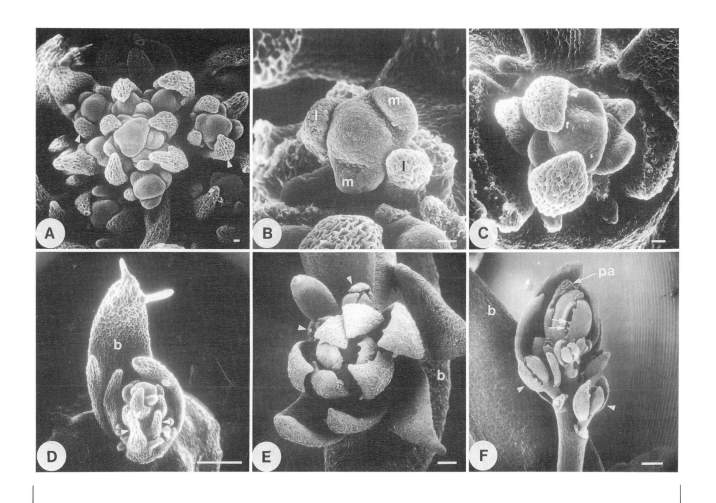

Plate 3.29

211

Plate 3.30
LEAFY and *APETALA1*:
expression patterns

(**A**) and (**B**) *In situ* hybridization of a *LEAFY* (*LFY*) anti-mRNA probe with a longitudinal section (8 μm) through a wild-type inflorescence. Note the inflorescence meristem (i), a floral anlagen (f), two stage 2 flowers (2), a stage 6 flower (6), and older flowers in which the floral organs have differentiated (se, sepal; st, stamen; c, carpel).

(**A**) Bright-field micrograph.

(**B**) Dark-field micrograph. Distribution of *LFY* transcripts is represented by silver grains, which appear as white dots.

LFY mRNA is first detected at a low level in the flanks of the inflorescence meristem in cells constituting floral anlagen. *LFY* expression is not detectable in the central region of the inflorescence meristem. *LFY* is uniformly expressed at a high level throughout stage 1 and stage 2 flower primordia (Weigel et al., 1992). In stage 4 flowers *LFY* mRNA becomes reduced in the inner whorls relative to the outer whorl of the wild-type *Arabidopsis* flower. After stage 5, *LFY* expression is reduced in the first-whorl sepals, but is detected in petals, stamen filaments, and carpels. Additionally, a low level of *LFY* expression is detectable in cauline leaf primordia.

LFY expression is largely normal in an *apetala1* mutant background, but is greatly reduced in an *apetala1 cauliflower* background (Weigel et al., 1992; Bowman et al., 1993).

(**C**) and (**D**) *In situ* hybridization of an *APETALA1* (*AP1*) anti-mRNA probe with a longitudinal section (8 μm) through a wild-type inflorescence. Note the inflorescence meristem (i), two stage 2 flowers (2), a stage 4 flower (4), a stage 7 flower (7), and parts of older flowers in which the floral organs have differentiated (se, sepal; p, petal; st, stamen; c, carpel; pd, pedicel).

(**C**) Bright-field micrograph.

(**D**) Dark-field micrograph. Distribution of *AP1* transcripts is represented by silver grains, which appear as white dots.

AP1 mRNA is first detected in stage 1 flowers and is uniformly expressed at a high level throughout stage 1 and stage 2 flower primordia (Mandel et al., 1992b). However, in stage 3–4 flowers, *AP1* mRNA becomes restricted to the first and second whorls of the wild-type *Arabidopsis* flower, with no detectable expression in the presumptive third and fourth whorls. Expression is maintained until past stage 12, throughout the development of sepals and petals. No expression is detected in the stamens and carpels. In addition, *AP1* mRNA is present throughout the pedicel during flower development at least through stage 12. *AP1* mRNA is not detectable in the inflorescence meristem.

AP1 expression is largely normal in an *apetala1* mutant background, but is greatly reduced in an *apetala1 cauliflower* background (Bowman et al., 1993).

J.L. Bowman

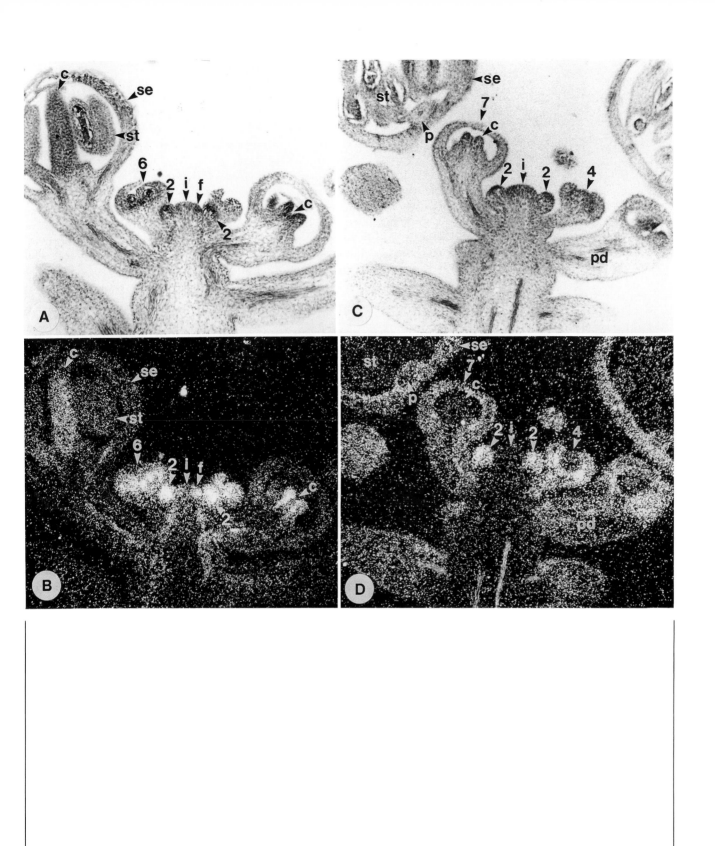

Plate 3.30

213

Plate 3.31
Inflorescence development in *clavata* **mutants**

(**A**) and (**B**) Homozygous *clavata1-4 ERECTA* (*clv1-4 ER*) plants.

(**A**) A dramatic defect in inflorescence meristem structure can be observed when *clavata1-4 erecta* plants are grown at 16°C. In this case, the inflorescence meristem, instead of becoming fasciated, develops into a huge, misshapen mass that dwarfs inflorescence meristems of wild-type plants. Despite this immense disruption in the apical meristem, it is still able to generate floral primordia at its base.

(**B**) When *clv1-4* plants are grown at 25°C, the primary inflorescence meristem is fasciated. The apical meristem grows as a line rather than a point.

(**C**) *clv1-4 leafy-1* double mutant plants grown at 25°C generate only 5–15 "flowers" after the transition from vegetative growth, and then produce hundreds of filamentous organs and bracts. The few flowers that develop are almost completely transformed into inflorescence-like shoots (not shown). All traces of a whorled pattern of organ initiation typical of flowers are absent, there is often internode elongation between the spirally initiated organs, and the flowers are indeterminate and continue to generate carpelloid and leaf-like organs.

(**D**) *hanaba taranu-1* (*han-1*) mutant plants grown at 25°C. *han-1* flowers have reduced second and third whorls, allowing for the formation of only a few petals and stamens, and occasionally, anther-less filaments. *han-1* inflorescences develop normally, with no fasciation of the stem occurring when grown at 25°C.

(**E**) *clv3-1 han-1* double mutant plants grown at 25°C produce very few flowers (less than 10) after the transition to flowering. The phenotype of the flowers varies acropetally, with the deviation from wild type greater in the apical positions. The most basal flowers have sepals and a gynoecium, while the partially formed flowers lack sepals and consist of a gynoecium with occasional sepaloid tissue. In contrast, the structures produced in apical positions have little remaining floral character, as many are simply filamentous organs tipped with stigmatic tissue. In this *clv3-1 han-1* plant, the inflorescence structure is extremely disrupted, and the inflorescence meristem has generated hundreds of filamentous organs and bracts that are tipped with stigmatic tissue.

Bar = 40 μm in **A**; 100 μm in **C**, **D**; 200 μm in **B**; 400 μm in **E**.

S.E. Clark, H. Sakai, and E.M. Meyerowitz

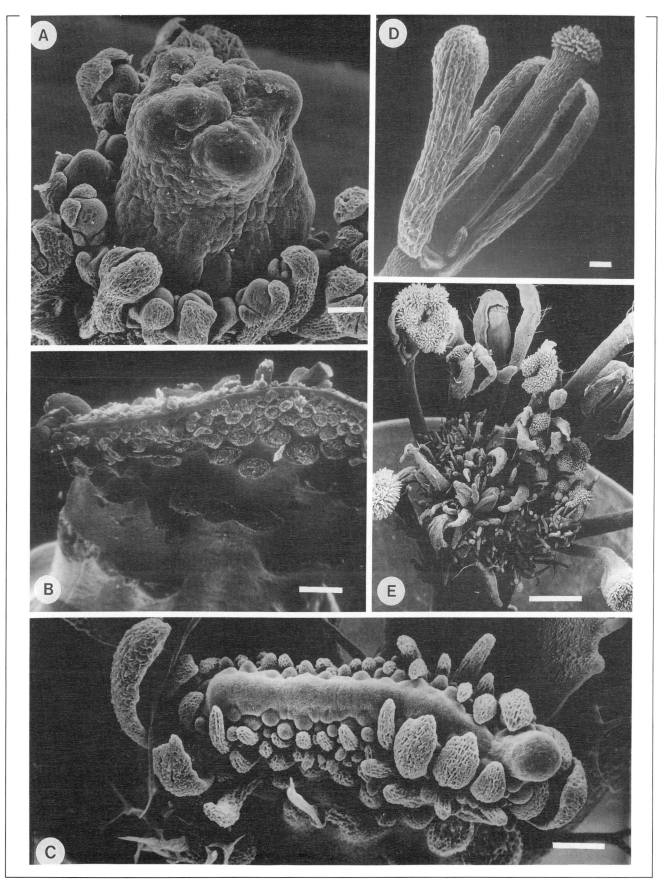

Plate 3.31

215

Plate 3.32
The floral homeotic gene
AGAMOUS: **phenotype of**
agamous **mutants**

Mutations in the *AGAMOUS* (*AG*) gene cause homeotic alterations of organ identity in the third and fourth whorls of the *Arabidopsis* flower (Bowman et al., 1989; Bowman et al., 1991b). Flowers of plants homozygous for the recessive *ag-1*, *ag-2*, or *ag-3* mutations consist of an indeterminate number of whorls of sepals and petals, with no staminoid or carpelloid tissue. Observations on developing flowers make clear the basis of the *agamous* phenotype. The sepals develop normally in the first whorl; the primordia of the second and third whorls also form in their wild-type positions. However, while the second-whorl primordia develop into petals as in wild type, the third-whorl primordia do not develop into the wild-type stamens. Instead they develop into petals indistinguishable from those of the second whorl. In addition, the cells that would normally develop into the gynoecium instead behave as if they constituted another floral meristem. This internal floral meristem reiterates the developmental program of the three outer whorls. This process of reiterative development continues indeterminately, but the precise positions of internal whorls of organ primordia are irregular, and many of the internal organs are mosaics of both petal and sepal tissue. Thus, the phenotype of *agamous* flowers can be thought of as (sepals, petals, petals)$_n$ with respect to organ identity, and (1, 2, 3)$_n$ with respect to organ number and position. The *ag-2* flowers exhibit marked internode elongation between successive internal flowers while *ag-1* and *ag-3* flowers, which are indistinguishable from each other, do not. This is due to differences at the *ERECTA* locus; *ag-1* and *ag-3* are in the Landsberg *erecta* background while *ag-2* is in the Wassilewskija background. Similar mutations, probably allelic, have been described previously in *Arabidopsis* (Braun, 1873; Conrad, 1971).

The *AGAMOUS* gene has been cloned (Yanofsky et al., 1990) and is a member of the MADS-box family of genes. The *ag-1* mutation is the result of a splice-site acceptor mutation 3′ to the MADS box, and the *ag-2* mutation is the result of a 35 kilobase T-DNA insertion in an intron just 3′ to the MADS box.

(**A**) Cross-section of an *agamous-1* flower. *ag-1* flowers are characterized by a large number of whorls of sepals and petals in the pattern: (sepals, petals, petals)$_n$, produced by an indeterminate floral meristem.

(**B**) Mosaic organs of distinct sectors of wild-type sepal (se) and wild-type petal (p) tissue are common in positions of internal whorls of *ag-1* flowers. The transition between the two types of tissue is usually abrupt, with a zone of one to three cells of intermediate type.

(**C**) *ag-4* flower. While most alleles of *agamous* have no stamens or carpels, the flowers of *ag-4*, a presumed "weak" partial loss-of-function allele, plants consist of an outer whorl of sepals (se), a second whorl of petals (p), and a third whorl of slightly petaloid stamens (st). Functional pollen can develop in the third-whorl stamens. Development interior to the third whorl generally consists of indeterminate growth, with the production of alternate whorls of carpelloid sepals (cs) and petaloid stamens (*arrow*). Thus, the phenotype of *ag-4* flowers can be summarized as: sepals, petals, petaloid stamens, (carpelloid sepals, petaloid stamens)$_n$ (John Alvarez, unpublished).

Bar = 20 μm in **B**; 100 μm in **A**, **C**.

J.L. Bowman

A reproduced from Bowman and Meyerowitz (1991) with permission from The Company of Biologists Ltd. **B** reproduced from Bowman et al. (1989) with permission from the American Society of Plant Physiologists.

Plate 3.32

Plate 3.33
The floral homeotic gene
***AGAMOUS*: expression**
pattern and development
of *agamous* mutants

(**A**) *agamous-1* (*ag-1*) flower bud at stage 6. The development of the outer two whorls of organ primordia is indistinguishable from wild type. The third-whorl organ primordia (w3) are in the correct positions, but they appear smaller relative to second-whorl primordia (w2) than do wild-type third-whorl organ primordia. The floral meristem (f) interior to the third-whorl organ primordia is morphologically indistinguishable from that of wild type. Three of the first-whorl sepals have been removed.

(**B**) *ag-1* flower bud at approximately stage 8. The cells that would ordinarily give rise to the gynoecium in wild-type flowers have instead behaved as if they constituted another flower primordium, producing a fourth whorl (w4) of four organ primordia which will differentiate into sepals. Interior to the fourth whorl, supernumerary whorls of organ primordia will be produced in a manner reminiscent of the pattern of second- and third-whorl primordia. Both the second- and third-whorl organ primordia have begun to differentiate into petals. Three first-whorl sepals have been removed.

(**C**) Older *ag-1* flower bud that has not yet opened. Second (w2) and third (w3) petals are visible as well as the fourth whorl of sepals (w4), interior to which an indeterminate number of whorls of organs are developing. Three first-whorl sepals and a third-whorl petal have been removed.

(**D**) Floral meristem of an *ag-1* flower that has opened. After the first few iterations of the sequence (sepals, petals, petals)$_n$, the pattern in which the floral meristem produces organ primordia becomes irregular. Many of these primordia will give rise to mosaic organs comprised of distinct sectors of sepal and petal tissue.

(**E**) and (**F**) *In situ* hybridization of an *AG* anti-mRNA probe with a longitudinal section (8 μm) through a wild-type inflorescence. The inflorescence meristem (i), a stage 2 flower (2), a stage 3 flower (3), and an older flower in which the floral organs have differentiated, are visible.

(**E**) Bright-field micrograph.

(**F**) Dark-field micrograph. Distribution of *AG* transcripts is represented by silver grains which appear as white dots. *AG* is expressed specifically in the third and fourth whorls of the flower. *AG* mRNA is first detectable in stage 3 flowers, at which time only the first-whorl organ primordia have formed in those cells that will give rise to the third- and fourth-whorl organ primordia (Drews et al., 1991). No *AG* mRNA is detected in the inflorescence meristem, or in stage 1 or stage 2 flowers. When the third- and fourth-whorl organ primordia arise, *AG* is uniformly expressed throughout the organ primordia. No *AG* expression is detected in the first- and second-whorl organ primordia, or mature sepals (se) and petals. Later, when the third- and fourth-whorl organ primordia begin to differentiate into stamens and carpels, respectively, *AG* mRNA becomes restricted to certain cell types in the developing stamens (st) and carpels (c) (Bowman et al., 1991a). Specifically, in the stamens, *AG* is expressed in the connective and endothecium, but is notably absent from the microspore cell lineage from the time meiosis takes place. Within the carpel, *AG* is expressed in the stigmatic tissue and is initially expressed uniformly throughout ovule primordia, but later becomes restricted to the endothelium of the ovule, and is notably absent from the embryo sac. *AG* mRNA is also detected in the nectaries.

The early expression pattern of *AG* is altered in an *apetala2-2* background, with the domain of *AG* expression expanding to include all four whorls of

Bar = 20 μm in **A**, **B**, **D**; 50 μm in **C**.

(*Text continued on p. 221*)

Flowers: Floral Organ Identity Mutants

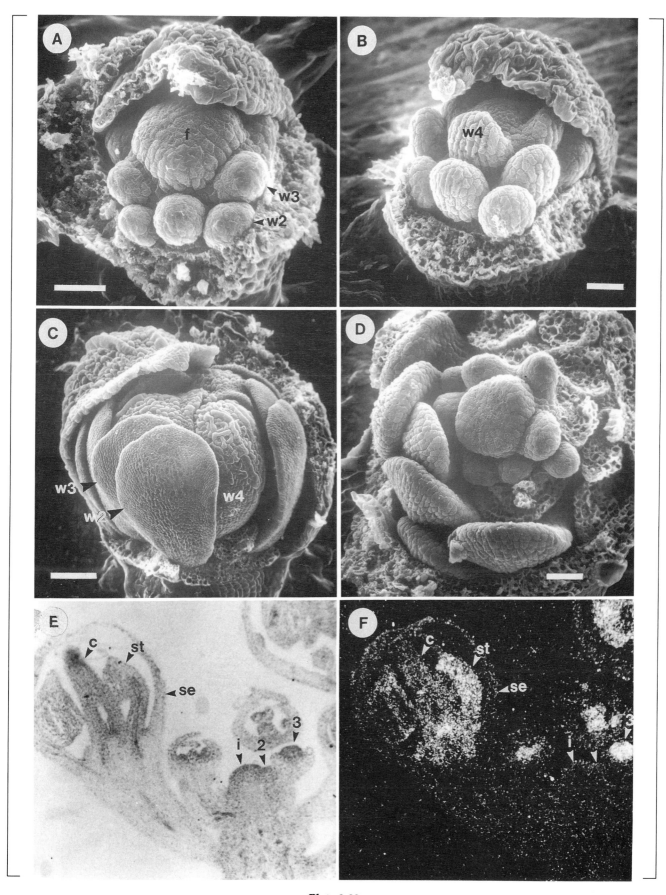

Plate 3.33

219

Plate 3.33
The floral homeotic gene
***AGAMOUS*: expression**
pattern and development
of *agamous* mutants
(*continued*)

stage 3 *ap2-2* flowers (Drews et al., 1991). This indicates that the *AP2* gene product is a negative regulator of *AG* mRNA accumulation in the first and second whorls of wild-type flowers. Furthermore, constitutive ectopic expression of *AG* in all whorls of transgenic flowers results in an *ap2*-like phenotype (Mandel et al., 1992a; Mizuhami and Ma, 1992), suggesting that the *AG* gene product can negatively regulate AP2 activity as well. The later cell-type specific *AG* expression patterns are not altered in *ap2* mutants, even in the ectopic first whorl carpels of *ap2-2* flowers (Bowman et al., 1991a). However, constitutive ectopic expression of *AG* also leads to developmental abnormalities other than floral organ identity changes: specifically, it results in male-sterility and abnormal ovule development (Mandel et al., 1992a; Mizuhami and Ma, 1992).

J.L. Bowman

A, **B**, **C** reproduced from Bowman et al. (1989) with permission from American Society of Plant Physiologists.

Plate 3.34

The floral homeotic gene *AGAMOUS*: floral histology of *agamous-1* and *clavata1-1* mutant flowers

(**A–C**) *agamous-1* (*ag-1*) flowers.

(**A**) Tangential longitudinal section of a stage 5 *ag-1* flower. *Arrows* indicate L2 periclinal divisions in initiating third-whorl organs primordia. L2 elongated cells in initiating second-whorl primordia are indicated (*above* ∗). At this stage, the development of *ag-1* flowers is indistinguishable from that of wild type.

(**B**) Tangential longitudinal section of stage 6 *ag-1* flower. *Arrow* indicates L3 periclinal division in the initiation of a fourth-whorl (w4) organ primordium. The third-whorl organ primordia of *ag-1* flowers initiate with a pattern similar to that observed for wild-type third-whorl stamens. However, the third-whorl organ primordia of *ag-1* flowers differentiate into petals (w3, third-whorl organ primordia; p, second-whorl petal primordia).

(**C**) Diagonal longitudinal section of a slightly older *ag-1* flower in which the third-whorl organs (w3) have begun to develop petaloid features. The fourth-whorl organ primordia (w4) are initiated in a manner similar to wild-type fourth-whorl organs but on the flanks of the remaining floral meristem (Crone, 1992). The fourth-whorl organ primordia rapidly develop characteristics of sepals. The floral meristem continues to produce additional whorls of organ primordia interior to the fourth whorl.

(**D–G**) *clavata1-1* (*clv1-1*) flowers. While the identities of floral organs are not affected, organ numbers in all whorls are altered in *clv1-1* flowers, most notably in the gynoecium which is usually composed of four carpels. Under the growth conditions used, the floral apices of wild-type and *clv1-1* flowers appears similar before gynoecial initiation (Crone, 1992).

(**D**) Radial longitudinal section of a late stage 5 *clv1-1* flower prior to gynoecial initiation. This stage appears to be prolonged in the mutant as compared to the wild type, resulting in an increase in cell number in the apex before carpel initiation (fa, floral apex; st, stamen).

(**E**) Tangential longitudinal section of a stage 6 *clv1-1* flower. The carpels in *clv1-1* flowers initiate on the flanks of the floral apex, in a manner similar to the fourth-whorl organs of *ag-1* flowers (g, gynoecium; ms, medial stamen).

(**F**) Scanning electron micrograph of an approximately axial view of a *clv1-1* flower just after fourth-whorl gynoecial initiation. Note the gynoecial cylinder in *clv1-1* flowers is larger than that of wild type (g, gynoecium; st, stamen; se, sepal).

(**G**) Transverse section of a relatively mature *clv1-1* flower. Note the four carpels and seven stamens (g, gynoecium; st, stamen; p, petal; se, sepal).

Bar = 50 μm in **A**, **B**, **C**, **D**, **E**, **F**; 200 μm in **G**.

W. Crone and E.M. Lord

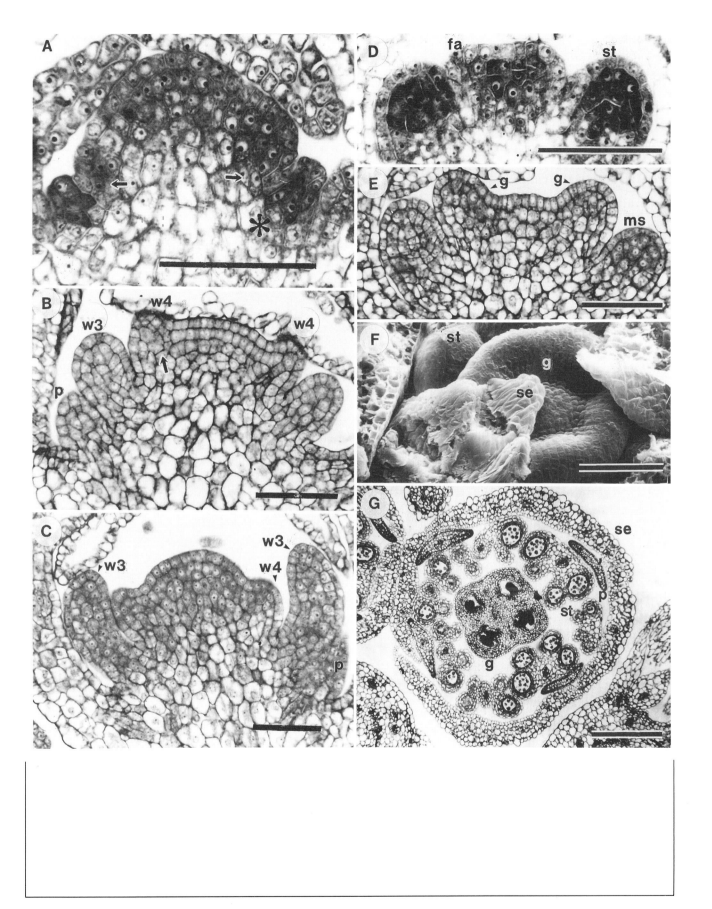

Plate 3.34

223

Plate 3.35
The floral homeotic gene
AGAMOUS: **nectaries**

Characteristics of nectarial tissue in *agamous-1* (*ag-1*) flowers (Landsberg *erecta* ecotype). Plants were grown under constant illumination and temperature (22°C).

(**A**) Open *ag-1* flower possessing sepals and petals but lacking stamens and carpels. A small petaloid appendage arising from a sepal (*open arrow*) and nectarial tissue (*white arrow*) is indicated.

(**B**) Close-up, from a medial-lateral perspective, of nectarial tissue identified in **A**, showing two prominent outgrowths (*arrows*) each surrounded by the bases of three petals: a second-whorl petal (p), a lateral third-whorl stamen (p-ls), and a medial third-whorl petal (p-ms) (mse, medial sepal).

(**C**) Higher magnification image of the foremost nectarial tissue of **B**, showing three modified stomata (*arrowheads*) on its surface. Note the cuticular patterning of the epidermal cells on the nectarial tissue, and its virtual absence on the guard cells.

(**D**) Open *ag-1* flower, older than that of **A** and from the same plant, showing abscission zones of the first two series (six whorls) of organs (sepals, petals, petals)$_n$, with a successive series of sepals and petals above them. One sepaloid petal has a serrated edge (*open arrow*). Persistent nectarial tissue is demarcated (*arrows*) in the lowest abscission zone.

(**E**) Close-up of the lowermost abscission zone shown in **D**, illustrating the two outgrowths of nectarial tissue each bearing several modified stomata (*arrowheads*). Bulbous cells of the abscission zones (az) occupy positions previously held by petals and sepals.

(**F**) Higher magnification of the leftmost nectarial tissue of **E**. Note the open pores of the modified stomata. The guard cells generally lack the prominent cuticular pattern of adjacent epidermal cells.

Bar = 5 μm in **F**; 20 μm in **C**; 50 μm in **E**; 100 μm in **B**; 200 μm in **D**; 500 μm in **A**.

A.R. Davis

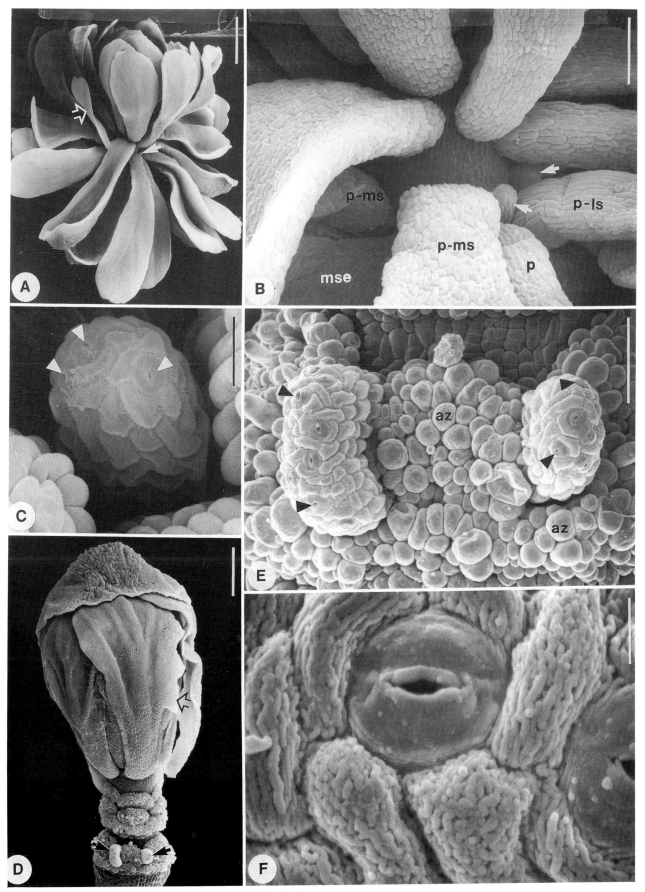

Plate 3.35

Plate 3.36
The floral homeotic gene
APETALA2: phenotype
and development of
apetala2 mutants

Mutations in the *APETALA2* (*AP2*) gene cause homeotic alterations of organ identity in the first and second whorls of the *Arabidopsis* flower (Pruitt et al., 1987; Bowman et al., 1988; Komaki et al., 1988; Bowman et al., 1989; Kunst et al., 1989b; Meyerowitz et al., 1989; Bowman et al., 1991b). Several recessive *ap2* mutant alleles have been isolated and can be arranged into an allelic series. Each of the mutant *ap2* alleles is sensitive to environmental conditions, suggesting that the underlying process in which the wild-type *AP2* gene product is involved is affected by environmental conditions (Komaki et al., 1988; Bowman et al., 1989; Bowman et al., 1991b). Additionally, the phenotypes of *ap2* flowers vary acropetally such that the first flowers produced are closer in phenotype to the wild type than are later produced flowers.

At one extreme of the allelic series are *ap2-2* and *ap2-8* ("strong" alleles) (Bowman et al., 1991b) in which the medial first-whorl organs develop as carpels, the lateral first-whorl organs are either leaf-like or fail to develop, and no second-whorl organs develop. The third- and fourth-whorl organs are affected with respect to number and position, but their identity is largely unaltered. The third-whorl stamens are greatly reduced in number (averaging less than one stamen per flower when grown at 25°C), and the overall morphology of the fourth-whorl gynoecium is slightly abnormal. At the other extreme, the "weakest" allele thus far isolated is *ap2-1* (Koornneef et al., 1983; Bowman et al., 1989). The first whorl of *ap2-1* flowers is occupied by leaf-like organs that may have some carpelloid characteristics. The second-whorl organs of *ap2-1* flowers range from stamens to staminoid petals to petals to leaf-like organs, depending on growth conditions. The third and fourth whorls of *ap2-1* flowers are largely normal. Other alleles, such as *ap2-3*, *ap2-4*, *ap2-5*, *ap2-6*, and *ap2-9*, have phenotypes between these two extremes (Komaki et al., 1988; Kunst et al., 1989; Bowman et al., 1991b). The floral phenotype of the *ap2* allelic series may thus be summarized as follows. The first-whorl organs range from sepals (wild type) to leaf-like organs (*ap2-1*) to carpels (*ap2-2*), and the second-whorl organs range from petals (wild type) to staminoid organs (*ap2-1*) to a failure of organs to develop (*ap2-2*). Additionally, the medial and lateral first-whorl organ primordia have distinctly different fates. Lateral first-whorl organ primordia arise lower on the flank of the floral meristem, and either abort or differentiate into leaf-like organs, while the medial first-whorl organ primordia differentiate into carpels.

Mutations in the *AP2* gene are pleiotropic in that their phenotypic effects are not confined to homeotic conversions of floral organ identity and floral organ number. For example, the seed coat morphology of *ap2* seeds is altered, as well as the seed size and shape. In addition, the size of the apical meristem appears to be smaller in *ap2* mutants as compared to wild type. The *AP2* gene has been cloned (Okamuro et al., 1993) and is expressed in vegetative organs as well as flowers.

(**A**) Developing *ap2-2* flowers (Bowman et al., 1991b). Two medial (m) and one lateral (l) first-whorl organ primordia are indicated on the stage 3 flower. The lateral and medial first-whorl organ primordia have different developmental fates. Usually, four first-whorl organ primordia arise. The medial first-whorl organ primordia develop into carpels, while the lateral either abort or develop into leaf-like organs. In *ap2-2* flowers, lateral first-whorl organ primordia arise lower on the pedicel than do medial first-whorl organ primordia. The remaining floral meristem may give rise to a small number of stamens and a central

Bar = 20 μm in **A, C, E**; 100 μm in **B, D, F**.

(*Text continued on p. 229*)

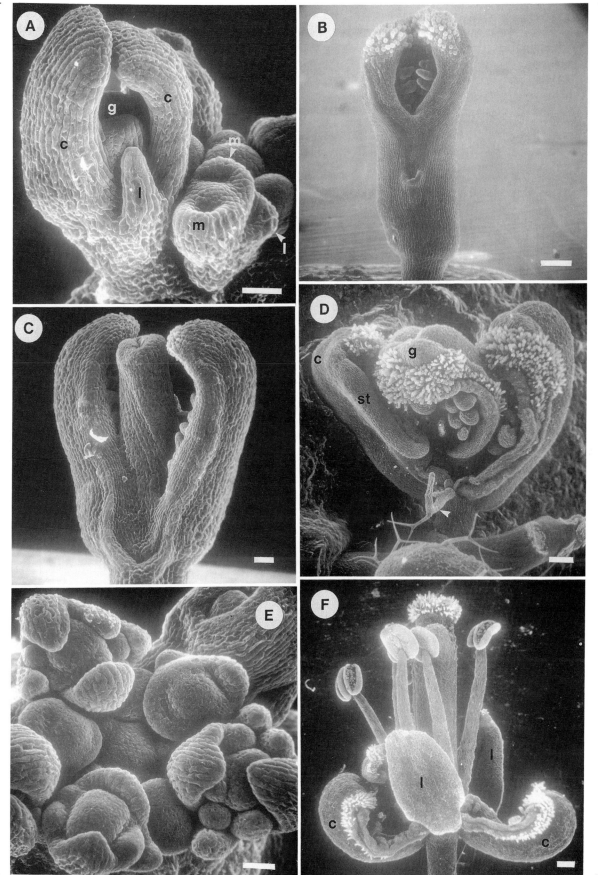

Plate 3.36

227

Plate 3.36
The floral homeotic gene
APETALA2: phenotype
and development of
apetala2 mutants
(*continued*)

gynoecium. The lack of second-whorl organs, and the reduction in number of third-whorl organs is due to a failure to initiate organ primordia. In the older, approximately stage 8 flower, two medial first-whorl carpels (c) and one lateral first-whorl leaf-like organ (l) are developing, with the developing fourth-whorl gynoecial cylinder (g) also visible.

(**B**) An *ap2-8* flower consisting of a single abnormal gynoecium of two partially fused carpels. It is not clear whether the two carpels developed from the first or the fourth whorls. This phenotype, a single gynoecium, either fused or unfused, is common in *ap2-2* and *ap2-8* flowers developing in unfavorable growth conditions (e.g., 29°C).

(**C**) An *ap2-2* flower in which the central gynoecium is twisted and the two medial first-whorl carpels are partially fused to the central gynoecium.

(**D**) Mature *ap2-2* flower in which the medial first-whorl organs are solitary carpels, and the central gynoecium (g) has failed to fuse properly. One of the medial first-whorl organs is a mosaic of marginal stamen (st) and central carpel (c) tissue, a common feature of *ap2-2* and *ap2-8* flowers. A lateral first-whorl position is occupied by a filamentous organ with a stellate trichome (*arrowhead*).

(**E**) Inflorescence meristem and young floral buds of an *ap2-9* plant grown at 16°C. Under these growth conditions, all first-whorl organs often fuse together to form a tube, constricting the inner whorls of floral organs. Note the pattern of floral organ primordia is irregular. Therefore, it is difficult to trace a primordium's origin to a specific whorl.

(**F**) Post-dehiscence *ap2-1/ap2-2* trans-heterozygote flower with a phenotype intermediate between *ap2-1* and *ap2-2* homozygotes. In this case, the medial first-whorl organs are solitary carpels (c), the lateral first-whorl organs are leaf-like (l), no second-whorl organs have formed, and the third and fourth whorls are largely normal. All tested *ap2* trans-heterozygotes exhibit phenotypes intermediate between the two respective homozygotes.

J.L. Bowman

A, C, E, F reproduced from Bowman et al. (1991b), and D reproduced from Bowman and Meyerowitz (1991) with permission from The Company of Biologists Ltd.

Plate 3.37
The floral homeotic gene
APETALA2: **phenotype**
and development of
apetala2-1 **flowers**

Flowers of plants homozygous for the *apetala2-1* (*ap2-1*) mutation have pattern defects in the first and second whorls of the flower (Bowman et al., 1989). The phenotype is temperature sensitive and varies acropetally such that apical flowers exhibit more severe defects than do basal flowers. When grown under favorable conditions (e.g., 16°C), *ap2-1* flowers consist of a first whorl of leaf-like organs, while second-whorl organs vary from leaf-like organs to nearly wild-type petals. In contrast, when grown under less favorable conditions (e.g., 29°C), the first-whorl organs of *ap2-1* flowers are carpelloid leaves, and the second whorl is occupied by staminoid organs, or second-whorl organs fail to develop. Under both growth regimes, the third and fourth whorls are largely normal with respect to floral organ identity.

(A–G) Phenotypes of organs observed in second-whorl positions of *ap2-1* flowers. Petals and leaf-like organs, or mosaics with sectors of petal and leaf-like tissue are common (over 90% of positions) at 16°C. Staminoid petals are common (about 60% of positions) at 25°C, while staminoid organs (about 10% of positions) or an absence of organs (over 70% of positions) occur at 29°C. Scoring the phenotypes of second-whorl organs in temperature-shift experiments indicates that the wild-type *AP2* gene product functions during stages 2 through 4 to specifying, in part, the fate of the second-whorl organs. Abaxial views shown in **A**, **B**, and **C**; adaxial views in **D**, **E**, **F**, and **G**.

(A) Leaf-like organ.
(B) Mosaic organ with distinct sectors of petal and leaf-like tissue.
(C) Mostly white, petal-shaped organ with trichomes and stomata.
(D) Nearly morphologically wild-type petal.
(E) Petal-shaped organ with rudimentary locules.
(F) Stamen-shaped organ with petaloid tissue at its apex. (Note: Staminoid petals in the second whorl of *ap2-1* flowers may either be mosaics of sectors of wild-type stamen and wild-type petal tissues, or have sectors that appear intermediate between petal and stamen tissue.)
(G) Morphologically normal stamen.
(H) *ap2-1* inflorescence meristem and flower buds up to stage 5.
(I) Approximately stage 8 *ap2-1* flower. Development of first-whorl organs is characterized by the presence of stipules (sp), a feature of wild-type leaves but not of wild-type sepals. Additionally, developing trichomes (t) can be seen on the abaxial surface of the first-whorl organs. At this time in development, the second-whorl organ primordia (w2), which will likely develop into staminoid petals, are indistinguishable from wild-type second-whorl organ primordia.
(J) Mature *ap2-1* flower. The first-whorl organs are leaf-like, with numerous stellate trichomes, a feature of wild-type leaves but not of wild-type sepals. Three staminoid petals (*arrowheads*) and a petaloid organ (p) occupy the second whorl. Five third-whorl stamens surround the gynoecium.
(K) Stigmatic tissue at the tip of a first whorl organ as can be seen in **J**. Carpelloid features such as stigmatic tissue at the tip, or rudimentary ovules along the margins occur most frequently on medial first-whorl organs.
(L) Close-up of a staminoid petal showing that the epidermal cells have features of both petals (general shape of cells) and stamens (irregularly sized and interspersed with stomata, *arrowhead*).

Bar = 20 μm in **H**, **I**, **K**, **L**; 100 μm in **A**, **B**, **C**, **D**, **E**, **F**, **G**, **J**.

J.L. Bowman

A–I reproduced from Bowman et al. (1989) with permission from American Society of Plant Physiologists. J reproduced from Bowman and Meyerowitz (1991) with permission from The Company of Biologists Ltd.

Flowers: Floral Organ Identity Mutants

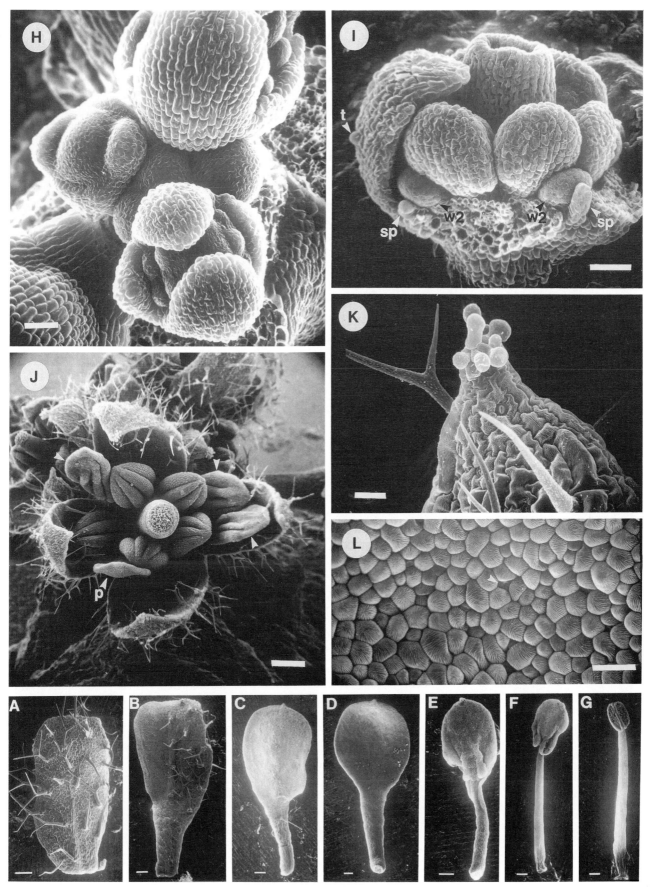

Plate 3.37

231

Plate 3.38
The floral homeotic gene
***APETALA2*: histology of**
***apetala2-1* mutant**
flowers

Patterns of cell division during the initiation of the first- and second-whorl organ primordia (when present) of *ap2-1* flowers are similar to those of the corresponding organ primordia in wild type (Crone, 1992).

(**A**) Radial longitudinal section of a stage 4 *ap2-1* flower. A periclinal L2 division during the initiation of the first whorl organ primordia is indicated (*arrow*) (ab, abaxial first-whorl organ).

(**B**) Radial longitudinal section of a stage 5 *ap2-1* flower. *Arrow* indicates L2 periclinal division in initiation third-whorl (stamen) primordium. L2 anticlinally elongated cells (*above* *) indicate an initiating second-whorl primordium. This is similar to the initiation of second-whorl petal primordia in wild type.

(**C**) Tangential longitudinal section of a stage 6 *ap2-1* flower showing a third-whorl stamen (st) primordium and a second-whorl organ primordium (w2). No distinction can be made between the wild-type petal and the *ap2-1* second-whorl organ until approximately stage 8 (Bowman et al., 1989).

(**D**) Radial longitudinal section of an *ap2-1* flower at approximately stage 8. Stigmatic papillae (*arrowhead*) at the tip of a first-whorl organ are indicated. Thus, the first- and second-whorl organs of *ap2-1* flowers achieve their altered phenotype (into carpelloid leaves and staminoid petals, respectively) during later differentiation, and not in the early initiation processes (g, gynoecium; ab, abaxial first-whorl organ; ad, adaxial first-whorl organ; ms, medial stamen).

Bar = 25 μm in **C**; 50 μm in **A**, **B**, **D**.

W. Crone and E.M. Lord

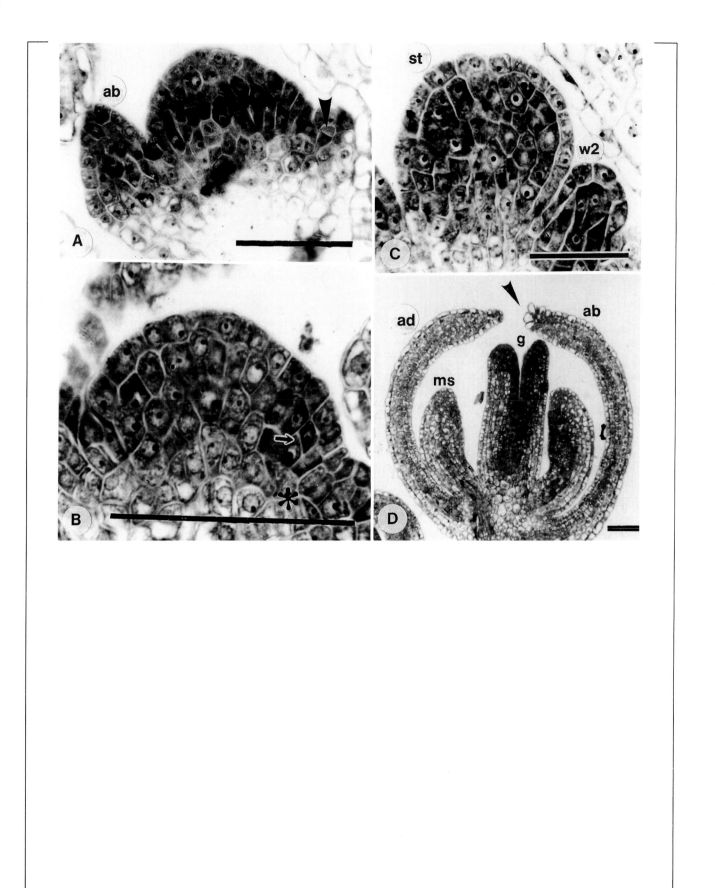

Plate 3.38

233

Plate 3.39
The floral homeotic gene
PISTILLATA: phenotype
and development of
pistillata mutants

Mutations in the *PISTILLATA* (*PI*) gene cause homeotic alterations of organ identity in the second and third whorls of the flower (Bowman et al., 1989; Hill and Lord, 1989; Bowman et al., 1991b). Homeotic conversions in *pi* flowers are petals to sepals in the second whorl, and stamens to carpels in the third whorl, transformations that are also seen in *apetala3* flowers. Three alleles of *pi* (*pi-1*, *pi-2*, *pi-3*) have been analyzed and can be arranged into an allelic series (Bowman et al., 1991b). Of these, *pi-1* flowers display the greatest phenotypic deviation from wild type. Second-whorl organs of *pi-1* flowers are completely transformed into sepals. Cells that would in wild type form the organs of the third whorl appear to be largely incorporated into the gynoecium which, as a consequence, is abnormal in size and irregular in structure. In the "weakest" allele, *pi-3*, the second-whorl organs develop as sepals, and while the third-whorl organs are generally distinct from the fourth-whorl gynoecium, they develop into solitary carpels, carpelloid stamens, or filamentous organs. Flowers of *pi-2* plants exhibit a phenotype intermediate between that of *pi-1* and *pi-3*. The allelic series of *pi* mutants parallels the allelic series of *apetala3* mutants.

(**A**) Stage 5 *pi-1* floral bud. Development of *pi-1* flowers is indistinguishable from that of wild type until the time of appearance of the second- and third-whorl organ primordia. The second-whorl organ primordia (*arrowheads*) arise in the correct number and position, and at this stage of development are indistinguishable from wild-type second-whorl organ primordia. By stage 5 in wild-type flowers, the third-whorl organ primordia have become distinct from the cells that will give rise to the gynoecium. In contrast, in *pi-1* flowers, third-whorl organ primordia seldom become distinct from those cells that would normally comprise the fourth-whorl organ primordia.

(**B**) Approximately stage 7–8 *pi-1* flower. The second-whorl organ primordia (*arrowhead*) are still indistinguishable from those of wild type. The cells that normally develop into third-whorl organ primordia appear to be incorporated into the developing central gynoecium (g). Gynoecial growth proceeds with characteristic rapid vertical growth at the periphery of the central dome of cells of the flower bud, but the diameter of the cylinder is larger than in wild type. Sometimes, rather than behaving as a single cylinder of tissue, somewhat distinct but fused third- and fourth-whorl organs may be evident.

(**C**) Older *pi-1* flower showing that the cells of the third and fourth whorls are developing as a single central gynoecium. The second-whorl organs (*arrowhead*) are clearly developing as sepals rather than petals, even though they develop on the delayed time course of wild-type petals.

(**D**) Mature *pi-1* flower with a gynoecium of abnormal morphology. Gynoecia of *pi-1* flowers are usually composed of more than two carpels (average approximately 2.7). In addition, filamentous structures (*arrowhead*) may develop from the side of the abnormal gynoecium.

(**E**) Mature *pi-1* flower showing that the epidermal morphology (elongated abaxial cells and a margin of smaller cells) of the second-whorl sepals (w2) is indistinguishable from that of the first-whorl sepals (w1). A nectary (*arrowhead*) is visible at the base of the flower, indicating that nectary development is independent of stamen development.

Bar = 20 μm in **A**, **B**; 50 μm in **C**; 100 μm in **D**, **E**.

J.L. Bowman

A and **B** reproduced from Bowman et al. (1989) with permission form the American Society of Plant Physiologists; **E** reproduced from Bowman and Meyerowitz (1991) with permission from The Company of Biologists Ltd.

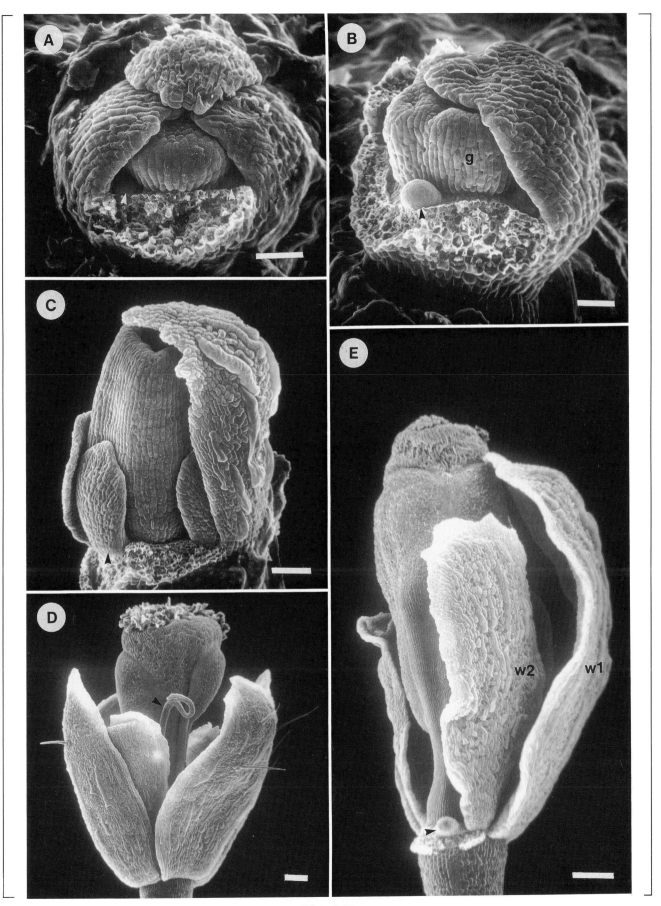

Plate 3.39

Plate 3.40

The floral homeotic gene PISTILLATA: histology of pistillata-1 mutant flowers

(A) SEM view of a *pi-1* flower with two first-whorl sepals (se) and two second-whorl sepals (w2) removed. The morphology of the gynoecium (g) is extremely variable in *pi-1* flowers (sg, stigma).

(B) Radial section of *pi-1* flower at an early stage of abaxial (ab) and adaxial (ad) sepal development. The structure of wild-type and *pi-1* flowers is similar at this stage (fa, floral apex).

(C) An oblique section of a *pi-1* flower near the time of stamen initiation in wild-type flowers. The floral meristem has an abnormal shape and the pattern of cell division initiated by periclinal divisions in the L2 and L3 distributed along the flanks of the floral meristem (*) is not restricted to the sites of stamen initiation, as in the wild type. A primordial buttress resembling a wild-type stamen primordia is visible (*small arrow*) above a developing second-whorl organ primordium (w2) (fa, floral apex; la, lateral sepal).

(D) In contrast to the wild type, distinct stamen and gynoecial primordia do not develop in *pi-1* flowers. Rather, all of the remaining floral meristem interior to the second-whorl primordia gives rise to gynoecial tissue, with the regions of apical growth (*arrows*) no longer conforming to the wild-type pattern.

(E) Oblique section of a *pi-1* flower. Following the production of second whorl primordia (w2), there is an excessive number of periclinal divisions (*) along the flanks of the floral meristem. No third-whorl primordia are evident in this section, but the second-whorl primordia are comparable to those of wild type.

Bar = 1 mm in **A**.

J.P. Hill and E.M. Lord

Reproduced from Hill and Lord (1989) with permission from the National Research Council of Canada.

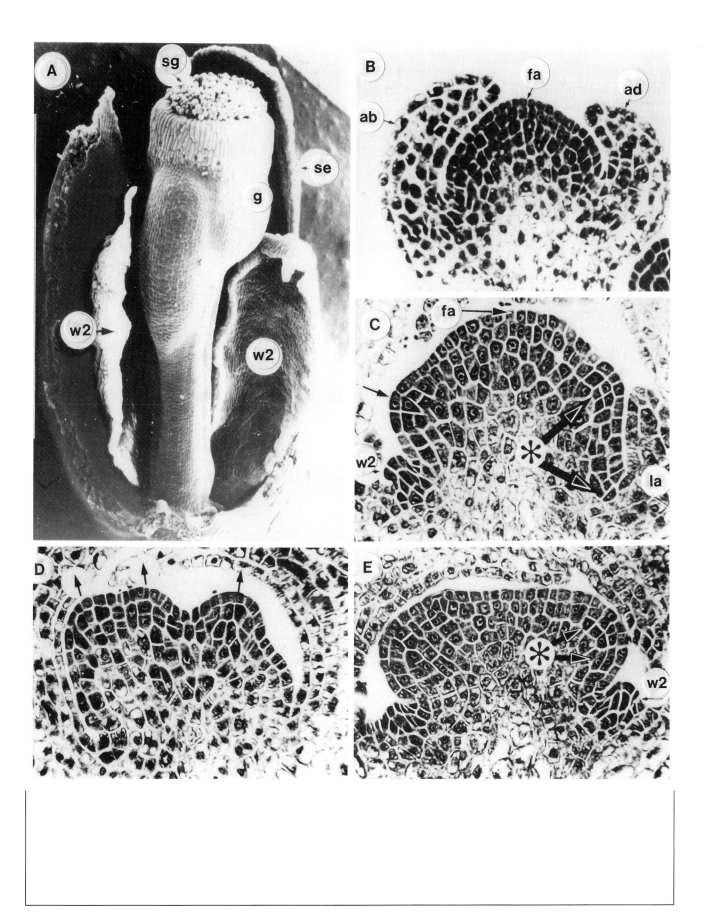

Plate 3.40

237

Plate 3.41
The floral homeotic gene *PISTILLATA*: histology of *pistillata-1* mutant flowers

(**A**) Second-whorl primordia (w2) initiation in *pistillata-1* (*pi-1*) flower. The overall position, size, and pattern of cell division is similar to wild-type second-whorl primordia initiation. The *arrow* (*) indicates a recent periclinal division of an L2 cell.

(**B–D**) Longitudinal sections of *pi-1* second-whorl organs.
(**B**) *pi-1* second-whorl organ (w2) 40 μm in length. The organ is indistinguishable from that of wild type at this stage.
(**C**) *pi-1* second-whorl organ (w2) 90 μm in length. The apical regions (*arrow*) of the organ are differentiating at this stage when the wild-type petal would still show active cell division at its apex.
(**D**) *pi-1* second-whorl organ (w2) 160 μm in length. At this stage it is clearly different in histology from the wild-type second-whorl petal, and is taking on sepal characteristics.

(**E**) A *pi-1* flower at a stage prior to ovule initiation. The terminal portion of the floral meristem has produced two zones of growth reminiscent of wild-type gynoecial development (g?), but also has abnormal gynoecial tissue (ag) along the flanks of the developing central gynoecial primordium.

J.P. Hill and E.M. Lord

Reproduced from Hill and Lord (1989) with permission from the National Research Council of Canada.

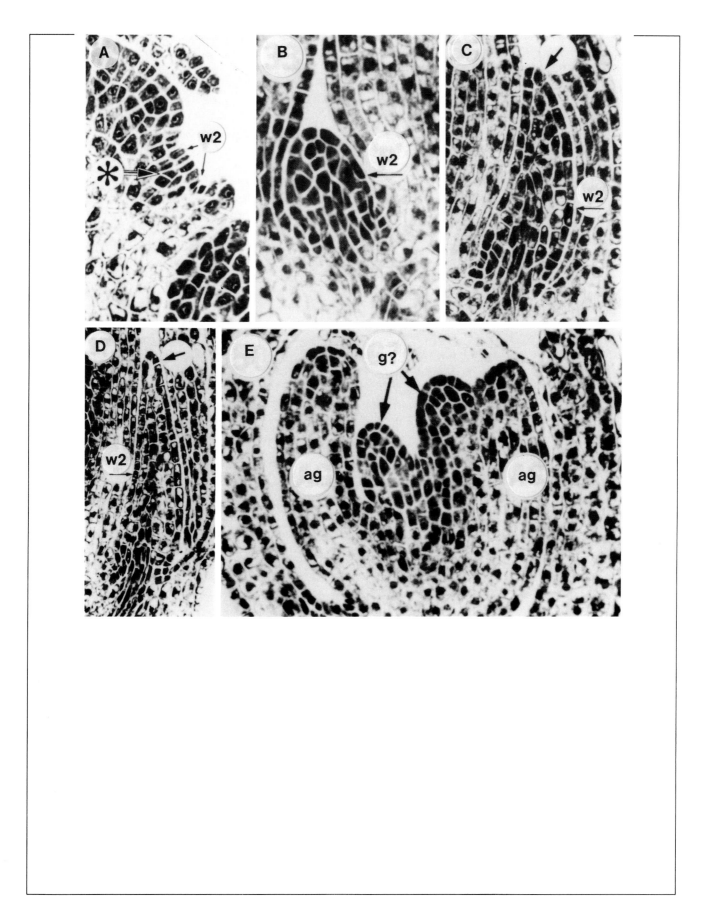

Plate 3.41

239

Plate 3.42
The floral homeotic gene
***APETALA3*: expression**
pattern and phenotype
and development of
***apetala3-1* mutants**

Recessive *apetala3* (*ap3*) mutations result in homeotic alterations of organ identity in the second and third whorls of the *Arabidopsis* flower (Bowman et al., 1989; Jack et al., 1992). Homeotic conversions observed in *ap3* flowers are petals to sepals and stamens to carpelloid organs, transformations that are also seen in *pistillata* flowers. The three mutant alleles of *ap3* can be arranged into an allelic series that parallels that of *pistillata*. *ap3-3* flowers display the greatest deviation from wild type, with complete transformations occurring in both affected whorls. In contrast, the phenotype of the weakest allele, *ap3-1*, is variable and temperature sensitive. When grown at 16°C, *ap3-1* flowers are nearly wild-type, except that the petals are slightly sepaloid. At 30°C, complete transformations of both the second- and third-whorl organs are observed; intermediate growth temperatures result in intermediate phenotypes. The phenotype of *ap3-1* flowers also varies acropetally such that later-produced flowers exhibit more severe alterations than do earlier-produced flowers. The *ap3-1* phenotype can be summarized as (whorls 1, 2, 3, 4) sepals, sepaloid petals, stamens, carpels, at 16°C; sepals, sepals, carpelloid stamens, carpels, at 25°C; and sepals, sepals, carpels, carpel, at 30°C.

(**A–E**) Phenotypes of organs observed in the third-whorl positions of *ap3-1* flowers. The fates of the lateral and medial third-whorl organs differ in that the lateral third-whorl organs are consistently less affected (less carpelloid) than the medial third-whorl organs. For example, at 29°C over 70% of medial organs are carpels, while over 70% of lateral organs are staminoid. In addition, the third-whorl organs in basal flowers are less carpelloid than those of more apical flowers. Morphologically wild-type stamens (**A**) are present in all positions at 16°C. Stamens capped with stigmatic tissue are common in the basal flowers of plants grown at 25°C. Carpelloid stamens (see **B**; organs shaped like stamens, but possessing ovules and stigmatic tissue) and staminoid carpels (see **C**; organs shaped like carpels, but possessing rudimentary locules) are common in the medial position at 25°C and lateral positions at 29°C. Filamentous organs (**D**) may be present in both lateral and medial positions at 29°C, and carpels (**E**) are common in medial positions at both 25°C and 29°C.

Scoring the phenotypes of third-whorl organs in temperature-shift experiments indicates that wild-type *AP3* gene product activity is necessary through stage 6 (after the third-whorl primordia have arisen, but before they begin to differentiate) to specify, in part, the identity of the third-whorl organs.

(**F**) and (**G**) *ap3-1* flowers from plants grown at 25°C.

(**F**) Development of *ap3-1* flowers parallels that of wild-type until about stage 7–8, when the third-whorl organ primordia begin to differentiate. This stage 7 *ap3-1* flower (three first-whorl sepals have been removed) is indistinguishable from wild-type, with all whorls of organ primordia in the normal numbers and positions. However, the second-whorl organs differentiate into sepals, although they develop on a time course of wild-type petals. The epidermal surface of developing third-whorl organs is characterized by vertical cell files, typical of wild-type carpels, but not of wild-type stamens.

(**G**) Mature *ap3-1* flower with some first- and second-whorl organs removed. When *ap3-1* plants are grown at 25°C, third-whorl organs are typically mosaic organs comprised of sectors of wild-type stamen and wild-type carpel tissue. A solitary carpel (c) and carpelloid stamens (*arrowheads*) occupy third-whorl positions in this flower. Third-whorl organs developing in adjacent positions can

Bar = 20 μm in **F**; 30 μm in **D**; 100 μm in **A**, **B**, **C**, **E**, **G**.

(*Text continued on p. 243*)

Flowers: Floral Organ Identity Mutants

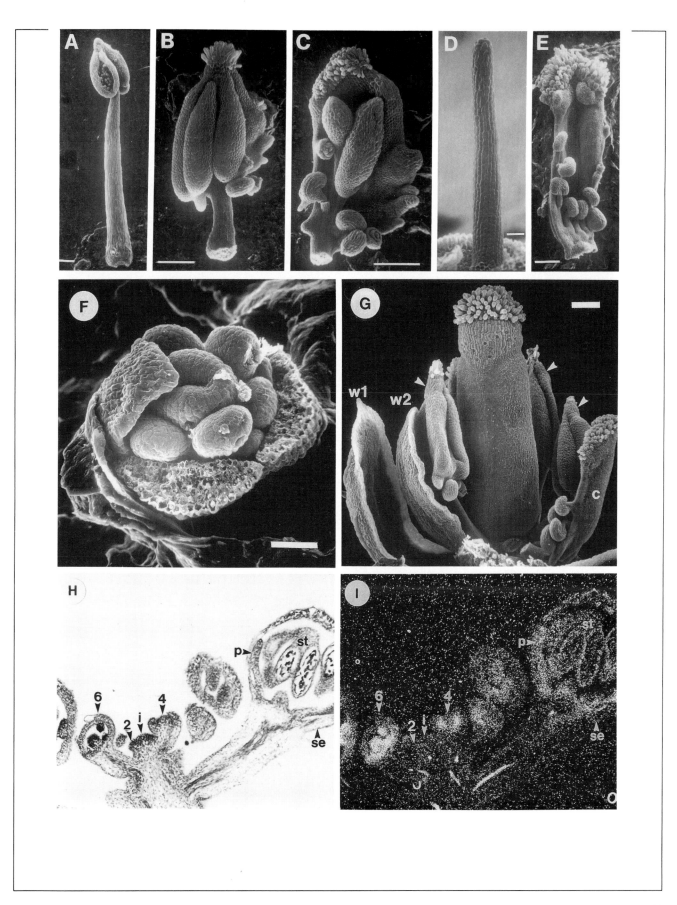

Plate 3.42

241

Plate 3.42
The floral homeotic gene
APETALA3: **expression**
pattern and phenotype
and development of
apetala3-1 **mutants**
(*continued*)

have significantly different developmental fates. One first-whorl sepal (w1) and one second-whorl sepal (w2) are also visible.

(**H**) and (**I**) *In situ* hybridization of an *AP3* anti-mRNA probe with a longitudinal section (8 μm) through a wild-type inflorescence. Note the inflorescence meristem (i), a stage 2 flower (2), a stage 4 flower (4), a stage 6 flower (6), and an older flower in which the floral organs have differentiated (se, sepal; p, petal; st, stamen).

(**H**) Bright-field micrograph.

(**I**) Dark-field micrograph. Distribution of *AP3* transcripts is represented by silver grains which appear as white dots. *AP3* is expressed specifically in the second and third whorls of the flower (Jack et al., 1992). *AP3* mRNA is first detectable in stage 3 flowers, at which point only the first-whorl organ primordia have formed in those cells that will give rise to the second- and third-whorl organ primordia (Jack et al., 1992), as in the stage 4 flower shown. No *AP3* mRNA is detected in the inflorescence meristem, stage 1 flowers, or stage 2 flowers. When the second and third-whorl organ primordia arise, *AP3* is uniformly expressed throughout the organ primordia, as in the stage 6 flower. No *AP3* expression is detected in the first- and fourth-whorl organ primordia, or mature sepals and carpels. *AP3* continues to be expressed throughout the petals and stamens until at least stage 12.

J.L. Bowman

A–E, G reproduced from Bowman et al. (1989) with permission form the American Society of Plant Physiologists.

Plate 3.43
Genetic interactions
among floral homeotic
genes: double mutants

Double mutant combinations have been constructed between many of the floral homeotic mutations described in *Arabidopsis* (Bowman et al., 1989; Bowman et al., 1991b). Two general classes of interactions are observed: combinations in which the double-mutant phenotype is an addition of the single-mutant phenotypes, and combinations in which the double-mutant phenotype has features not observed in either of the single mutants. Essentially additive interactions are observed in *agamous apetala3, agamous pistillata, apetala2 apetala3,* and *apetala2 pistillata* double mutants, while non-additive interactions occur in *apetala2 agamous* double mutants. The phenotype of *apetala3 pistillata* double mutants is indistinguishable from that of the single mutants.

(**A**) and (**B**) *apetala2-2 agamous-1* (*ap2-2 ag-1*) flowers. The medial first-whorl positions in *ap2-2 ag-1* flowers are occupied by carpelloid leaf-like organs (in contrast to the solitary carpels observed in *ap2-2* single mutants). These organs often have stigmatic tissue at their tips, rudimentary ovules along their margins (features of carpels), and stellate trichomes and basal stipules (characteristics of leaves). The lateral first-whorl positions may be occupied by leaf-like organs, or no fully developed organ may be present. The second- and third-whorl organs are intermediate between stamens and petals. As in *ag-1* flowers, after production of the first three whorls of floral organs, the floral meristem of the doubly-mutant flowers reiterates the developmental process of the first three whorls, with the result that *ap2 ag* flowers consist of an indeterminate number of whorls of organs in the pattern: (carpelloid leaves, petaloid stamens, petaloid stamens)$_n$.

(**A**) Approximately stage 8 *ap2-2 ag-1* flower. The number of second- and third-whorl primordia (8) is intermediate between that of wild-type (and *ag-1*) (10) and *ap2-2* single mutants (<1). Although the number of second- and third-whorl primordia is close to wild type, their spatial pattern may be altered. Four fourth-whorl organ primordia (4) are visible as well. Two medial first-whorl organs (1) have been removed.

(**B**) Mature *ap2-2 ag-1* flower with carpelloid leaves in the first whorl (1), and petaloid stamens in the second and third whorls (*arrowheads*). Thus, the *ag-1* mutation has phenotypic effects in the first and second whorls in an *ap2-2* background, an effect not seen in *ag-1* single mutants. Conversely, the *ap2-2* mutation affects organ identity in the third and fourth whorls in an *ag-1* background. This has led to the idea that the wild-type *AP2* and *AG* gene products act in an antagonistic manner to specify, in part, the identity of the floral organs. Additionally, in an *ag-1* mutant background, heterozygosity of *ap2-2* can be detected as occasional carpelloid sepals and staminoid petals.

(**C**) and (**D**) *agamous-1 pistillata-1* (*ag-1 pi-1*) flowers (as well as *ag-1 ap3-1*) consist of an indeterminate number of whorls of sepals.

(**C**) Approximately stage 8 *ag-1 pi-1* flower. The four first- and four second-whorl primordia initiate and differentiate as they do in *pi-1* flowers. However, interior to the second whorl (*arrowheads*), the remaining floral meristem reiterates the developmental program of the first two whorls such that the pattern of organ primordia of *ag-1 pi-1* flowers may be summarized as: (whorl 1, whorl 2)$_n$. In combinations involving *pi* (or *ap3*) alleles in which the third-whorl primordia are distinct from those of the fourth whorl, the pattern of organ primordia in the double mutants is: $(1, 2, 3)_n$.

(**D**) Mature *ag-1 pi-1* flower.

Bar = 20 μm in **A, C, E**; 200 μm in **B, D, F**.

(*Text continued on p. 247*)

Flowers: Floral Organ Identity Mutants

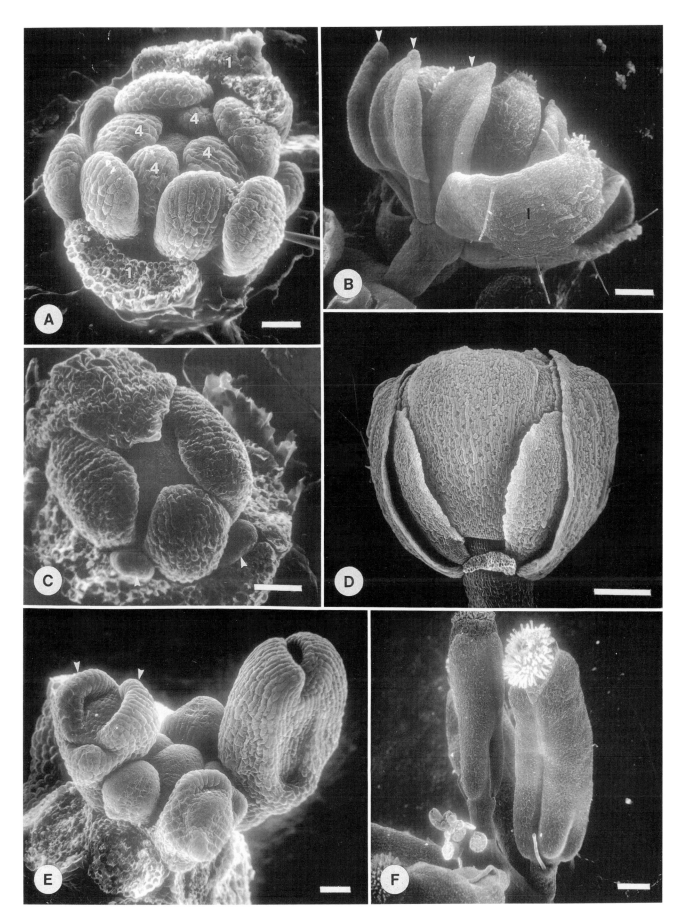

Plate 3.43

Plate 3.43
Genetic interactions among floral homeotic genes: double mutants
(continued)

(E) and (F) *apetala2-2 pistillata-1* flowers consist of a single gynoecium, usually comprised of four carpels and an occasional lateral first-whorl leaf-like organ.

(E) Inflorescence meristem and young flower primordia. The medial first-whorl organ primordia (*arrowheads*) are visible on the approximately stage 4 flower. These appear to fuse later in development with the cells that would normally comprise the fourth whorl, resulting in a single gynoecium (as in oldest flower in the apex). The second- and third-whorl organ primordia are not formed (as in *ap2-2* and *pi-1* single mutants, respectively).

(F) Mature *ap2-2 pi-1* flower comprised of four fused carpels.

J.L. Bowman

B–F reproduced from Bowman et al. (1991b) with permission from The Company of Biologists Ltd.

Plate 3.44
Genetic interactions
among floral homeotic
genes: triple mutants

Triple-mutant combinations have been constructed between floral homeotic mutations of *Arabidopsis* (Bowman et al., 1991b). The triple-mutant combinations *apetala2 agamous pistillata* and *apetala2 agamous apetala3* have flowers comprised of an indeterminate number of carpelloid leaf-like organs. The extent of carpellody of the organs is dependent on the allele of *apetala2* used.

(**A–C**) *apetala2-2 agamous-1 pistillata-1* (*ap2-2 ag-1 pi-1*) flowers. This triple-mutant combination involves the strongest mutant alleles of each of the genes isolated to date. Flowers of these triple homozygotes are composed of an indeterminate number of carpelloid leaf-like organs similar to those observed in the medial first-whorl positions of *ap2-2 ag-1* flowers. The carpelloid leaf-like organs are characterized by stigmatic tissue at their tips, rudimentary ovules on their margins, stellate trichomes on their abaxial surface, and occasional stipules at their base. The inner organs appear more carpelloid than the outer, and the lateral first-whorl organs are the most leaf-like.

(**A**) Inflorescence meristem and young flower primordia. On the oldest flower bud, which is approximately stage 4, two lateral (l) and two medial (m) first-whorl organ primordia are visible, with the lateral arising lower on the pedicel than the medial.

(**B**) Two medial first-whorl organs (1) as well as a lateral organ interior to the first whorl (l) have been removed from this *ap2-2 ag-1 pi-1* flower bud. Subsequent to the production of the first-whorl organ primordia, the remaining floral meristem produces two more organ primordia in lateral positions (l), directly inside the lateral first-whorl organs (when present), and interior to all the first-whorl organ primordia. This is followed by the production of two additional organ primordia in medial positions (m), interior to all the previously produced organ primordia. This pattern of paired lateral, then medial organ primordia formation repeats indefinitely. These pairs of organs could be interpreted as the repeating pattern: (medial first whorl organs)$_n$, if each successive internal flower is rotated 90° with respect to the next outer one.

(**C**) Mature *ap2-2 ag-1 pi-1* flower with all organs except the lateral first-whorl leaf-like organ (l) differentiating into carpelloid leaves. Note stigmatic tissue and fusion of the organs, both characteristics of carpels, as well as stellate trichomes, a feature of leaves.

(**D**) and (**E**) *ap2-1 ag-1 pi-1* and *ap2-1 ag-1 ap3-1* flowers consist of an indeterminate number of whorls of leaf-like organs, with basal stipules and abaxial stellate trichomes. The leaf-like organs of the triple mutant senesce on the time course of leaves rather than sepals which senesce soon after anthesis.

(**D**) Young flower bud with the first-whorl organs removed. Stipules (*arrowheads*) are visible at the base of the first-whorl organs. The pattern of organ primordia is similar to that of *ag-1 pi-1* flowers: (1, 2)$_n$.

(**E**) Cross-section of a mature *ap2-1 ag-1 pi-1* flower. All organs are leaf-like as evidenced by basal stipules (*arrowheads*) and numerous stellate trichomes.

(**F**) Mature *ap2-1 ag-1 ap3-1* flower in which all organs are leaf-like, with numerous stellate trichomes.

Bar = 20 μm in **A**, **B**, **D**; 200 μm in **C**, **E**, **F**.

J.L. Bowman

Reproduced from Bowman et al. (1991b) with permission from The Company of Biologists Ltd.

Flowers: Floral Organ Identity Mutants

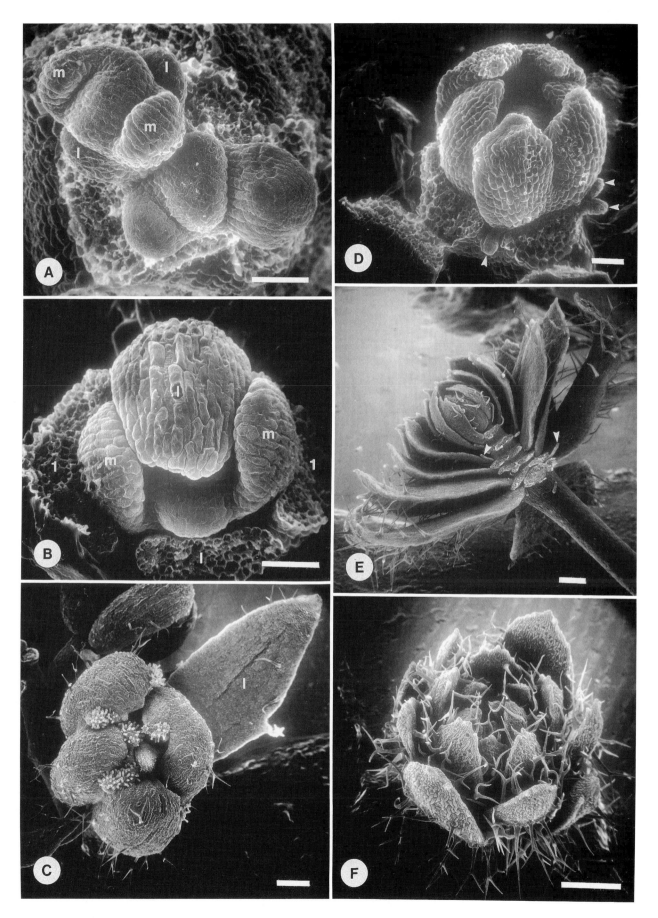

Plate 3.44

Plate 3.45
**The flower patterning
gene *SUPERMAN*:
phenotype and
development of *superman*
mutants**

Mutations in the *SUPERMAN* (allelic with *FLO10*) locus of *Arabidopsis* result in extra stamens developing at the expense of the central gynoecium (Schultz et al., 1991; Bowman et al., 1992). The outer three whorls of *superman* (*sup*) flowers are morphologically normal sepals, petals, and stamens. However, interior to the third whorl, cells that would ordinarily give rise to the fourth-whorl gynoecium instead develop into supernumerary rings of stamens. The number of extra stamens that develop varies, with more than 20 extra stamens in some flowers. Carpelloid tissue may develop interior to the extra stamens, but there is seldom a functional gynoecium. That the ectopic first-whorl carpels of *apetala2* (*ap2*) mutants are not converted into stamens in *sup ap2* flowers indicates that the *SUP* gene product is not required for proper carpel development, but rather is involved with delineating the boundary between the third and fourth whorls of the flower. Specifically, *SUP* appears to be a regulator of the floral homeotic genes *APETALA3* and *PISTILLATA*. Consistent with this is the observation that the initial domain of *AP3* expression is altered in *sup-1* flowers (Bowman et al., 1992). In wild type, *AP3* expression is confined to the second and third whorls of the flower. In contrast, in *sup-1* flowers, expression of *AP3* expands into those cells that would in wild type give rise to the gynoecium, but in *sup-1* flowers give rise to extra rings of stamens. However, a narrow band of cells in the center of the floral meristem of *sup-1* flowers does not express *AP3*. Thus, mutations in the *SUP* gene lead to an increase in stamen number with a concomitant loss of the central gynoecium, and also result in a slight loss of determinacy in the flower as extra rings of organs are produced.

(**A–F**) *sup-1* flowers.
(**A**) Stage 7–8 *sup-1* flower. The initiation and development of the first three whorls of floral organs of *sup-1* flowers are indistinguishable from wild type. After production of the third-whorl organ primordia, instead of the production of the gynoecium as in wild type, another ring of organ primordia is produced on the flank of the remaining floral meristem. In this flower the extra ring contains four organ primordia (*arrowheads*), each of which will develop into a stamen. The number of organ primordia in the extra rings does not exceed six.
(**B**) Stage 9–10 *sup-1* flower. Interior to the six third-whorl stamens, a ring of six stamens (*arrowheads*) and another ring of three organ primordia are visible.
(**C**) Approximately stage 11 *sup-1* flower. Seven stamens interior to the six third-whorl stamens are present.
(**D**) Mature *sup-1* flower. Ten stamens and a central gynoecium lacking most of the ovary tissue are visible. The carpelloid tissue at the center of *sup-1* flowers ranges from a complete absence of carpelloid tissue, to mosaic organs composed of stamen and carpel tissue, to nearly normal gynoecia.
(**E**) A filamentous structure capped with stigmatic papillae occupies the center of this flower. Nectary tissue (*arrowheads*) is present at the base of the third-whorl stamens as well as the inner stamens.
(**F**) Mosaic organs consisting of large stamen and carpel sectors often occupy the center of *sup-1* flowers.

Bar = 20 μm in **A**, **B**; 100 μm in **C**, **E**, **F**; 200 μm in **D**.

J.L. Bowman

C–F reproduced from Bowman et al. (1992) with permission from The Company of Biologists Ltd.

Flowers: Floral Organ Number Mutants

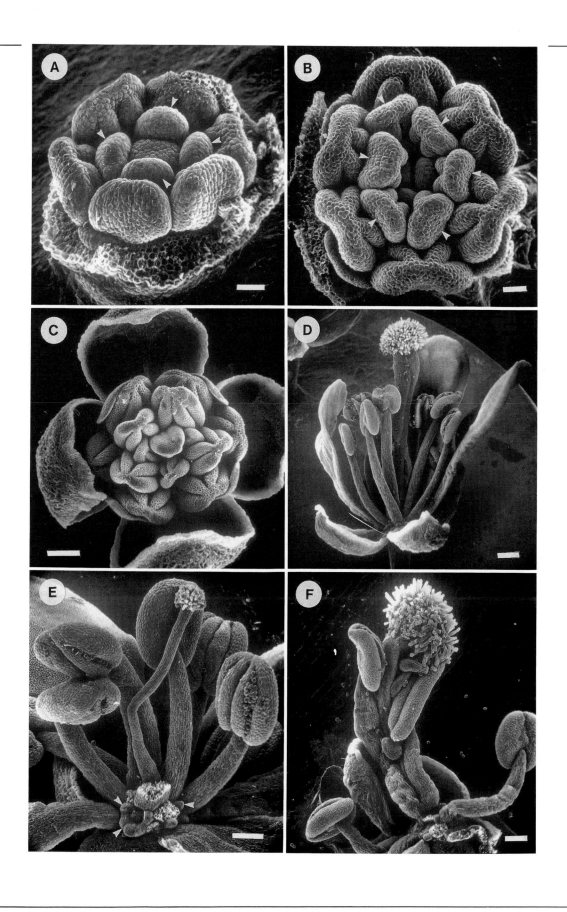

Plate 3.45

Plate 3.46
The flower patterning
gene *SUPERMAN*:
phenotype and
development of *superman*
double mutants

Double-mutant combinations between floral homeotic mutants and *superman* (allelic with *flo10*) have been constructed (Schultz et al., 1991; Bowman et al., 1992). Two classes of interaction are observed: those that are additive, and those that are epistatic. In terms of floral organ identity, *superman apetala2* and *superman agamous* double mutants exhibit essentially additive phenotypes, while both *pistillata* and *apetala3* mutations are largely epistatic to *superman* mutations. However, in terms of determinacy, *superman* and *agamous* mutations act synergistically to increase indeterminacy of the floral meristem.

(**A–D**) *superman-1 agamous-1* (*sup-1 ag-1*) flowers consist of a first whorl of sepals, and an indeterminate number of petals interior to the first-whorl sepals. In contrast to *ag-1* flowers, which exhibit internode elongation between successive internal flowers when in an *ERECTA* background, no internode elongation between whorls of organs occurs in *sup-1 ag-1 ERECTA* flowers (Bowman et al., 1992).

(**A**) Development of *sup-1 ag-1* flowers is indistinguishable from that of *ag-1* flowers until the initiation of the fourth-whorl primordia. In *ag-1* flowers, four fourth-whorl primordia develop, while in *sup-1 ag-1* flowers usually more than four develop. In this flower, six fourth-whorl primordia can be seen interior to the six third-whorl primordia. The remaining floral meristem (f) is larger than in either *ag-1* or *sup-1* flowers. The four first-whorl sepals have been removed.

(**B**) After the fourth-whorl organ primordia develop, the floral meristem continues to produce irregular numbers of organ primordia from its flank, and the size of the floral meristem (f) continues to enlarge. All of the organ primordia interior to the first-whorl sepals develops into petals.

(**C**) Mature *sup-1 ag-1* flower. After several rings of organ primordia are produced, the floral meristem of *sup-1 ag-1* flowers becomes fasciated.

(**D**) Close up of **C**. The fasciated floral meristem (f) is continuing to produce organ primordia from its entire perimeter. Each organ primordia develops into a petal, although occasional organs have some sepaloid characteristics.

(**E**) and (**F**) *superman-1 apetala3-1* flowers. Strong *apetala3* and *pistillata* mutations, such as *ap3-3* and *pi-1*, are epistatic to *sup* mutations, such that the double mutants are indistinguishable from the *ap3* or *pi* single mutant. In contrast, *sup* mutations enhance weak *ap3* and *pi* mutations, such as *ap3-1* and *pi-3*, such that the phenotype of the double mutant resembles closely that of strong *ap3* or *pi* alleles. *sup ap3* and *sup pi* double mutants consist of an outer two whorls of sepals and an inner two whorls of carpels. Thus, *sup* mutations have a phenotypic effect in the third whorl in *ap3-1* and *pi-3* backgrounds, but not in a wild-type background.

(**E**) Approximately stage 7 *sup-1 ap3-1* flower. The cells that normally constitute the medial third-whorl organ primordia (m) are congenitally fused with some fourth-whorl tissue. The second-whorl primordia (*arrowheads*) develop into sepals. This pattern of development resembles that of *ap3-3* flowers.

(**F**) Mature *sup-1 ap3-1* flower. The fused medial third-whorl organs retain some characteristics of stamens such as the filamentous base (*double arrowhead*), but are mostly carpelloid and have fused with fourth-whorl tissue. Nectary tissue (*arrowhead*) is visible at the base of the third-whorl organs.

Bar = 20 μm in **A**, **E**; 100 μm in **B**, **D**; 200 μm in **C**, **F**.

J.L. Bowman

B, C, D, and F reproduced from Bowman et al. (1992) with permission from The Company of Biologists Ltd.

Flowers: Floral Organ Number Mutants

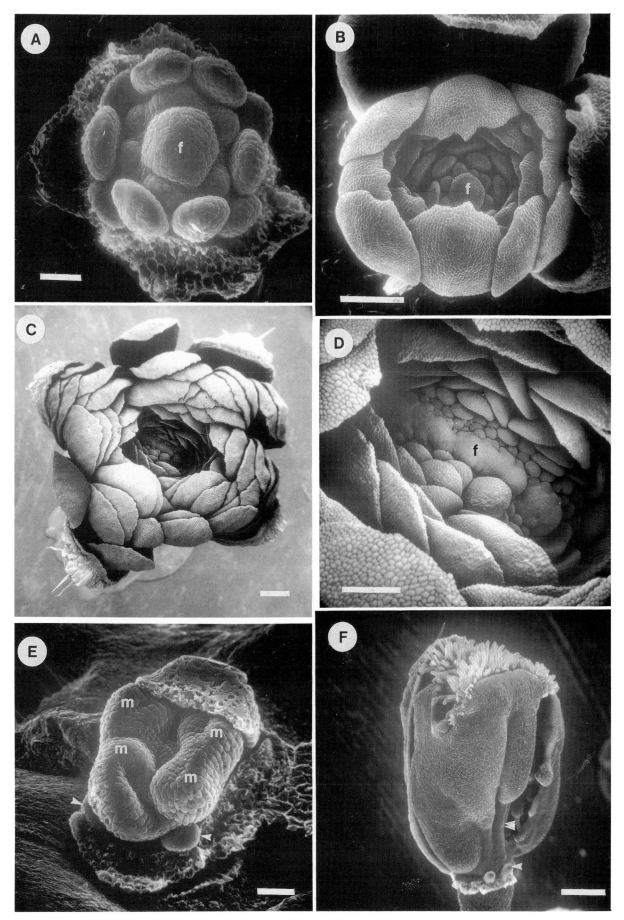

Plate 3.46

Plate 3.47
Flower development in
***clavata3*, a mutation that**
produces enlarged floral
meristems

The vegetative, inflorescence, and flower meristems of plants homozygous for recessive mutations in three loci, *CLAVATA1* (McKelvie, 1962), *CLAVATA2* (Koornneef et al., 1983), and CLAVATA3 (M. Griffith, J. Alvarez, S. Clark and D. R. Smyth), are enlarged and often distorted in shape. A greater number of rosette leaves is produced compared with wild type. An increase in the number of cauline leaves with associated axillary meristems is also observed. In the flowers of *clavata1-1* (*clv1-1*) and *clavata2-1* (*clv2-1*) mutant plants, the most obvious effect of a larger flower meristem is the formation of super-numerary carpels and the characteristic club-shaped siliques, after which these mutants were named (Koornneef et al., 1983). The flowers of *clv1-1* and *clv2-1* may also have additional organs in any of the other organ whorls (Leyser and Furner, 1992).

The *clv3-2* mutation produces an even greater enlargement in vegetative meristems (see Plate 1.4) as well as inflorescence and flower meristems. Concomitantly it has a greater number of additional floral organs. The numbers of different organs in each whorl was scored for the first ten flowers of ten *clv3-2* mutant plants. The average number of sepals was 5.3 ($\pm$0.4, range 4–7), petals 5.0 ($\pm$0.3, range 4–6), stamens 8.7 ($\pm$0.6, range 7–11), and carpels 6.7 ($\pm$0.5, range 5–8). Additional filamentous structures that appear to be antherless filaments also develop in the third whorl. In the *clv3-2* plants scored, approximately one of these structures is produced per flower (0.9 $\pm$ 0.4, range 0–2).

The development of *clv3-2* flowers is similar to that of wild type but with some notable differences. The *clv3-2* stage 2 flowers appear as a large, almost hemispherical primordium of greater height than in wild type. Another difference is seen at the initiation of gynoecial development in stage 6. The *clv3-2* gynoecium develops as a ring of carpel tissue encircling an internal dome of undifferentiated cells. A similar dome of cells, of smaller size, is also present in flowers of plants mutant at the *CLV1* and *CLV2* loci, but is absent in wild type. The dome of tissue continues to grow inside the carpels as they elongate and is encased by the carpels at around stages 11 and 12. Alternatively, the carpels may never fully encapsulate these internal structures and a small hole in the stigmatic surface may result. Analysis of sections reveals that additional carpels may be produced from the flanks of this expanding dome inside the fourth whorl. After stage 12, the central dome continues to grow in its own right. As a consequence, in flowers of stages 17–19, tissue from the proliferating dome may rupture the gynoecium. This mass of tissue has the appearance of callus, but the surface often has small pockets of carpelloid tissue and stigmatic papillae.

(**A–F**) Scanning electron micrographs of the inflorescence and flowers of *clv3-2*. The *clv3-2* mutation was induced in the Landsberg *erecta* ecotype.

(**A**) Lateral view of a stage 4 *clv3-2* flower. Note the domed flower meristem interior to the sepals which is relatively higher than in wild type.

(**B**) Side view of a stage 8 *clv3-2* flower. Seven sepals have been dissected from the flower bud. Many stamens (9) are in evidence around a developing six-carpelled gynoecium. Note that the gynoecium encircles a large internal dome of cells (*arrowhead*).

(**C**) Vertical view of a stage 11 *clv3-2* flower. Note the increased numbers of flower organs, 6 sepals, 5 petals (hidden), 8 stamens, and the enlarged gynoecium. Stigmatic papillae are developing on top of the gynoecium, but the

Bar = 20 μm in **A**; 50 μm in **B**; 100 μm in **C**, **D**, **F**; 500 μm in **E**.

(*Text continued on p. 257*)

Flowers: Floral Organ Number Mutants

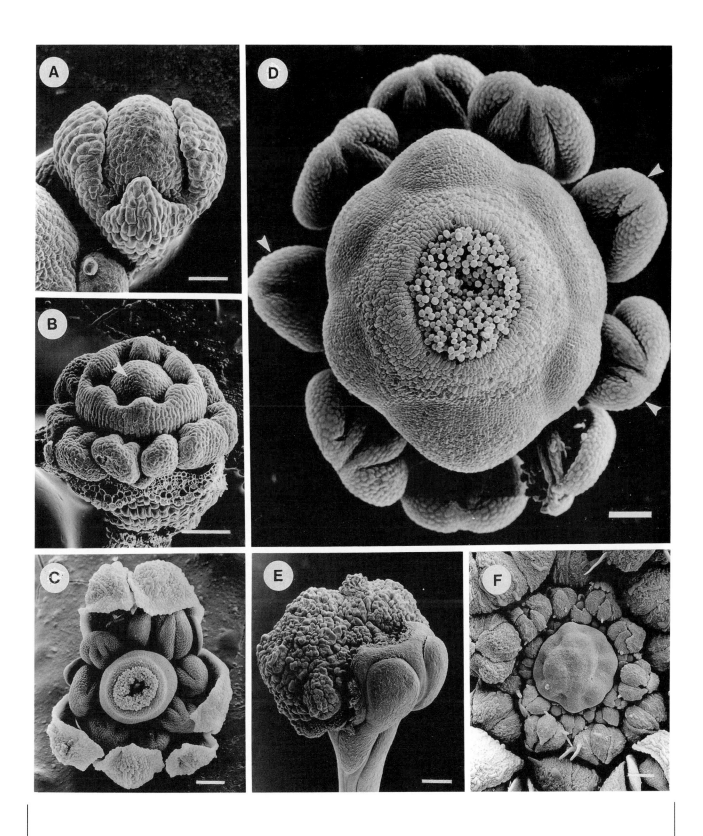

Plate 3.47

255

Plate 3.47
Flower development in
clavata3, **a mutation that**
produces enlarged floral
meristems
(*continued*)

stigmatic surface is incomplete as the fourth-whorl carpels fail to cover the central dome of cells and internally developing carpels (see above).

(**D**) Vertical view of the third and fourth whorls of a stage 13 *clv3-2* flower. Six sepals and six petals have been dissected from the flower. The third whorl is composed of 9 stamens, 2 lateral (ls) and 7 medial. Interestingly, three of the medial stamens have abnormal anthers (*arrowheads*). It is possible that these stamens represent additional third-whorl structures. One of the medial stamens has dehisced, and pollen has been released. The gynoecium is composed of 7 carpels that have grown to enclose the central dome of tissue. Additional carpels develop from the flanks of this dome (see above). This gives the gynoecium its radially expanded appearance.

(**E**) Side view of a stage 17 *clv3-2* gynoecium. The central dome of tissue has proliferated to a point where it has "erupted" from the developing *clv3-2* silique. Note the undifferentiated, callus-like appearance of this tissue. In some flowers, small regions of carpelloid tissue may be seen in this mass.

(**F**) Vertical view of a 15-day-old *clv3-2* inflorescence apex. Note the large mass of apical tissue and the young flowers produced from the flanks of this dome. The inflorescence meristem becomes increasingly enlarged as it develops, the phyllotaxis is disrupted, and internodal elongation is reduced. Consequently, apical flowers are "bunched" together.

J. Alvarez and D.R. Smyth

Plate 3.48
Phenotypic analysis of
unusual floral organs
mutants

A mutant plant exhibiting an abnormal floral phenotype was isolated from an EMS mutagenized population. The mutant, due to a recessive allele of a single nuclear gene, was given the name *unusual floral organs* (*ufo*) due to the high degree of phenotypic variability among floral organs.

(**A**) and (**B**) Typical *ufo* flowers, with one sepal removed for clarity. The first whorl of *ufo* flowers is generally not affected, with four sepals most common, although five sepals have been observed occasionally. The second and third whorls show a variety of homeotic transformations (either partial or complete); cell types typical of every floral organ have been observed on second- and third-whorl organs of *ufo* flowers. The fourth whorl is composed of 2–4 morphologically normal carpels fused to form a fully fertile gynoecium.

(**C**) and (**D**) Most second whorl organs are mosaics of cell types representative of two or more types of floral organs. The enlarged portion of each figure reveal the distinct boundaries which are often evident between component cell types.
(**C**) A second-whorl mosaic organ of stamen (st) and sepal (se) cell types.
(**D**) A second-whorl mosaic organ of petal (p) and sepal (se) cells types.

(**E**) Like those of the second whorl, third-whorl organs are typically mosaic in nature, although complete transformations of stamen to carpel or stamen to petal have been observed at low frequency. An example of a mosaic stamen–carpel organ is shown. At the tip of this organ, papillae (p) have developed. A locule, typical of wild-type anthers, is also present.
(**F**) The most common organ type in both the second and third whorls is a filament with no obvious homology to any wild-type floral organ. Second-whorl filaments (2) are generally shorter and more slender than third-whorl filaments (3). The cell types along the length of a filament are typical of stamen filament cells, however, the cells at the tip are ambiguous in identity, having features in common with anther, stylar, and nectary cell types.
(**G**) In addition to the abnormalities in floral shoots described above, the primary shoot of *ufo* mutant plants terminates in a pistil-like structure. The organs which make up this structure are composed of both carpel (c) and sepal (se) cell types. This phenotype suggests that the *UFO* gene product is also required to maintain indeterminacy in the inflorescence.
(**H**) In wild-type plants, the several basal lateral shoots develop as secondary inflorescences subtended by cauline leaves. When *ufo* plants are grown in short days (10 hours light/14 hours dark), more secondary inflorescences are produced as compared to wild-type plants grown in the same conditions. In addition, after initiation of the first few flowers, secondary inflorescence-like structures (co) may be produced. The apical meristem lies to the upper right of the figure. Two secondary inflorescence-like structures (co) are visible, distinguished by the production of cauline leaf-like organs and tertiary meristems. The more apical of these is preceded by two *ufo* flowers (*arrows*), indicating that the secondary inflorescence-like structure is in a position that would normally be occupied by a flower.

M. Wilkinson and G.W. Haughn

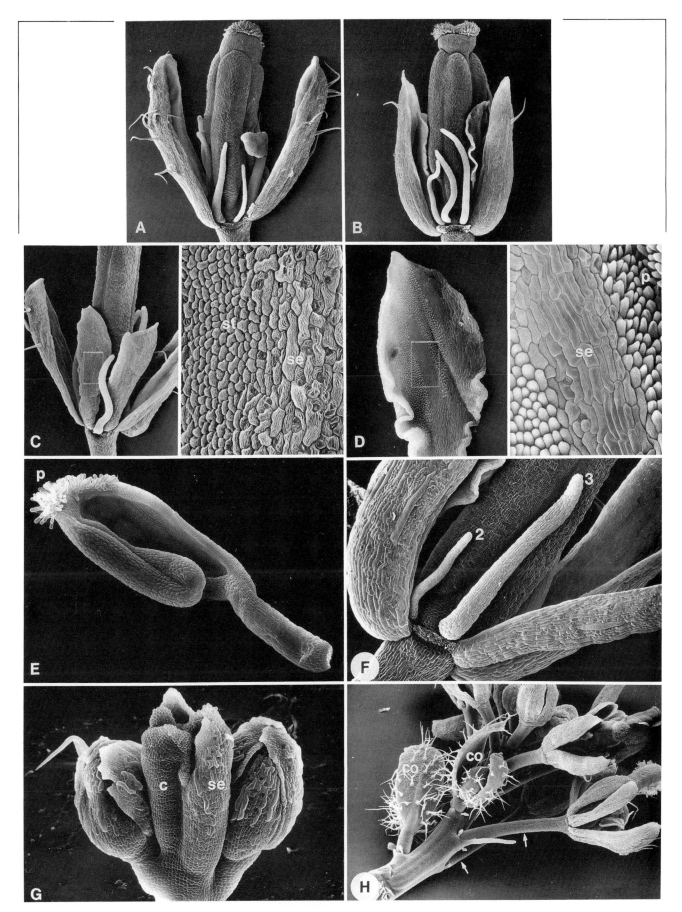

Plate 3.48

Plate 3.49
The *PETAL LOSS* gene

The *petal loss-1* (*ptl-1*) mutation is recessive and produces characteristic abnormalities in the flower structure. It has no obvious effects on vegetative or inflorescence development. The most striking effect is a reduction in petal number. This increases acropetally such that first-formed flowers may have four petals, later-arising flowers have fewer than four, and the most apical flowers lack petals completely. The petals that are produced are often reduced in size, "buckled," tubular in form, and occasionally staminoid. The region in which the nectaries appear in wild type flowers also appears to be more extensive in *ptl-1* flowers. In addition, the first whorl is affected. Sepals are more boat-shaped in appearance than their wild-type counterparts, and they often have carpelloid outgrowths from their margins, including ovules. In more apical flowers, petal tissue may be found along the margins of sepals, especially those in medial positions. The third (stamen) whorl is largely unaffected except for the rare occurrence of an additional stamen. The gynoecium appears normal.

(A–E) Scanning electron micrographs of the developing *ptl-1* flowers. The mutation was induced in the Columbia ecotype.

(A) Vertical view of stage 2, 3, and 5 flowers developing from the *ptl-1* inflorescence apex. No differences are obvious between these early *ptl-1* flowers and those of wild type at equivalent stages.

(B) Stage 5 *ptl-1* flower viewed from a lateral orientation. One lateral and two medial sepals have been removed. Medial stamen primordia are apparent, but a lateral stamen primordium has failed to develop. At this stage, petal primordia are evident in wild-type flowers but are not present in this *ptl-1* flower.

(C) Oblique lateral view of a stage 8 *ptl-1* flower. One lateral and two medial sepals have been removed. There is no evidence of petal development in the region internal and alternate to the sepals (*arrowhead*).

(D) Medial view of a late-stage 11 or early-stage 12 *ptl-1* flower. One lateral and two medial sepals have been removed. The third-whorl stamens and fourth-whorl gynoecium appear normal in number and structure. Petals have failed to develop in the second whorl.

(E) Side view of a stage 13 *ptl-1* flower. The petals have failed to develop, but the numbers of all other organs is the same as in wild type. One of the lateral stamens appears slightly abnormal. Note the "boat-shaped" sepals, one of which has an outgrowth on its lateral margin (*arrowhead*).

(F–H) Characteristic structures associated with the *ptl-1* phenotype.

(F) Side view of the receptacle of a stage 12 *ptl-1* flower from which sepals have been removed. Nectary tissue has expanded to encompass the region between and interior to the sepals from which a petal would normally develop in wild type (*arrowhead*).

(G) Lateral view of a stage 11 *ptl-1* flower with sepals removed. A tubular petaloid structure (*arrowhead*) is developing in the location where a petal normally develops.

(H) Lateral view of a stage 12 *ptl-1* flower from which one lateral sepal has been removed to expose stamens and the gynoecium. A mid-stage 12 ovule (Robinson-Beers et al., 1992; *arrowhead* and inset) is developing on the margin of one of the medial sepals.

Bar = 10 μm in **B**; 50 μm in **A, C, D, E, F, G**; 100 μm in **H**.

J. Alvarez and M. Griffith

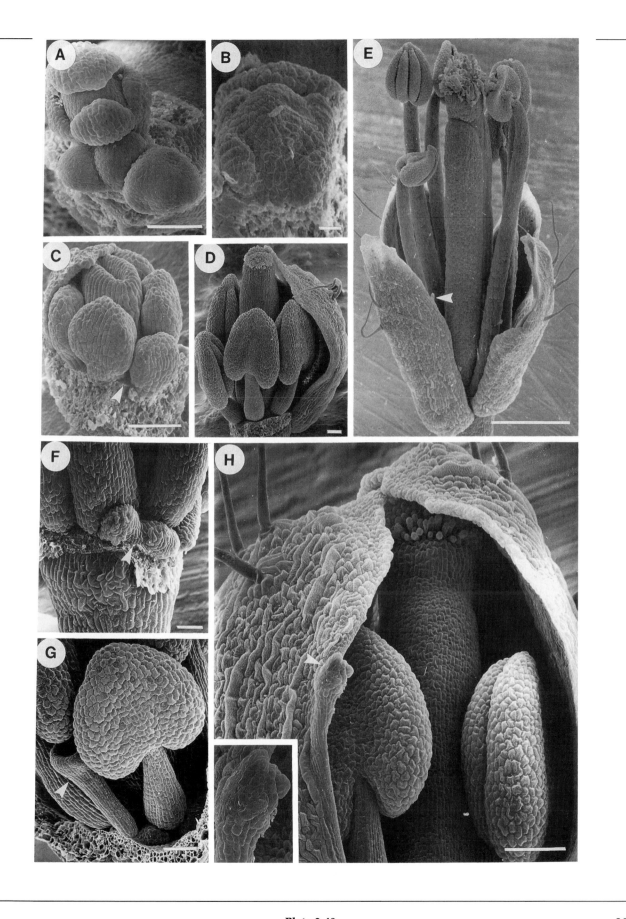

Plate 3.49

261

Plate 3.50
Structural male sterility
in the *antherless* mutant

(A) In *antherless* mutants, the development of the anther is severely modified (*arrow*) although the development of the stamen filaments is largely normal.

(B) Close-up of the structure that forms at the tip of the stamen filaments in *antherless* mutant flowers. The pattern of epidermal cells of the apical structure does not resemble that found in wild-type anthers. Instead, it resembles those found on sepals. Allelism tests have not been performed between *antherless* and *fl54*, *unusual floral organs*, or *spitzen*.

S. Craig and A. Chaudhury

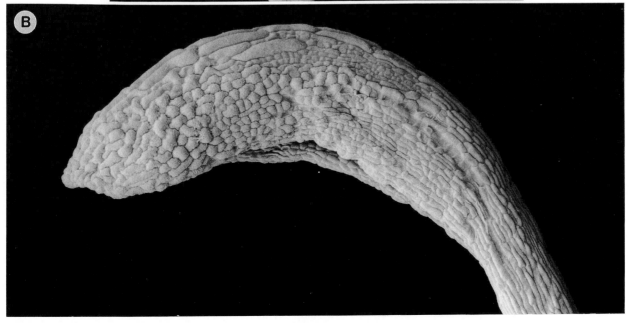

Plate 3.50

Plate 3.51
The CRABS CLAW gene

Recessive mutations in the *CRABS CLAW* (*CRC*) gene cause characteristic abnormalities in the developing *Arabidopsis* gynoecium, but do not affect the development of other flower organs. The gynoecium of *crc-1* flowers appears to develop normally up to stage 6 (Smyth et al, 1990). During stage 7 in wild-type plants, the gynoecium begins to grow as a uniform cylinder. In contrast, the carpels of a *crc-1* mutant gynoecium, while congenitally fused at the base, grow separately at the apex, and often appear to be slightly twisted. The mutant gynoecium is noticeably wider than wild type in the lateral plane. In a small number of *crc-1* flowers, a third carpel (or a reduced carpelloid structure) develops medially between the two lateral carpels. From stages 8 through 12, the *crc-1* gynoecium continues to develop into a structure wider and shorter than the wild-type gynoecium. During this period in the wild type, the upper gynoecium becomes tapered and the carpels join at the apex to enclose the locules of the ovary. In the *crc-1* gynoecium, the outer, lateral edges of the carpels continue to grow separately upwards. The carpel apices and presumptive precursors of style and stigma curve toward the center of the floral axis, and are angled as much as 40° from the vertical. Following the differentiation of the style and stigmatic papillae at stages 11 and 12, the apex of the *crc-1* gynoecium is essentially bilobed, with 2 stigmatic surfaces that angle into each other so that the stigmatic papillae intermingle. On fertilization, the ovary of the *crc-1* gynoecium begins to expand. The style and stigma of the carpel lobes incline further inwards such that the respective stigmatic surfaces oppose each other and the apical cleft between the carpels is further enlarged. The apex of the *crc-1* silique thus has the appearance of the opposing pincers of a crab's claw.

(**A–F**) Scanning electron micrographs of developing *crc-1* flowers. *crc-1* was induced in the Landsberg *erecta* ecotype.

(**A**) Vertical view of a stage 5 *crc-1* flower. Two medial sepals have been removed. Development at this stage is indistinguishable from wild type.

(**B**) Vertical view of a stage 7 *crc-1* flower. Four sepals have been dissected from the flower bud. The gynoecial cylinder (g) is noticeably wider than wild type in the lateral plane.

(**C**) Vertical view of a stage 8 *crc-1* flower. Four sepals have been dissected from the flower bud, and one lateral stamen has failed to develop. An apical cleft is apparent between the tops of the growing carpels, which are separated and slightly twisted.

(**D**) Vertical view of a late stage 9 *crc-1* flower. Four sepals have been dissected from the flower bud. A third carpelloid structure (*arrowhead*) is developing between the two lateral carpels.

(**E**) Medial side view of a stage 12 *crc-1* flower. Three sepals, two petals, and two medial stamens have been removed from the flower. The gynoecium appears as a short, squat structure which is wider laterally than in wild type. Note that the gynoecium is bilobed at its apex. The unfused carpels are angled inward so that their respective stigmatic surfaces abut (*arrowhead*).

(**F**) Medial side view of a fertilized *crc-1* gynoecium. The organs of the outer three whorls have senesced and fallen from the flower. The expansion of the fertilized ovary has apparently resulted in the style and stigma of the two unfused carpels bending further inward so that the stigmatic surfaces directly oppose each other.

Bar = 10 μm in **A**, **B**; 50 μm in **C**, **D**; 100 μm in **E**, **F**.

J. Alvarez and D.R. Smyth

Flowers: Floral Organ Mutants

Plate 3.51

265

Plate 3.52
The SPATULA gene

Recessive mutations in the *SPATULA* (*SPT*) gene specifically disrupt the structure of the gynoecium. The rest of the flower remains unaffected. In the strongest known mutant allele (*spt-2*), the gynoecium develops normally up to stage 7 of flower development. During stages 8 and 9 it starts to show developmental irregularities. Vertical growth of the two carpels is usually faster in the lateral regions than in the medial regions where they are congenitally fused. The trend continues during stages 10 and 11 of flower morphogenesis and, as a consequence, the carpels are not always fused at the top of the gynoecium. In those flowers where the carpels are fully fused, a central hole in the stigmatic surface is frequently present and associated with a reduction in the transmitting tissue of the style. In addition, the stigmatic papillae that develop around stage 11 in *spt-2* mutant flowers are not as abundant as in wild type flowers. Also the false septum which normally separates the two locules is usually absent in the upper regions of the *spt-2* gynoecium. The *spt-2* flowers are only partially female fertile, possibly due to defects in the stigma, transmitting tissue, and septum. The mature *spt-2* silique is often flattened laterally, especially in the upper half. This gives it a spatula-like appearance after which the mutation was named.

(**A–F**) Scanning electron micrographs of *spt-2* flowers. *spt-2* was induced in the Landsberg *erecta* ecotype.

(**A**) Lateral view of a developing stage 6 *spt-2* flower. Three sepals have been removed. At this stage the pattern of development is indistinguishable from that of wild type.

(**B**) Vertical view of a stage 8 *spt-2* gynoecium in the same orientation as in **A**. Growth in a region where the two carpels are congenitally fused (*arrowhead*) is lagging behind that of the other regions of the gynoecial cylinder.

(**C**) Vertical view of a stage 10 *spt-2* gynoecium in the same orientation as in **A** and **B**. The two carpels, at the top and bottom of the growing cylinder in this view, are beginning to grow together apically, but the growth is asymmetrical. The lower carpel is twisting away and a cleft is evident at the presumptive fusion point between the carpels (*arrowhead*).

(**D**) A lateral/medial side view of a stage 11 *spt-2* flower. Three sepals have been removed. The carpels are not joined along one edge at the top of the gynoecium.

(**E**) Vertical view of a stage 12 *spt-2* gynoecium shown in an orientation 90° to that in **A**, **B**, and **C**. The carpels in this flower have grown together normally but there is a large central hole above the region where the transmitting tissue of the style would normally develop.

(**F**) Medial view of a stage 13 *spt-2* flower. Note the gap along the carpel fusion margin in the gynoecium (*arrowhead*), and the reduced number of stigmatic papillae.

Bar = 10 μm in **A**, **B**, **C**; 50 μm in **E**; 100 μm in **D**, **F**.

J. Alvarez and D.R. Smyth

Plate 3.52

267

Plate 3.53
The *ETTIN* Gene

Recessive mutations in the flower gene *ETTIN* (*ETT*) disrupt normal flower development (Sessions et al., 1993). The most obvious effect involves the gynoecium. In plants homozygous for strong alleles, this appears as a distorted, sterile, apically bilobed structure with disrupted anatomy and abnormal cell surface histology (Sessions et al., 1993). In addition, the numbers of sepals, petals, and stamens are also altered in *ettin* mutants, and male fertility may be reduced.

(A–F) Scanning electron micrographs of the developing flowers of *ett-3* mutant plants. The *ett-3* mutant was induced in the Landsberg *erecta* ecotype.

(A) Medial view of an early stage 6 *ett-3* flower. At this time this particular flower is indistinguishable from wild type. Petal (p) and stamen (s) primordia can be seen. The gynoecium has not yet developed from the central dome of cells. Some *ettin* flowers are abnormal at this age because they have an altered number of organs in the first three whorls.

(B) Medial/lateral view of a late stage 6 *ett-3* flower. The gynoecium is developing abnormally. Cells at the rim of the central dome are growing to produce the gynoecial tube but they have begun to "funnel" outwards much more than in wild-type flowers (*arrowhead*).

(C) Medial/lateral view of a stage 7 *ett-3* flower. The gynoecium continues to develop aberrantly as it grows to produce a strikingly funnel-shaped structure.

(D) Vertical view of a stage 11 *ett-3* gynoecium. Cells at the rim of the gynoecium are growing irregularly and have failed to enclose the gynoecium. Developing ovules (o) are exposed. Developing stigmatic papillae can be seen on the more apical gynoecial outgrowths.

(E) Lateral view of a stage 12 *ett-3* flower. Note the exaggerated internode between the third and fourth whorls, the vertical "flange" of tissue that appears to be associated with the carpel margins (*arrowhead*), and the extensive cap of stigmatic papillae.

(F) Medial view of severely affected mature *ett-3* gynoecium. Note the bilobed structure of the gynoecium, the irregular occurrence of stigmatic papillae, the absence of a style, and the abnormal cell surface structure of the carpels. An enlargement of the carpel surface is seen (inset).

Bar = 10 μm in **A**, **B**, **C**; 50 μm in **D**; 100 μm in **E**, **F**.

J. Alvarez

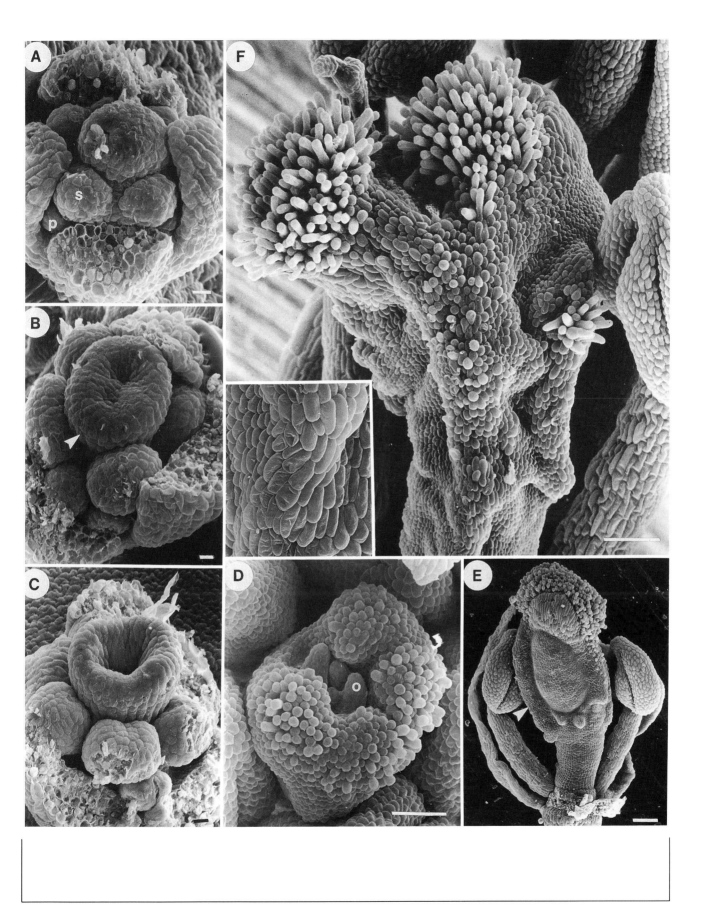

Plate 3.53

269

Plate 3.54
The *FL82* gene

(A) *fl82* mutant flowers have extraordinarily large pistils (Komaki et al.,1988; Okada et al., 1989), as exemplified in these two *fl82* pistils.

(B) In some flowers two distinct pistils develop.

(C) *fl82* inflorescence. Due to the large size of the gynoecial structures, floral buds are not fully enclosed by the sepals.

(D) Anatomical analysis shows that the pistil is composed of from three to ten carpels. Unlike the *clavata* mutants, the supernumerary carpels of the gynoecium are not positioned symmetrically.

(E) The carpels are often not completely fused, as seen in this transverse section of a *fl82* gynoecium (see also **A**). In addition, as seen in this flower, anthers with pollen sacs are sometimes attached at the margin of the un-fused carpels. Fertilization frequency of the pistil is generally low and the pistils composed of unfused carpels are completely sterile. The number and shape of other floral organs are usually normal, except that mutant flowers can occasionally have up to five sepals, five petals, and eight stamens. The *fl82* mutation is recessive.

K. Okada and Y. Shimura

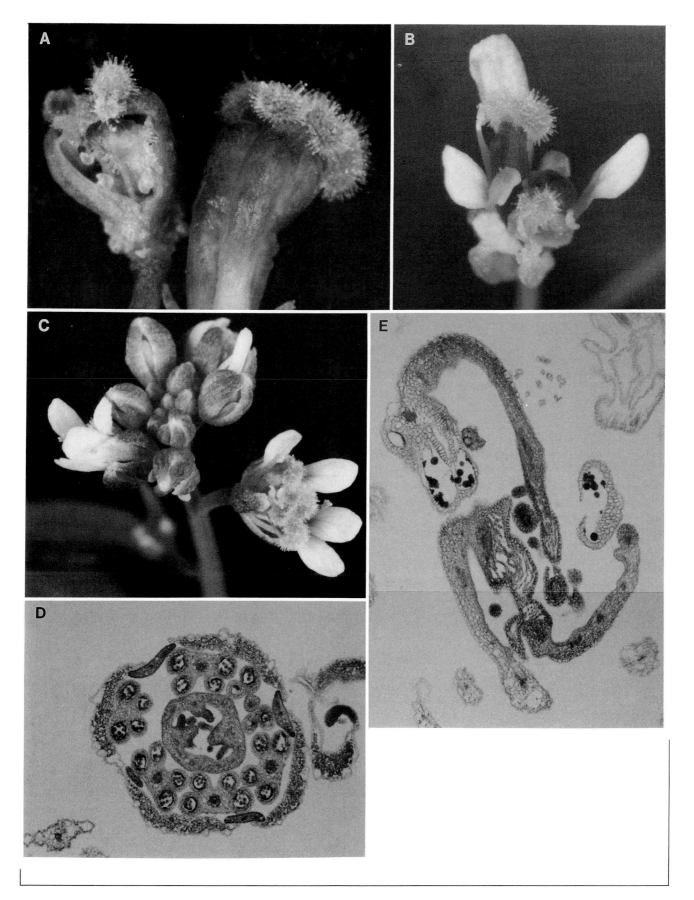

Plate 3.54

Plate 3.55
The *FL89* gene

The pistils, petals, and sepals of *fl89* mutants are aberrant in structure (Komaki et al., 1988; Okada et al., 1989). The *fl89* mutation is recessive and maps to chromosome 4 (M. K. Komaki, K. Okada, Y. Shimura, unpublished).

(**A**) and (**B**) The pistil of *fl89* mutants has two clumps of stigmatic papillae, and two horn-shaped projections at its apex. The horn-shaped structures are located at the top of the carpels, whereas the clumps of stigmatic tissue reside on the seamline of the carpels.

(**C**) At an early stage of pistil development, when the apices of the two carpels are fusing at the top of the wild-type pistil, four swellings are observed at the top of the mutant pistil.

(**D**) and (**E**) Inside the ovary, the septal tissues are not fused at the top of the gynoecium. In some pistils, the two carpels are not fused at the top. In addition, the median vascular bundle of the carpels is often separated into two distinct parts. These features indicate that the process of fusion of the two carpels does not proceed normally in the mutant. This causes the seed set by normal self-fertilization to be low in *fl89* mutants (**D**). The number of the other floral organs is apparently normal, but the sepals and the petals are slightly more slender than those of wild type.

Bar = 100 μm in **C**.

K. Okada and Y. Shimura

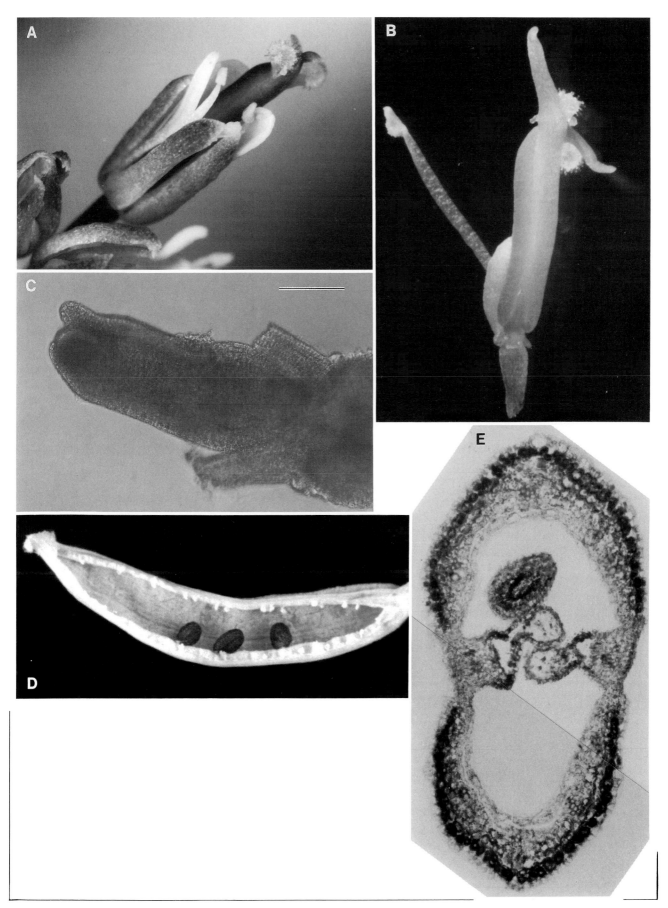

Plate 3.55

4
Pollen

Introduction

Pollen development in *Arabidopsis* has been described from the appearance of the initial archesporial cells through dehiscence, and divided into a number of stages (Misra, 1962; Regan and Moffatt, 1990). Table 1 summarizes some of the landmark events and attempts to relate them to defined stages of flower development (Müller, 1961; Smyth et al., 1990). However, since the data on pollen development were amalgamated from several sources, it is likely that some refinement of the temporal correlations in Table 1 will be necessary.

Pollen development has its origins in the hypodermal archesporial cells which are located at each of the four corners of immature anthers. The archesporial cells undergo periclinal divisions to give rise to the primary parietal and primary sporogenous layers (Misra, 1962). The primary parietal layer undergoes two more periclinal divisions such that three anther wall layers are eventually formed. The outermost of the three layers develops into the endothecium, the innermost (adjacent to the sporogenous tissue) develops into the tapetum, and the middle one is appropriately termed the middle layer. The tapetum, which completely encompasses the locule, is derived from both the

Table 1. Stages of pollen and stamen development.

Approximate floral stage[a]	Pollen development stage[b]	Stamen morphology[c]
7	Archesporial cells divide to give rise to primary parietal and sporogenous cells	Filament and anther regions distinct
8	Microsporocytes conspicuous	Anther region becomes lobed on adaxial side
9	Pollen mother cells (PMCs) become separated from each other and from tapetum by a callose wall	The three anther wall layers (endothecium, middle, tapetum) are evident
	PMCs undergo meiosis to form tetrads (isobilateral and tetrahedral) of microspores	
10	Microspores separate from each other after breakdown of the callose wall, and lie freely in pollen sac	Filaments begin to elongate
	Microspores round up; walls thicken due to formation of the exine; bacula of exine visible	
11–12	First mitotic division of microspores follows resorption of prominent vacuole	Tapetum degenerating
	Second mitotic division of microspores; storage bodies visible in microspores	
	Desiccation of pollen grains	
13		Dehiscence
14		Fertilization

[a] Floral stages refer to those of Müller (1961), Smyth et al. (1990), and Bowman et al. (1991a).
[b] Stages of pollen development summarized from Misra (1962) and Regan and Moffatt (1990).
[c] Stages of stamen development summarized from Misra (1962), Smyth et al. (1990), and Bowman et al. (1991a).

primary parietal cells (adaxially) and from connective tissue (abaxially). The tapetal cells, whose primary function is nutritive, are initially uninucleate but before meiosis of the pollen mother cells, become binucleate (Misra, 1962). The tapetal layer starts to degenerate about the time that the microspores become separated due to the degradation of surrounding callose, and the degeneration is complete after the first mitotic division of the microspores (Misra, 1962). The middle layer degenerates soon after its formation. The endothecium differentiates relatively late, developing secondary wall thickenings just prior to dehiscence. Thus, at dehiscence, only the endothecium and some withering epidermal cells comprise the wall of the pollen sac.

The primary sporogenous cells develop into pollen mother cells (PMCs) which become isolated from the tapetum and each other by surrounding callose walls. Each PMC undergoes two meiotic divisions, characterized by simultaneous cytokinesis, resulting in the formation of a tetranucleate PMC (Misra, 1962). The callose surrounding the tetrad of microspores (tetrahedral or isobilateral) is broken down after meiosis, releasing individual microspores (Misra, 1962; Regan and Moffatt, 1990). The microspores then round up, and their walls thicken due to exine deposition (Regan and Moffatt, 1990). A large vacuole forms in each microspore causing a rapid increase in size, and displacing the nucleus to one side of the microspore (Regan and Moffatt, 1990). The vacuole is resorbed and soon after, the first mitotic division occurs, resulting in a large vegetative cell and a small generative cell (Misra, 1962; Regan and Moffatt, 1990). Subsequently, the second mitotic division occurs, the generative cell divides, and the cytoplasm of the vegetative cell accumulates numerous storage bodies (Misra, 1962; Regan and Moffatt, 1990). Throughout the later stages of microspore development, the complex cell wall of the pollen grain is laid down. Briefly, it consists of an inner polysaccharide intine layer derived from the gametophyte, and an outer sporopollenin exine layer derived from both the gametophyte and sporophyte (Esau, 1977). The exine layer is highly resistant to chemicals, high temperature, and decay, and is intricately sculptured. Maturation is accompanied by dehydration of both anther tissues and pollen grains. The mature pollen grains are tricolpate and tricellular.

Development of Anthers and Pollen in Male Sterile Mutants of *Arabidopsis thaliana*

Recessive mutations causing male sterility have been reported in many plant species (Gottschalk and Kaul, 1974; Kaul, 1988). The biology of male sterility has been investigated in numerous species including maize (Beadle, 1932; Albertsen and Phillips, 1981), tomato (Rick, 1948; Rick and Butler, 1956; Lapushner and Frankel, 1967; Sawhney and Bhahula, 1988), *Brassica* species (Nieuwhof, 1968; Theis and Röbbelen, 1990), and soybean (Albertsen and Palmer, 1979; Palmer et al., 1978; Skorupska and Palmer, 1989). These studies have provided a wide spectrum of male sterile mutants for comparative work in other species. Male sterile mutants provide a potential starting point for the genetic and molecular investigation of anther and pollen development in higher plants.

The following plates compare stages in anther and pollen development in male fertile wild-type *Arabidopsis* (Landsberg *erecta* ecotype), in five recently isolated *male sterile* (*ms*) mutants (Dawson et al., 1993), and in a previously isolated mutant, *ms1* (van der Veen and Wirtz, 1968). Details of ethyl methane sulfonate (EMS) and X-ray mutagenesis of *Arabidopsis* (Koornneef et al., 1982a), microscopical methods, including a staining procedure for callose (Regan and

Moffatt, 1990), and a procedure for pollen germination (Pickert, 1988) can be found in Dawson et al. (1993). Five of the *ms* mutants were from EMS M2 populations (*msK, msW, msY, msZ,* and *ms1*) and one (*msH*) was from an X-ray M2 population. The *ms* mutations described represent different loci. Seed set on all manually pollinated (pollen derived from wild-type plants) *ms* lines was normal. Buds from fertile wild-type *Arabidopsis* plants (Landsberg *erecta* ecotype) and the male sterile lines (*ms* homozygotes) were fixed, sectioned, and analyzed by light microscopy as described by Dawson et al. (1993). Heterozygotes (*MS/ms*) showed anther and pollen characteristics of wild type, indicating the recessive nature of each of the *ms* mutations.

J.L. Bowman, J. Dawson, Z.A. Wilson, L.G. Briarty, and B.J. Mulligan

Plate 4.1
Wild-type
Microgametogenesis

Male development in *Arabidopsis thaliana* involves the initiation and elaboration of the male structural organs, the development of normal gametes inside the locules of the anthers, and the interaction of mature pollen with the stigma, leading to pollen tube growth. Male fertility is the result of the normal expression of a large number of genes; inactivation of any of these genes may lead to male sterility.

(**A**) Mature pollen grains are formed inside anther locules by a process known as microsporogenesis. Initially, a group of archesporial cells generate four thecae with pollen mother cells surrounded by a nutritive tapetum (t), endothecium, and epidermis. Pollen mother cells undergo meiosis to form haploid tetrad microspores (m).

(**B**) Enzyme activity releases individual microspores within the anther locules where they undergo mitosis to form vegetative and generative nuclei, and develop their characteristic wall sculpturing.

(**C**) Finally, the anthers dehisce, releasing functional pollen.

S. Craig and A. Chaudhury

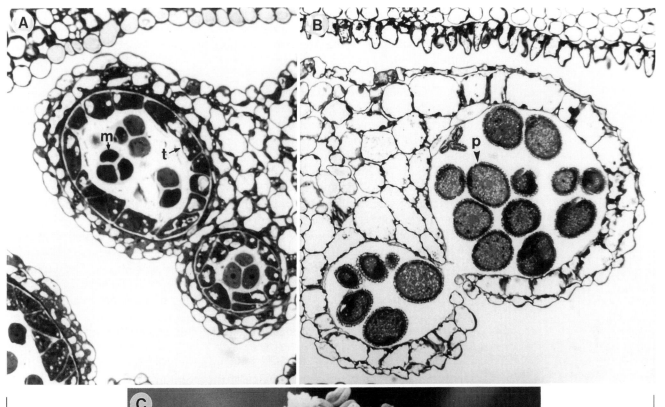

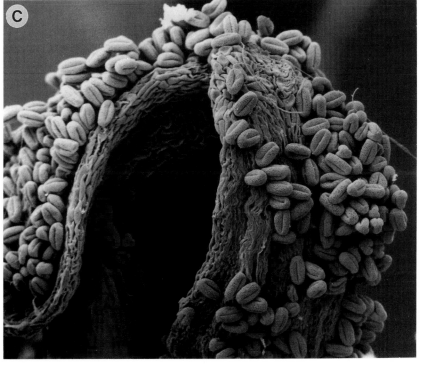

**Plate 4.2
Wild-type
microgametogenesis**

This plate shows stages in pollen development in a single (**A–F**) or in paired (**G–H**) locules of a wild-type (Landsberg *erecta* ecotype) anther. A detailed microscopical study of fertile *Arabidopsis* anthers has been described by Regan and Moffatt (1990).

(**A**) Pollen mother cells (PMCs) (p). The central mass of sporogenous tissue is surrounded by the cell layers of the anther, tapetum (tp), middle layer (m), endothecium (en), and epidermis (ep).

(**B**) PMCs in meiosis. Chromosomes are visible (*arrow*) and PMCs are surrounded by a layer of callose.

(**C**) Tetrads (t). Individual microspores are separated by callose.

(**D**) Microspores. Following callose dissolution and microspore release, the cytoplasm of young microspores is densely stained and the sculptured nature of the microspore cell wall is apparent.

(**E**) Vacuolate microspores. At this point in development, which probably corresponds to the vacuolate microspore stage described by Regan and Moffatt (1990), microspores tend to collapse during fixation procedures. The tapetum also collapses, though tapetal cells remain intact.

(**F**) Microspores at a later stage. Possible storage bodies and small vacuoles are apparent. Tapetal cells have a granular appearance and show signs of degeneration. Pollen mitosis probably occurs at this stage (Regan and Moffatt, 1990).

(**G**) Mature pollen. Pollen grains are enclosed within the anther locules. The tapetum has degenerated and cleavage of the stomium (s) and the septum (se) between the pair of locules has commenced.

(**H**) Dehisced anther. After cleavage of the septum and stomium, the anther wall springs back from the point of cleavage and pollen is released.

Bars = 10 μm.

J. Dawson, Z.A. Wilson, L.G. Briarty, and B.J. Mulligan

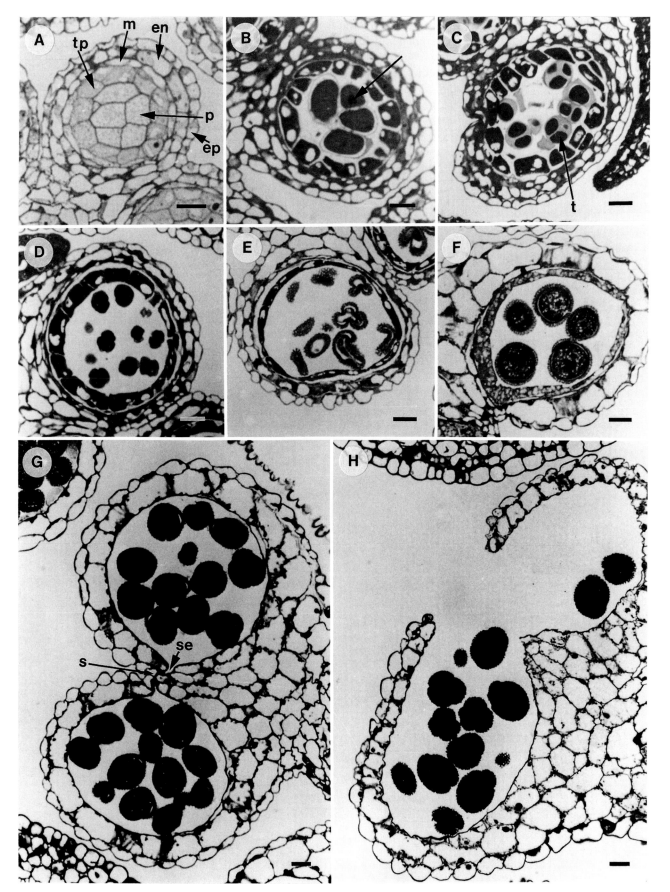

Plate 4.2

283

Plate 4.3
Male sterile mutants: *ms1*

Homozygous *male sterile1* (*ms1*) plants (van der Veen and Wirtz, 1968) show normal meiosis and callose dissolution. Pollen development is blocked just after microspore release from tetrads. Abnormalities are also apparent in tapetal cells at the point when abnormalities in pollen development are apparent. Tapetal abnormalities have frequently been observed in male sterile mutants of other plant species (e.g., Rick, 1948; Palmer et al., 1980; Kaul, 1988; Sawhney and Bhadula, 1988) and in cytoplasmic male sterile plants (e.g., Laser and Lersten, 1972; Kaul, 1988).

(A) Pollen mother cells (PMCs) at the onset of meiosis in a homozygous *ms1* plant. This stage of development appears normal.

(B) Newly released microspores of the *ms1* mutant appear very similar to those of the wild type. Immediately after release, microspores develop a granular, vacuolate cytoplasm, and tapetal cells become abnormally vacuolate.

(C) Microspores adhere, collapse, and the cytoplasm degenerates.

(D) Tapetal cells and microspores become enlarged as the cytoplasm continues to degenerate.

(E) Microspores and tapetal cells degenerate to form an undifferentiated mass. Eventually, both microspores and tapetal cells degenerate completely, leaving an empty anther locule.

Bars = 10 μm.

J. Dawson, Z.A. Wilson, L.G. Briarty, and B.J. Mulligan

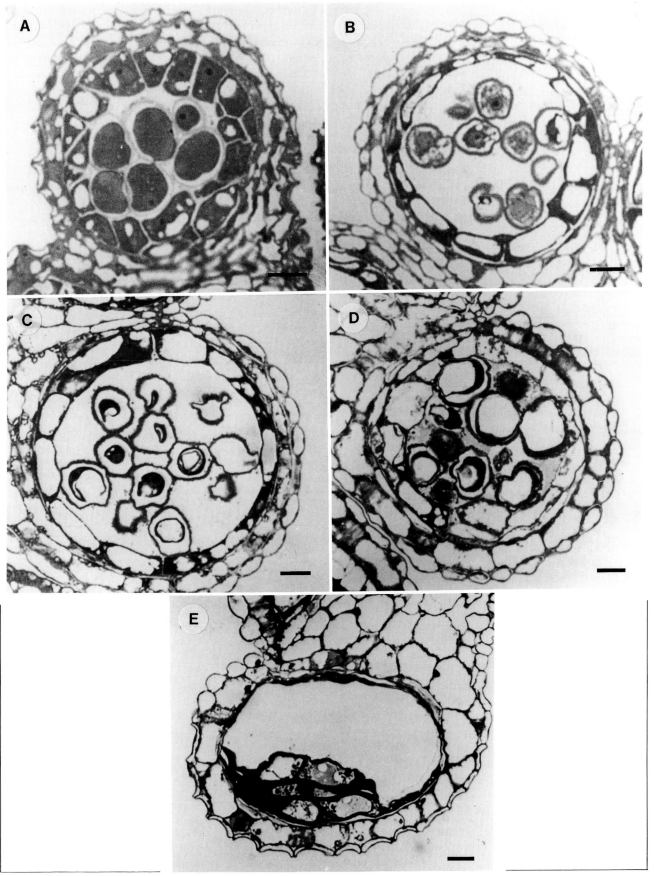

Plate 4.3

Plate 4.4
Male sterile mutants: *msZ*

A characteristic feature of fertile flowers of *Arabidopsis* is the rapid elongation of the stamen filaments just prior to bud opening (Müller, 1961; Smyth et al., 1990). This does not occur in homozygous *msZ* mutants, in which pollen development fails at a late stage, probably about the time of pollen mitosis. A similar phenotype results from the *ms2* mutation in cotton (Richmond and Kohel, 1961). The importance of the stamen filament in anther development has been discussed by Schmid (1976).

(A) In *msZ* homozygotes, pollen development is normal until the stage at which stamen elongation would normally occur; at this point, pollen contains numerous small vacuoles or storage bodies. A similar stage occurs in the development of normal pollen (see **F** of wild-type gametogenesis, Plate 4.2). Tapetal cells in *msZ* homozygotes are shrunken, however, and almost devoid of cytoplasm. In fertile anthers, tapetal cells are filled with granular cytoplasm (see **F** of wild-type gametogenesis, Plate 4.2).
(B) Tapetal cells swell and degenerate.
(C) The tapetum has degenerated but pollen still shows vacuoles and storage bodies.
(D) In mature buds of *msZ* plants, anthers remain positioned about midway up the gynoecium, well below the stigma. Rapid elongation of the stamen filaments, as seen in fertile flowers just prior to bud opening, does not occur in the *msZ* mutant. Dehiscence was very delayed and was observed only in open flowers. Pollen manually extruded from mature anthers does not germinate *in vitro* or when manually applied to receptive stigmas (not shown).

Bars = 10 μm.

J. Dawson, Z.A. Wilson, L.G. Briarty, and B.J. Mulligan

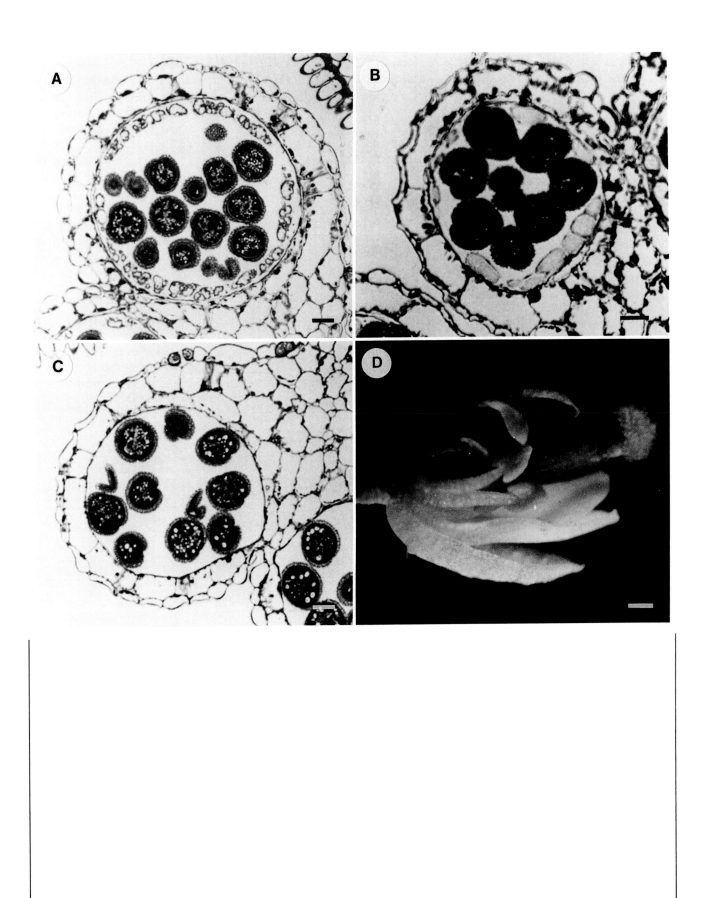

Plate 4.4

287

Plate 4.5
Male sterile mutants:
msH

The primary defect caused by the *msH* mutation appears to be a restraint on complete dehiscence, so that the anther walls do not pull back and open up the stomium after cleavage. Thus, fertile pollen cannot be released. Dehiscence is a multi-step process in which anther tissue desiccation probably plays an important part (Keijzer, 1987). The wild-type *MSH* gene product may therefore be required for the correct anatomical changes upon desiccation.

(**A**) Pollen development and the onset of tapetal degeneration prior to dehiscence is normal.

(**B**) Pollen development continues normally, and tapetal degeneration is completed as anther dehiscence begins. The septum separating the locules and the stomium are cleaved as expected. However, the anther walls do not retract from the cleaved stomium (s), and thus, pollen is not released. Examination with the dissecting microscope of unfixed open flowers indicated that no pollen emerges through the stomium.

(**C**) Pollen grains from manually disrupted anthers can be successfully germinated *in vitro*. Such pollen can be used to manually "self-fertilize" *msH* homozygotes; as expected, seed from the "selfed" *msH* cross produces all sterile progeny.

Bars = 10 μm.

J. Dawson, Z.A. Wilson, L.G. Briarty, and B.J. Mulligan

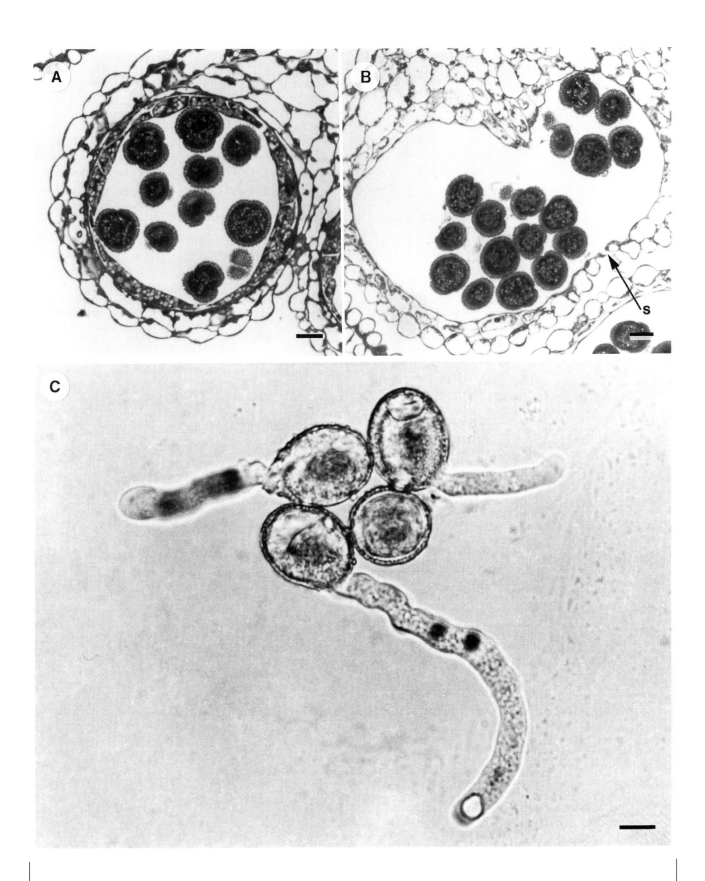

Plate 4.5

289

Plate 4.6
Male sterile mutants: *msY*

In the *msY* mutant, aberrations in anther tissues are apparent from the start of meiosis, and continue throughout anther development. A similar phenotype is produced by an *ms* mutation in *Curcurbita maxima* (Singh and Rhodes, 1961).

(**A**) Pollen mother cells (PMCs) prior to first division of meiosis are abnormally granular. Other anther tissues appear normal.

(**B**) At the start of meiosis, PMCs are abnormally granular, though occasional apparently normal PMCs can be seen. Callose is reduced, as judged by the lack of yellow fluorescence in aniline blue-stained, UV-illuminated sections. The tapetum and other anther tissues appear normal (see **B** of wild-type gametogenesis, Plate 4.2).

(**C**) The products of meiosis are heterogeneous in size and appearance, and include abnormal tetrad-like structures. Callose is usually absent. Occasional normal meiotic figures were observed in younger buds, but at a far lower frequency than in buds of the wild type.

(**D**) PMC-derived cells and tapetal cells become enlarged and both eventually degenerate.

(**E**) Later stage in the degeneration of the tapetal cells and meiotic products in *msY*. Eventually, tapetal and PMC-derived material is completely degraded, leaving the anther locules empty.

Bars = 10 μm.

J. Dawson, Z.A. Wilson, L.G. Briarty, and B.J. Mulligan

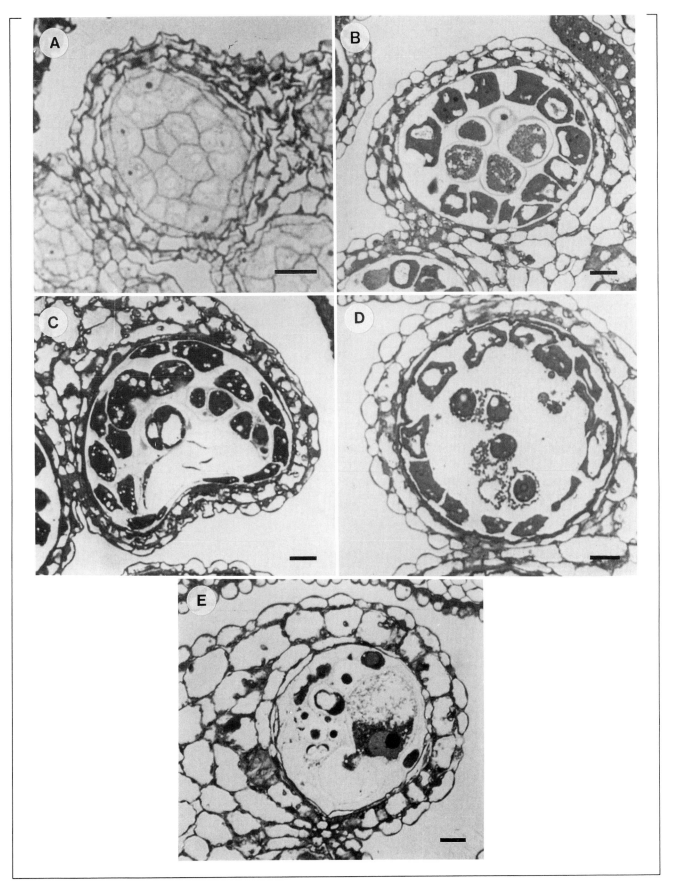

Plate 4.6   291

Plate 4.7
Male sterile mutants:
msW

The *msW* mutation causes a block in pollen development around the onset of meiosis. Also, callose is absent from pollen mother cells (PMCs) and tetrads.

(**A**) As meiosis begins, the cytoplasm of PMCs appears normal. However, callose is reduced or absent (confirmed by lack of UV-induced fluorescence in aniline blue-stained sections, not shown).

(**B**) Meiotic products are irregularly shaped, and abnormal tetrad-like cell groups are formed. Callose is absent.

(**C**) Meiotic products and tapetal cells become vacuolate and enlarged.

(**D**) Meiotic products and tapetal cells degenerate leaving an empty anther. No evidence of dehiscence was observed in sections of fixed material or in fresh anthers from opened flowers.

Bars = 10 μm.

J. Dawson, Z.A. Wilson, L.G. Briarty, and B.J. Mulligan

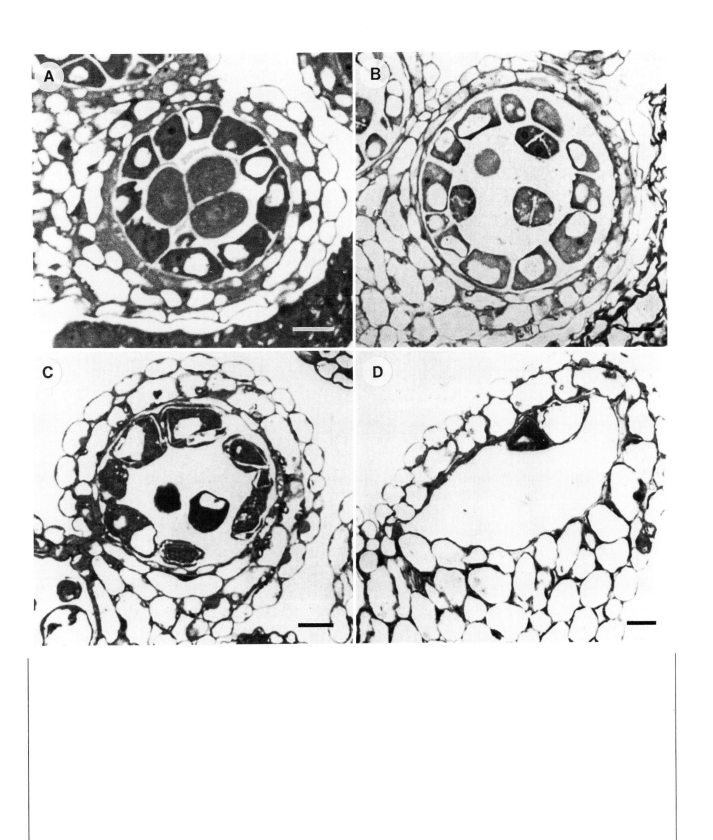

Plate 4.7

293

Plate 4.8
Male sterile mutants: *msK*

In the *msK* mutant, male sterility is accompanied by an unusual pattern of callose distribution and dissolution. The correct timing of callose dissolution has been shown to be important for normal pollen development. In studies of cytoplasmic male sterile *Petunia* (Izhar and Frankel, 1971), and of transgenic tobacco carrying a foreign β-1,-3-glucanase (callase) gene, abnormal expression of callose caused male sterility (Worrall et al., 1992).

(**A**) and (**D**) In wild-type *Arabidopsis* (Landsberg *erecta* ecotype), microspores making up a normal tetrad (t) are surrounded by callose, which fluoresces in UV-illuminated aniline blue-stained sections (**D**).

(**B**) and (**E**) In *msK* anthers, after meiosis, tetrads (t) and tapetal cells remain closely packed together. On the basis of the observed pattern of fluorescence, callose distribution around the tetrads appears normal (**E**), though, as shown in **B**, the cytoplasm of individual microspores is reduced.

(**C**) and (**F**) Microspore release from *msK* tetrads. Individual microspores remain densely packed in close proximity to the tapetum. Callose distribution is abnormal and callose dissolves slowly (**F**).

(**G**) Pollen development does not progress beyond abnormal microspores. The latter become swollen and eventually degenerate, though some abnormal microspore-derived material remains in the anther at dehiscence. Dehiscence appears to be normal, though no viable pollen is released.

Bars = 10 μm.

J. Dawson, Z.A. Wilson, L.G. Briarty, and B.J. Mulligan

Dawson, Wilson, Briarty, and Mulligan thank the UK AFRC for funding under the *Arabidopsis* Plant Molecular Biology Programme, and Mr. Brian Case for expert photographic assistance.

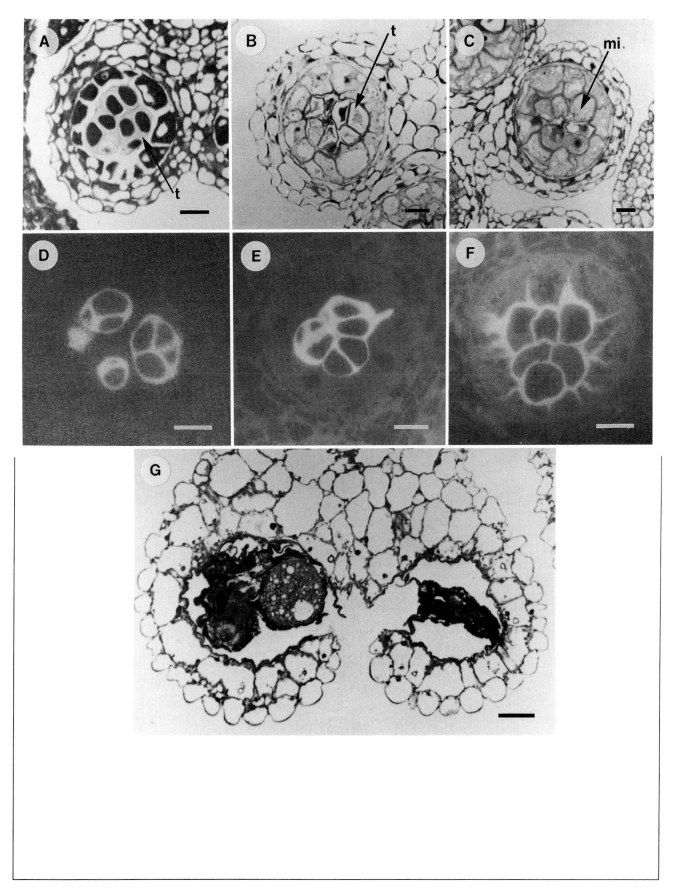

Plate 4.8

295

5
Ovules

Introduction

The gynoecium of *Arabidopsis* consists of two (or four in an alternate interpretation, see Flowers: Introduction) congenitally fused carpels that develop as a single cylinder. The ovary is divided into two locules by a false septum. The central region of the septum consists of transmitting tissue, while the tissue at the margins of the septum is fused with and is similar in appearance to the tissue of the inner layer of the ovary wall (Okada et al., 1989). Ovules arise from parietal placental tissue at the margins of fusion of the carpels, where the septum merges with the ovary wall. A total of 40–60 ovules are produced in four rows, with the two rows of ovules in each locule interdigitating.

The ovules of *Arabidopsis* are tenuinucellate (ovule has a small nucellus) and bitegmic (two integuments) (Misra, 1962). They have been characterized as having either an anatropous (Misra, 1962; Webb and Gunning, 1990; Mansfield et al., 1991) or amphitropous (Robinson-Beers et al., 1992) orientation. Detailed studies of ovule development, and in particular of embryo sac development (Webb and Gunning, 1990; Mansfield et al., 1991; Robinson-Beers et al., 1992) have allowed correlations between the stage of ovule and embryo sac development, and an external marker, the length of the gynoecium (Table 1).

The Embryo Sac

In *Arabidopsis* and most angiosperms, the megaspore mother cell undergoes meiotic divisions to form four megaspores. Three of these degenerate, and one undergoes mitotic divisions to form the cells of the embryo sac (monosporic development).

The embryo sac of *Arabidopsis* is curved and of the Polygonum type, consisting of seven cells that contain eight nuclei: the egg, two synergids, a single diploid central cell, and three antipodals (Misra, 1962; Polyakova, 1964; Webb and Gunning, 1990; Mansfield et al., 1991). The egg and synergids are intimately arranged in the micropylar chamber of the embryo sac, there forming what is known as the egg apparatus. The central cell separates the egg apparatus from the triad of antipodals in the chalazal portion of the embryo sac. Each of these cell types displays a different form of structural, and presumably also functional, specialization (Mansfield et al., 1991), and this, together with their precise disposition, probably relates to the cells' roles in early embryogenesis, and undoubtedly has implications for the process of fertilization.

Mutations Affecting Ovule Development

Selection on the basis of female sterility but morphologically normal carpels has led to the isolation of mutants that have altered ovule development (Robinson-Beers et al., 1992). In *bell* mutants, only a single integument-like structure is initiated and its growth is abnormal (organized cell layers do not develop) as it never envelops the nucellus (Robinson-Beers et al., 1992). In some cases, the cells that would normally give rise to the outer integument develop into structures resembling complete carpels (Plate 5.2). In addition, in *bell* ovules the funiculus is thickened compared to wild type, and embryo sac development is disrupted. Ovules of *short integument* (*sin*) mutants also exhibit

Table 1. Stages of embryo sac and ovule development.

Approximate floral stage[a]	Stage of embryo sac development[b]	Ovule morphology[c]	Pistil length[d] (mm)	Appearance of flower[e]
9		Ovule primordia arise	0.15–0.4	Small green bud, gynoecium slotted at top
10		Ovule primordia elongate	0.4–0.5	Small green bud, gynoecium starting to close
11	Megasporocyte	Inner and outer integuments initiated	0.5–1.0	Small green bud, anthers green, filaments not readily visible
	4 megaspores			Green bud, filaments beginning to elongate, petals green
	Functional megaspore (1-nucleate embryo sac)		1.0–1.5	Anthers green, petals white
early 12	2-nucleate embryo sac	Funiculus and nucellus curve; other integument exhibits assymmetric growth	1.5–2.0	Petals showing and anthers starting to turn yellow
mid 12	4-nucleate embryo sac	Integuments grow upward around nucellus		Bud starting to open, height of long anthers level with stigma, petals protruding well past sepals
late 12	8-nucleate embryo sac	Outer integument begins to cover both inner integument and nucellus		
13	Mature embryo sac (8-nucleate, 7-celled)	Integuments envelop nucellus; micropyle positioned near funiculus	2.0–2.5	Freshly opened flower, anthers dehiscing, petals bent outwards
14	Fertilization	Embryo sac becomes increasingly curved	2.5–3.0	Open flower, pistil extended past height of anthers, anthers dehisced, petals just beginning to straighten
15	Elongating zygote and developing endosperm		3.0–3.5	Open flower, ovary still green
15–16	Early embryo		3.5–4.0	Siliqua (fertilized pistil) extending past petals, sepals yellowing

Arabidopsis thaliana ecotype Columbia were grown under constant illumination at 18°C.

[a] Floral stages refer to those of Müller (1961), Smyth et al. (1990), Bowman et al. (1991), and Robinson-Beers et al. (1992).

[b] The stage of embryo sac was determined by clearing ovules (Jongedijk, 1987; Webb and Gunning, 1990).

[c] Stages of ovule development summarized from Robinson-Beers et al. (1992) where Landsberg *erecta* plants were grown under constant illumination at 25°C.

[d] Pistils were assigned to size classes which covered 0.5 mm increase in length, from the top of the stigma to the point of attachment to the receptacle.

[e] Although less precise than pistil length, this gives an estimation of floral appearance at a given stage of embryo sac development.

Table: M.C. Webb and J.L. Bowman.

altered embryo sac development. In addition, although both inner and outer integuments are initiated in *sin* mutants, they fail to enclose the nucellus, and continue to grow via cell division without the normal cell expansion (Robinson-Beers et al., 1992). Ovules of another female sterile mutant (*female gametophyte factor, gf*), which exhibits apparent megaspore selection, appear morphologically normal except for a low incidence of twinned embryo sacs (Rédei, 1965a).

S.G. Mansfield and J.L. Bowman

Plate 5.1
Ontogeny of the wild-type ovule

Morphogenesis of wild-type *Arabidopsis* ovules was examined from initiation to maturation by scanning electron microscopy. Stages are those of Smyth et al. (1990) and Robinson-Beers et al. (1992).

(**A**) During stage 9, two files of ovule primordia (op) are initiated on each of the two placentae. The positions of the primordia are staggered relative to those on the second placentae.

(**B**) As the ovule primordia enlarge during stage 10, and the central septum (cs) dividing the gynoecial cylinder forms, the two rows of ovule primordia in each locule display an interlocking pattern.

(**C**) The nucellus (nu) and funiculus (fu) begin to differentiate, followed by integument initiation marked by cell outgrowth (cog) at the base of the nucellar region during stage 11. Both inner and outer integuments are initiated through a series of cell divisions in the L1 a short distance behind the apex (Robinson-Beers et al., 1992). Within the nucellus, megasporogenesis results in formation of four megaspores (Webb and Gunning, 1990). *Arabidopsis* embryo sac morphogenesis is of the Polygonum type. Therefore only a single megaspore survives to form the embryo sac.

(**D**) By early stage 12, the inner and outer integument primordia (iip, oip) appear as two rings of tissue between the funiculus and nucellus.

(**E**) The outer integument primordia display asymmetrical growth through increased cell divisions on the side facing the central septum. Megagametogenesis is initiated (Robinson-Beers et al., 1992).

(**F**) Growth of inner and outer integuments (ii, oi) continues upward, enclosing the nucellus during mid-stage 12.

(**G**) At maturity (stage 13), the outer integument completely overgrows the inner integument and the nucellus, forming the micropylar opening (mp) where the pollen tube (pt) enters the ovule (ov). By late-stage 12 within the nucellus, megagametogenesis is completed with the formation of a seven-celled, eight-nucleate embryo sac (Mansfield et al., 1991).

(**H**) Following fertilization (stage 14), the ovules enlarge and the embryo sacs become highly curved, suggesting an amphitropous orientation. *Arabidopsis* ovules have been characterized as having either an anatropous (Misra, 1962; Webb and Gunning, 1990; Mansfield et al., 1991) or amphitropous (Robinson-Beers et al., 1992) orientation.

Z. Modrusan, L. Reiser, R.L. Fischer, and G.W. Haughn

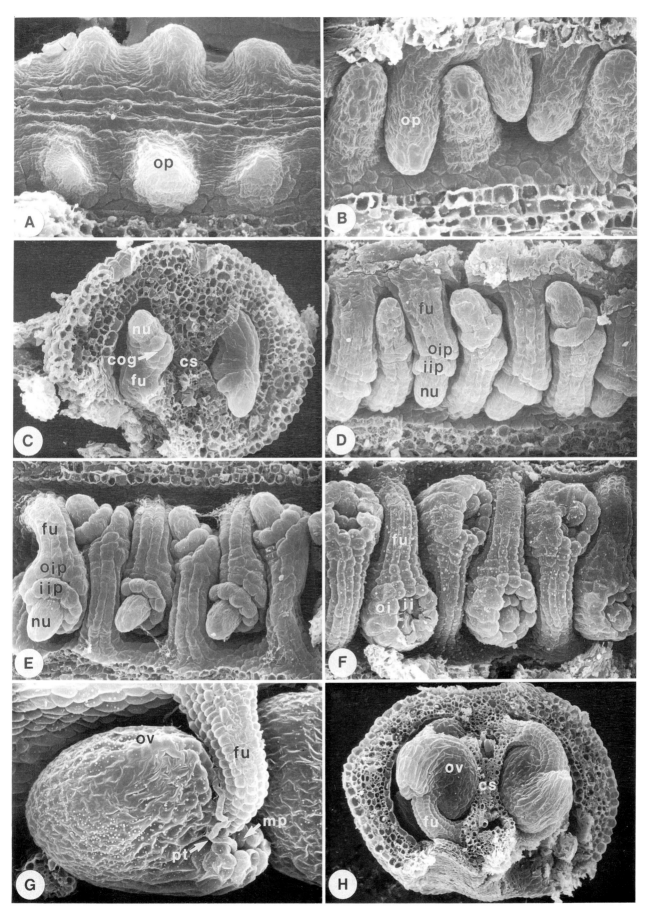

Plate 5.1

Plate 5.2

bell1 **mutants are defective in ovule morphogenesis**

Two mutants, originally designated as *fruitless* (*fts*) (Modrusan et al., 1992) and *female sterile mutant* (*fsm*) (Reiser and Fischer, 1991) have similar defects in ovule morphogenesis due to recessive mutant alleles of a previously identified nuclear gene BELL1 (BEL1) (Robinson-Beers et al., 1992). These two alleles are now designated *bel1-3* and *bel1-2*, respectively. All other aspects of morphogenesis appear to be normal in these two mutants.

Morphological and anatomical development of *bel1-3* mutant ovules from initiation to degeneration have been examined by light and scanning electron microscopy.

(**A**) Ovule primordia (op) are initiated from the placental tissue of *bel1-3* ovaries at the same time and in the same arrangement as in the wild type. As the primordia elongate, the appearance of a megasporocyte (ms) is an early indication that differentiation into nucellus (nu) and funiculus (fu) has been initiated.

(**B**) Differences between *bel1-3* mutant and wild-type ovule development are first observed at the time of integument initiation. The inner integument fails to develop, and an organized cluster of cells forms at the position of the outer integument. Subsequent cell growth generates an irregular thickening (it) at the base of the nucellus of mutant ovule.

(**C**) Within the nucellus of *bel1-3* mutant ovules, megasporogenesis is completed and the chalazal megaspore (cm) enlarges.

(**D**) and (**E**) Continued growth of the thickened region at the base of the nucellus of *bel1-3* mutants results in the formation of finger-like projections (integument-like structure, ils) which surround the nucellus and give a distinct morphology to the mutant ovule. The cells within the integument-like structures remain densely cytoplasmic, in contrast to wild-type integuments which have highly vacuolate cells.

(**F**) Within the nucellus the chalazal megaspore appears to undergo some of the mitotic divisions of gametogenesis, but a functional embryo sac (es) is never formed.

(**G**) Most of the deformed *bel1-3* mutant ovules degenerate as the flower senesces.

(**H**) In a small percentage of the *bel1-3* mutant ovules, the integument-like structures continue to develop and eventually differentiate carpel-like features giving the mutant ovules a pistil-like appearance. The carpel-like structures (cls) possess cell types typical of wild-type carpels, such as stigma (stg), style (sty), and ovaries (ova).

Z. Modrusan, L. Reiser, R.L. Fischer, K.A. Feldmann, and G.W. Haughn

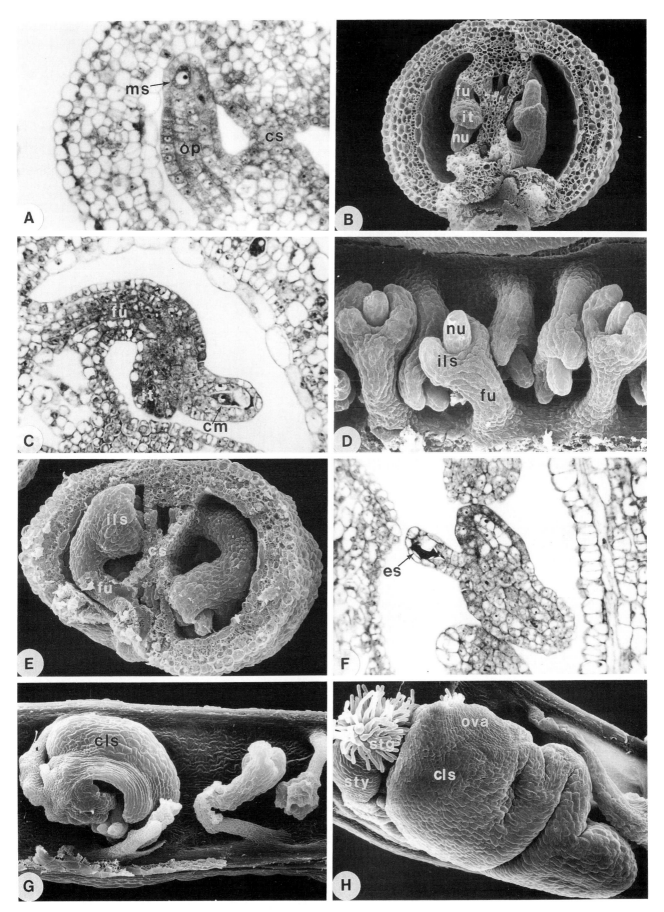

Plate 5.2

Plate 5.3
Morphology of the ovary and ovules

(**A**) Longitudinal section (1.5 μm, toluidine blue staining) of the ovary during megasporogenesis (ov, ovule; pl, placentae; w, ovary wall)

(**B**) Transverse section (2 μm) of the ovary containing mature embryo sacs. In **A** and **B**, the numerous ovules (ov) are variously oriented and attached to central placentae (pl) of a two-loculed ovary (w, ovary wall).

(**C–F**) Scanning electron micrographs showing stages of ovule development. The ovules are tenuinucellate (ovule has a small nucellus), bitegmic (two integuments), and are anatropous at maturity (Misra, 1962).

(**C**) During megasporogenesis the integument primordia (ii, inner integument primordia; oi, outer integument primordia), which encircle the base of the nucellus, begin to grow up unevenly around the nucellar epidermis (nu), a layer of cells immediately surrounding the developing archesporium and embryo sac (fu, funiculus).

(**D**) Ovule during early megagametogenesis, by which time the integuments have enclosed the nucellus (m, position of the micropyle).

(**E**) By maturity, the ovule has bent around such that the micropylc (m) is in close association with the funiculus (fu).

(**F**) Fertilized ovule showing the marked elongation which has occurred to accommodate the expanding embryo/endosperm inside.

Bar = 20μm.

M.C. Webb

A and **C** are reproduced from Webb and Gunning (1990) with permission from Springer-Verlag.

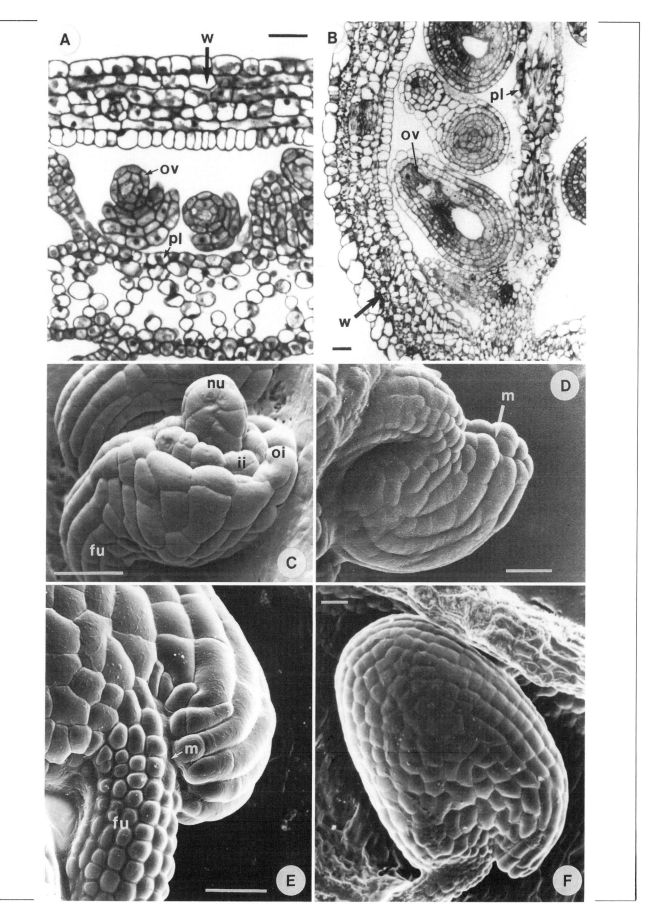

Plate 5.3

Plate 5.4
Megasporogenesis

Recent observations of megasporogenesis in *Arabidopsis thaliana* deviate from previous studies which report monosporic *Polygonum* type development with successive cytokinesis, giving a linear tetrad of megaspores (Misra, 1962; see also Rédei, 1970). Instead, the megasporocyte was found to undergo both meiotic divisions followed by simultaneous cytokinesis (i.e., without an intermediate dyad stage), to give a multiplanar (tetrahedral or decussate; Blackmore et al., 1987) tetrad (for details see Webb and Gunning, 1990).

(**A**) and (**B**) Equivalent longitudinal views of the megasporocyte. A hypodermal, one-celled archesporium functions directly as the megasporocyte (Misra, 1962), which enlarges more than threefold to be approximately 17 μm long and 10 μm wide before undergoing division (Webb and Gunning, 1990).

(**A**) Differential interference contrast (DIC) image of a cleared ovule showing the characteristic "in spireme" (Kausik, 1940) appearance of the megasporocyte nucleus.

(**B**) Longitudinal section (1.5 μm, toluidine blue staining) showing organelles dispersed throughout the cytoplasm. By this stage the chromatin has already condensed and the single prominent nucleolus stains intensely (*arrow*). The nuclear position is highly variable among ovules.

(**C**) and (**D**) Transverse sections (1.5 μm, toluidine blue staining) taken from a complete series through one ovule, showing (**C**) the nucleus with densely staining nucleolus toward the chalazal end, and (**D**) organelles which are concentrated at that end, site of the future functional megaspore, before meiosis begins.

(**E**) and (**F**) Approximately longitudinal sections (1.5 μm, toluidine blue staining) taken from a series through an ovule, showing late anaphase I of meiosis. The bivalents can be clearly seen approaching the (**E**) micropylar and (**F**) chalazal poles. Organelles are restricted to the outer cytoplasm.

(**G**) and (**H**) Longitudinally oriented, cleared ovule viewed with DIC optics, showing a tetrad directly after formation. The megaspores (m) are in a decussate arrangement. One focal plane (**G**) shows two of the megaspores, and (**H**) a lower plane of focus shows the other two tetrad members.

(**I–K**) Immediately after megaspore formation, the megaspore closest to the chalazal end of the ovule begins to enlarge while the other three rapidly degenerate. Expansion of the functional megaspore progressively displaces the degenerative ones. The final arrangement of degenerated cells is very flexible between ovules (Webb and Gunning, 1990).

(**I**) Longitudinal section (2 μm, DIC optics) showing two megaspores (*arrows*) beginning to degenerate micropylar to the active one. The fourth megaspore is not visible. Note that this could easily be interpreted as deriving from a linear tetrad.

(**J**) and (**K**) Oblique sections (1.5 μm, toluidine blue staining) taken from a complete sequence showing (**J**) the functional megaspore (f) and one degenerated cell (*arrow*), and (**K**) the other degenerating megaspores (*arrows*) extending below the level of the functional megaspore with respect to the micropylar end of the "embryo sac space."

Bars = 5 μm.

M.C. Webb

Reproduced from Webb and Gunning (1990) with permission from Springer-Verlag.

Ovules: Megasporogenesis

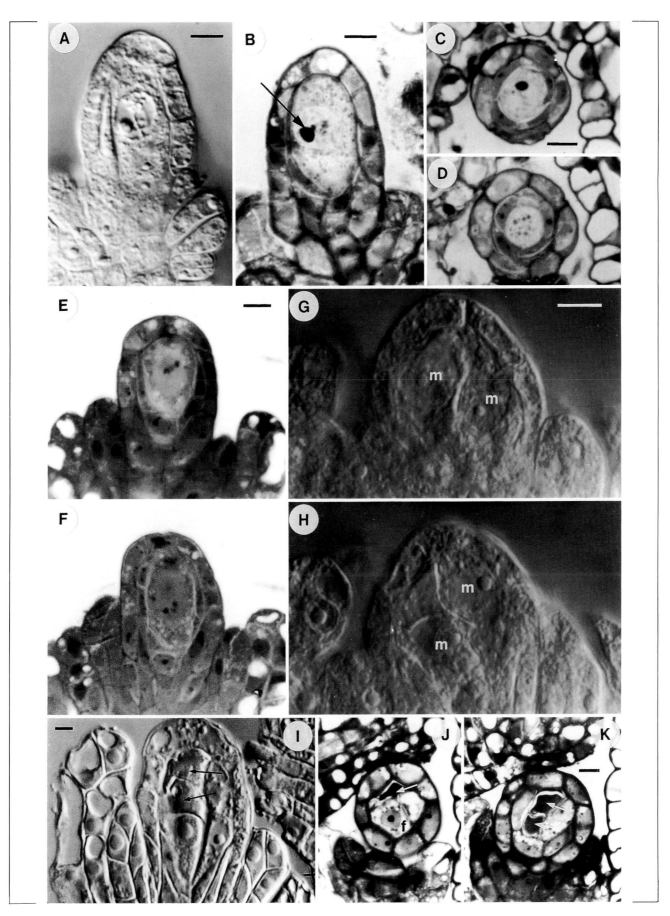

Plate 5.4

Plate 5.5
Megagametogenesis

The functional megaspore undergoes three rounds of mitosis. The resultant eight nuclei migrate to specific positions in the coenocytic embryo sac before cell wall formation occurs to give a seven-celled mature embryo sac (megagametophyte). In all parts of this plate the micropyle is to the left of the micrograph. Detailed studies of various aspects of megagametogenesis and the mature embryo sac in *Arabidopsis* include Polyakova (1964), Mansfield et al. (1991), and Webb (1991).

(**A**) and (**B**) Differential interference contrast (DIC) images of cleared ovules. The mitotic divisions of megagametogenesis give rise to the two-nucleate (**A**), the four-nucleate (**B**), and subsequently, the eight-nucleate stages of ovule development. Vacuoles (v) (particularly large central ones) are observed in the fully expanded phase of all these stages.

(**C**) through (**H**) Longitudinal views of the mature embryo sac.
(**C**) Section (2 μm, toluidine blue staining) through the micropylar end, showing the synergids (s), each with a filiform apparatus (fa), the egg (e), and central cell (cc). Three small antipodal cells are found at the chalazal end (not shown). The nucellar cells (nu) close to the micropylar end have started to degenerate. Further degeneration of these cells causes the innermost layer of the inner integument (ii) to be directly adjacent to the embryo sac at the micropylar end in mature ovules (Kapil and Tiwari, 1978; Robinson-Beers et al., 1992).
(**D**) Section (2 μm) stained with Hoechst dye to show the interesting phenomena of extremely weak staining of embryo sac nuclei (*arrows*) by fluorescence DNA stains compared to the brightly staining nuclei of the surrounding tissues (for discussion see Webb and Gunning, 1993).
(**E–H**) The component cells of the egg apparatus (i.e., the egg and synergids) shown from different angles to show their arrangement.
(**E**) and (**F**) Two sections (2 μm, toluidine blue staining) taken from a complete series through an ovule. The chalazal end of the embryo sac curves into the page. The two synergids (s) lie closest to the micropylar entrance and funicle (**E**), while the egg (e) is fastened "behind" them (**F**). This arrangement resembles that in barley (Engell, 1988). Note also the central cell (cc), the central part of which is occupied by a large vacuole (v).
(**G**) and (**H**) DIC image of a cleared ovule (**G**) and a section (**H**) of the egg apparatus viewed at right angles to the view in **E** and **F**. In both micrographs only one synergid (s) is visible, and the egg (e) is positioned "behind" it. In **H**, the synergids have direct contact with the micropyle (m). *Arrows* mark the extremities of the egg cell. Also note the positions of the inner (ii; 2–3 layers thick) and outer (oi; 2–4 layers thick) integuments which consist of well-defined parallel cells (Misra, 1962).

Bars = 10 μm.

M.C. Webb

A–D reproduced from Webb and Gunning (1993) with permission of Kluwer Academic Publishers.

Ovules: Megagametogenesis

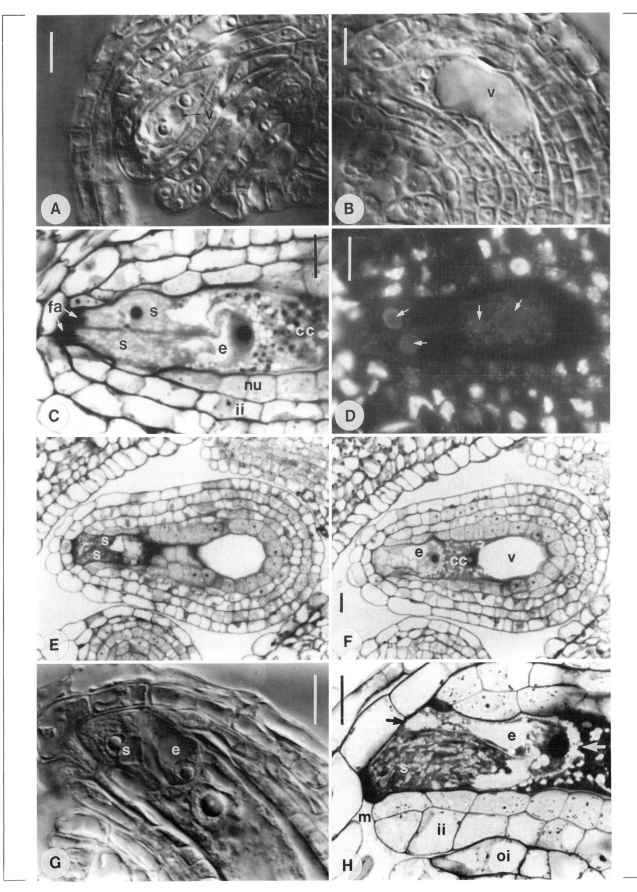

Plate 5.5

Plate 5.6
Zygote and early embryo development

Embryo development follows the *Capsella* variation of the Onagrad type (Misra, 1962). Aspects of normal embryogenesis in *Arabidopsis* have been studied in detail (Yakovlev and Alimova, 1976; Webb and Gunning, 1991; Mansfield and Briarty, 1991).

(**A–C**) Differential interference contrast images of cleared ovules. The micropyle is at the top in each micrograph.

(**A**) The primary endosperm nucleus divides immediately after fertilization while the zygote (z) remains quiescent for some time (c, chromosomes in metaphase configuration). The endosperm nuclei continue to divide without cell wall formation throughout the period of early embryogenesis.

(**B**) Telophase stage of division of nuclei in the coenocytic endosperm. Spindles (*arrows*) are visible between groups of chromosomes.

(**C**) A central vacuole occupies much of the space in the free-nuclear endosperm, confining the cytoplasm with nuclei to the periphery.

(**D**) Longitudinal section (2 μm, toluidine blue staining) through a fertilized ovule showing the zygote (z) which has elongated markedly. The nucleus and most of the cytoplasm are located at the chalazal end of the zygote (*arrow*), while a large vacuole occupies the micropylar end. Also note the highly vacuolate coenocytic endosperm (e). Approximately at anthesis, the innermost layer of the inner integument (*arrowhead*) differentiates into the integumentary tapetum, or endothelium, and is visible as an intensely staining layer of cells adjacent to the embryo sac (Kapil and Tiwari, 1978; Bowman et al., 1991a; Robinson-Beers et al., 1992). The endothelium appears to have a nutritive role with respect to the developing embryo (Kapil and Tiwari, 1978).

(**E**) and (**F**) The zygote undergoes a transverse division to form the apical cell (a) which is the precursor of the embryo proper except for the root (Jürgens et al., 1991), and a basal cell (b) which divides to form the root apex and suspensor. In this example, the divided zygote has been enzymatically isolated intact. (**E**) Differential interference contrast image. (**F**) Fluorescent Hoechst DNA-staining showing the nuclei.

(**G**) Longitudinal section (2 μm, toluidine blue staining) showing the quadrant stage embryo extending into the free-nuclear endosperm. The apical cell has divided longitudinally to give four cells, the second divisions at right angles to the first. The basal cell has divided transversely, creating a uniseriate suspensor.

(**H**) Enzymatically isolated early embryo in which the four cells of the quadrant embryo have undergone a transverse division to form the octant (eight-celled) embryo. Cellularization of the endosperm does not occur until the beginning of the heart-shaped embryo stage (e.g., Mansfield and Briarty, 1990b).

Bar = 10 μm.

M.C. Webb

A–D reproduced from Webb and Gunning (1993) with permission from Kluwer Academic Publishers.

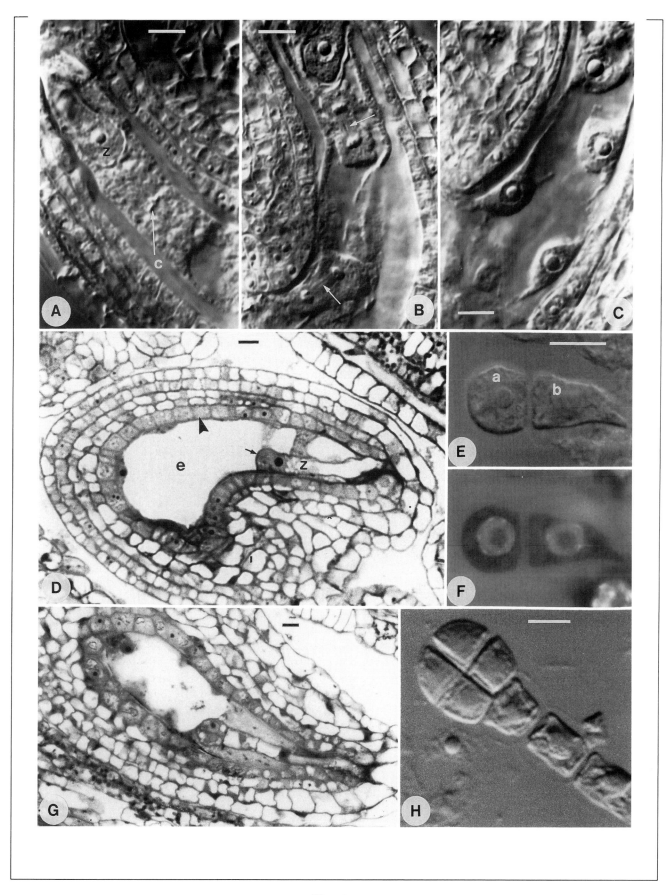

Plate 5.6

Plate 5.7
Microtubules and
megasporogenesis

The microtubule cytoskeleton is extensive throughout female reproductive development in *Arabidopsis thaliana* (Webb, 1991; Webb and Gunning, 1990; Webb and Gunning, 1991; Webb and Gunning, 1993). Many distinct and varied arrays occur and appear to reflect a sequence of disparate roles in the different cells and stages involved. The cells in this figure are all immunofluorescently labeled for their microtubules. All show cells which have been isolated enzymatically, except **J** and **K** which show a cryostat section (see Webb and Gunning, 1990).

(**A–D**) In all stages, from megasporocyte up to fertilization and including the coenocytic endosperm, the cells have very few cortical microtubules (MTs) but do have complex internal arrays. This example shows a one-nucleate embryo sac about to undergo mitosis.

(**A**) The shape of the cell (Differential, interference contrast, DIC).

(**B**) Hoechst staining showing the condensing chromosomes.

(**C**) and (**D**) Two focal planes showing MT distribution. (**C**) Focused at the nucleus' near face, (**D**) at the nucleus' mid-plane (∗, nuclear position).

(**E**) and (**F**) In contrast to the preceding stages, cortical arrays of MTs are well developed in the zygote and early embryo cells, and indicate that cell shape is internally determined in these stages (Webb and Gunning, 1991).

(**E**) Surface view of a young octant stage embryo (DIC).

(**F**) Microtubules are in a reticulate pattern through the cell cortices of the young embryo, proper, at interphase.

(**G–K**) Interphase populations of MTs which are closely associated with the nucleus occur widely throughout female reproductive cell development. In most cases, MTs appear to radiate from the nuclear envelope (**G**) and sometimes extend between sister nuclei (**H, I**) or to the plasma membrane (**J, K**) (∗, position of nuclei).

(**G**) Portion of the free-nuclear endosperm showing many MTs radiating from each nucleus.

(**H**) and (**I**) Chalazal end of a four-nucleate embryo sac showing (**H**) a prominent array of MTs extending between a pair of sister nuclei, and (**I**) the corresponding nuclei visualized with Hoechst DNA stain.

(**J**) and (**K**) Approximately longitudinal cryostat section (10 μm) showing (J) MTs in the functional megaspore concentrated around the nucleus (∗) with some spanning the cytoplasm to the plasma membrane (*arrowhead*). Microtubules are barely discernible in the degenerating cells. The DIC image of the same section (**K**) shows the cell boundaries (f, functional megaspore; *arrows*, degenerating megaspores).

(**L–O**) Microtubules contribute to typical mitotic configuration during karyokinesis. Phragmoplast arrays of MTs coincide with cell plate formation, and preprophase bands of MTs occur in unfertilized cell stages. This example shows, in two focal planes, division of the quadrant to form the octant embryo.

(**L**) Upper focal plane showing one nucleus in preprophase-prophase (pp), and two smaller nuclei which are the products of recent division (*arrows*). (Hoechst staining of the nuclei.)

(**M**) Lower plane showing two further daughter nuclei situated in the cell behind those shown in **L**, and chromosomes at metaphase (m) situated in the cell behind the preprophase-prophase cell of **L**. Suspensor nuclei are also visible (*arrowheads*).

Bar = 5 μm

(*Text continued on p. 317*)

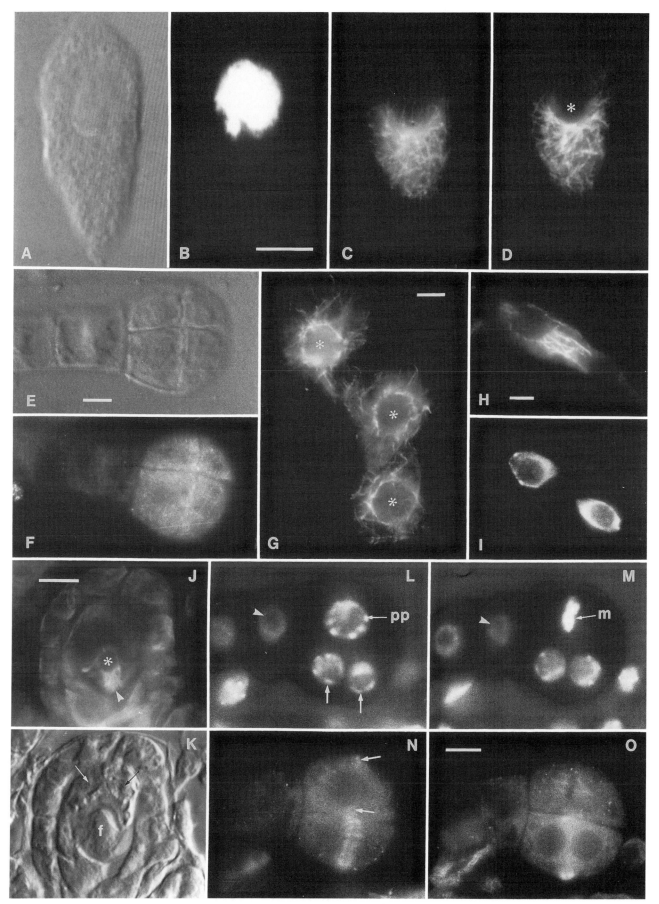

Plate 5.7

Plate 5.7
Microtubules and
megasporogenesis
(continued)

(**N**) and (**O**) Microtubules corresponding to the focal planes of **L** and **M**, respectively.

(**N**) A preprophase band of MTs (*arrows*) transversely girdles the cell in preprophase-prophase, and a parallel phragmoplast array has formed between daughter nuclei m the adjacent cell.

(**O**) A spindle array is visible in the metaphase cell and a late-stage phragmoplast array in the adjacent cell, with MTs persisting at the growing margins of the cell plate.

M.C. Webb

A–D, **J**, **K** reproduced from Webb and Gunning (1990) and **E**, **F**, **L–O** reproduced from Webb and Gunning (1991) with permission from Springer-Verlag. **H**, **I** reproduced from Webb and Gunning (1993) with permission from Kluwer Academic Publishers.

Plate 5.8
Microtubules and the zygote

Microtubules (MTs) undergo many changes in their configurations during zygote maturation, and this serves as a vivid reminder of the dynamic nature of the microtubular cytoskeleton in plant cells. All categories of MT population described occur in a single zygote cell.

(**A**) Summary of the changing MT distributions during zygote development. The micropylar end of the zygote is at the bottom of the drawing. (**1**) Initially, MTs are randomly oriented throughout the cytoplasm. (**2**) In the young zygote, MTs transversely encircle the cell at the distal end (with respect to the micropyle). The orientation of the cortical MTs gradually shifts to approaching longitudinal at the proximal end. (**3**) In the elongating zygote, MTs are predominantly cortical and transversely oriented with a distinct concentration at the growing end of the cell. (**4**) As the zygote reaches its maximum length, the concentration of MTs at the tip becomes less pronounced. (**5**) The mature zygote has an even distribution of cortical MTs along its length. (**6**) Entering preprophase. A broad preprophase band (PPB) of MTs girdles the cell at the putative site of cell plate formation. (**7–9**) Microtubules form standard spindle configurations during nuclear division. (**7**) Late prophase. (**8**) Metaphase. (**9**) Late anaphase. (**10**) A phragmoplast array of MTs is present in the interzone between daughter nuclei during telophase. A second population of MTs is closely associated with the nuclei.

(**B–H**) Representative micrographs of isolated and enzymatically isolated zygote cells of some of the stages described in **A**.
(**B**) and (**C**) Correspond to stage 1 of **A**.
(**B**) Fertilized embryo sac showing the newly formed zygote (z), the degenerating persistent synergid (ps), and the free-nuclear endosperm (e) (Differential interference contrast, DIC).
(**C**) Microtubules are found throughout the cytoplasm of the endosperm and zygote in random orientation and are particularly concentrated around the zygote nucleus (∗). Note the lack of fluorescence in the degenerating synergid.
(**D**) and (**E**) Correspond to stage 3 of **A**.
(**D**) DIC image.
(**E**) Microtubule distribution.
(**F**) Corresponds to stage 5 of **A**.
(**G**) Corresponds to stage 6 of **A**. Preprophase bands of MTs (for definition see Gunning and Hardham, 1982) are not present during megasporogenesis and subsequent embryo sac development (Hogan, 1987; Webb and Gunning, 1990; Webb, 1991), but are reinstated in the first zygotic division and are present consistently in somatic divisions thereafter during embryogenesis (Webb and Gunning, 1991).
(**H**) Corresponds to stage 8 of **A**.

Bar = 5 μm.

M.C. Webb

Reproduced from Webb and Gunning (1991) with permission from Springer-Verlag.

A

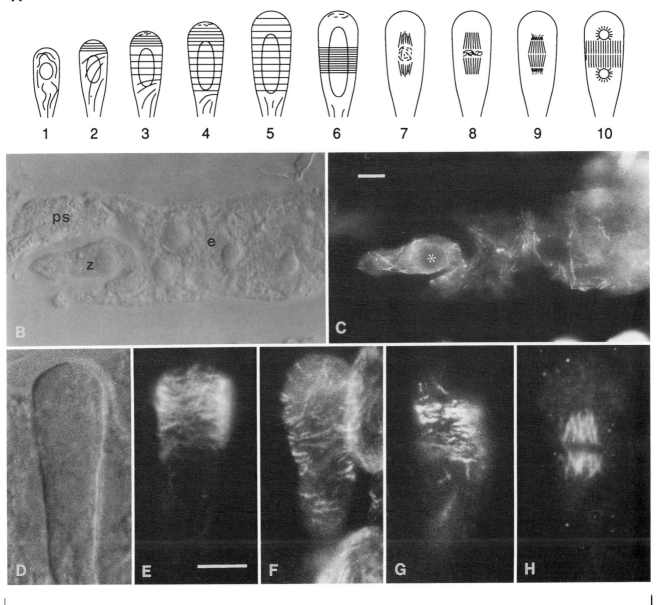

Plate 5.8

319

Plate 5.9
The mature embryo sac

(**A**) The mature flower of *Arabidopsis thaliana* at zero hours after flowering [stage 14, as defined by Müller (1961) and Smyth et al. (1990)], defined as the stage at which the length of the medial (long) stamens exceeds that of the gynoecium. One sepal (s) removed to show the arrangement of petals (p), anthers (a), and stigma (St). All descriptions of embryo development including embryo morphogenesis, endosperm formation, and storage reserve deposition (Chapter 7: Embryogenesis) have been related to the well-defined time scale Hours After Flowering (HAF).

(**B**) The mature embryo sac zero HAF. *Arabidopsis* has an anatropous bitegmic ovule, as seen in a median longitudinal section, and a monosporic, eight-nucleate, seven-celled, curved megagametophyte of the Polygonum type, typical of the Brassicaceae. The central cell (cc) lies between the egg apparatus and the antipodal cells (ac), the former consisting of two identical synergid cells (sc) and a singly highly vacuolate egg cell (ec) (f, funiculus; i, inner integument; mi, micropyle; n, nucellus; vb, vascular bundle).

Bar = μm in **B**; 1 mm in **A**.

Table 1. Cell dimensions[a] (μm) of embryo sac cells.

Cell type	Cell size	
	Length	Width
Central cell	6.5	25
Synergid cell	23	10.8 (long axis)
		5.5 (short axis)
Egg cell	21	13
Antipodal cell	7	19

[a] Values represent maximum profile measurements taken from a minimum of 10 micrographs.

S.G. Mansfield

Reproduced from Mansfield et al. (1991) with permission from the National Research Council of Canada.

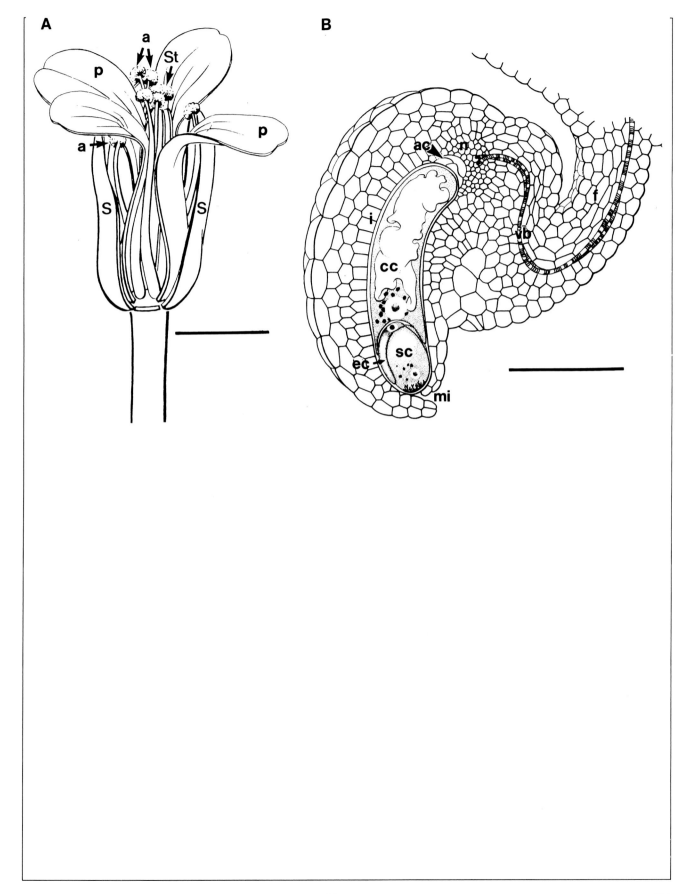

Plate 5.9

321

Plate 5.10

The egg cell zero hours after flowering (HAF)

(**A**) Light micrograph of the egg cell (ec) in longitudinal section. The chalazal nucleus of the egg cell (*arrowhead*) projects above the adjacent synergid cell (sc). A number of cytoplasmic strands (*arrows*) extend through the large chalazal vacuole.

(**B**) Light micrograph of the egg cell (ec) chalazal tip overlying the two synergid cells (sc). Large PAS-positive starch grains (*arrows*) are present around the egg cell nucleus (*arrowhead*) (PAS, periodic acid and Schiff's staining).

(**C**) Transmission electron micrograph of longitudinal section of the egg cell (ec). The egg is highly polarized with a chalazally situated nucleus (nu) regular in profile, and a single large micropylar vacuole (v) which may sometimes extend into the chalazal tip alongside the nucleus (not shown). Numerous nuclear pores (*arrows*) are evident in the nuclear membrane. The chalazal cytoplasm in this particular example contains two plastid profiles (pl).

(**D**) Enlargement of the chalazal cytoplasm adjacent to the nucleus (nu). The cytoplasm appears quiescent, ribosome density and plastid, mitochondria (m), endoplasmic reticulum (*arrowhead*), and dictyosome abundance is low. Note the absence of a cell wall at the boundary between the egg cell (ec) and central cells (cc). The gap between the egg and central cell plasmalemmas, which is widest at the extreme chalazal region, contains small masses of dense-staining material (*arrows*) which usually make contact with opposing areas of membrane.

Bar = 2 μm in **C**, **D**; 10 μm in **A**, **B**.

S.G. Mansfield

Reproduced from Mansfield et al. (1991) with permission from the National Research Council of Canada.

Ovules: The Mature Embryo Sac

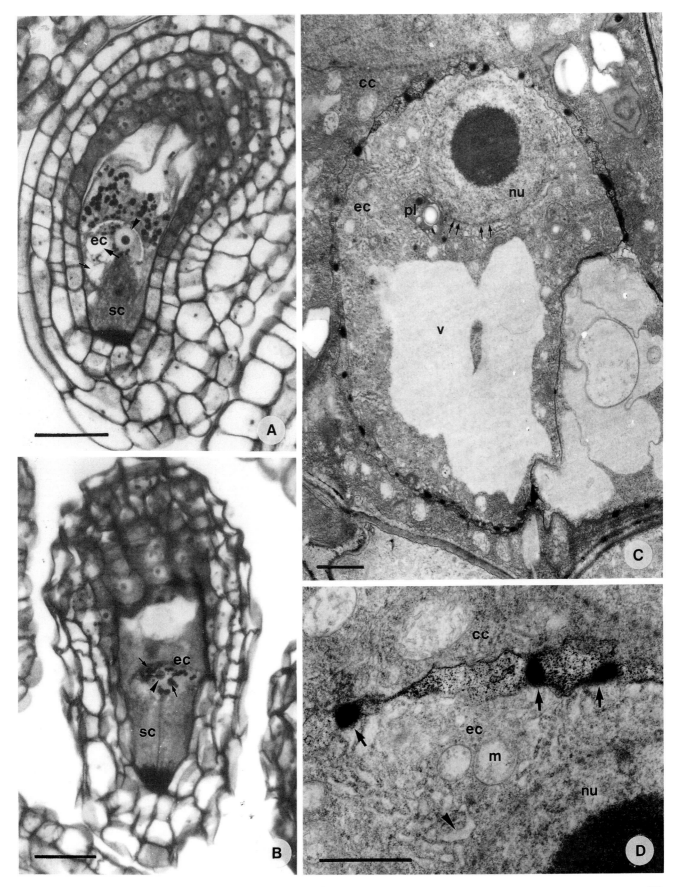

Plate 5.10

Plate 5.11
The synergid cells zero hours after flowering (HAF)

(**A**) and (**B**) Light micrographs of the two synergid cells (sc) in two different longitudinal planes. The cells are highly polarized with a mycropylar nucleus (*arrows*) and a single large chalazal vacuole (v). The filiform apparatus (*arrowheads*) at the apex of each cell stains intensely PAS-positive. Also note the accumulation of large starch bodies around the central cell nucleus in **B** (PAS, periodic acid-Schiff's staining).

(**C**) Micropylar half of a synergid cell in longitudinal section showing the large nucleus (nu) and nucleolus (nuc), and extensive wall proliferations of the filiform apparatus (fa). Plasmodesmatal connections are present in the common wall between the two synergid cells (*arrows*). The cytoplasm appears metabolically active; rough endoplasmic reticulum (rer), dictyosomes (*arrowheads*), and mitochondria (m) are ubiquitous, the latter accumulating in large numbers adjacent to the filiform apparatus.

Bar = 2 μm in **C**; 10 μm in **A**, **B**.

S.G. Mansfield

Reproduced from Mansfield et al. (1991) with permission from the National Research Council of Canada.

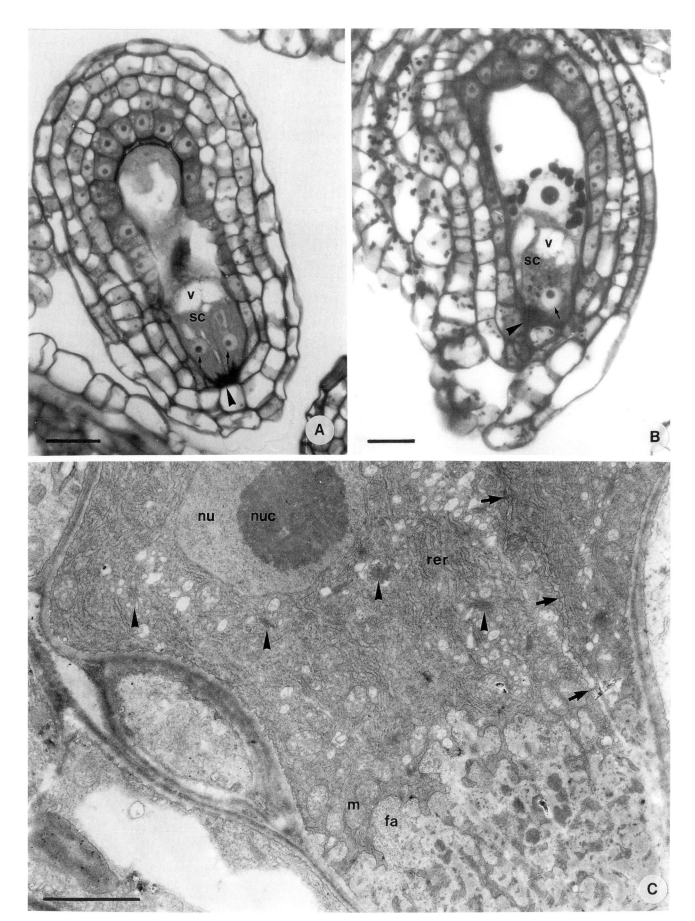

Plate 5.11

Plate 5.12

The synergid cells zero hours after flowering (HAF)

(A) The synergid cell cytoplasm in transverse section mid-way between the nucleus and chalazal vacuole. The rough endoplasmic reticulum (rer) is present in masses of parallel-stacked cisternae, and certain regions of the network show cisternal dilations (*arrowheads*). Mitochondria (m) are spherical to oval in profile and possess short cristae. Small wall crenellations are present along this region of the central cell-embryo sac wall (*large arrows*).

(B) Enlargement of the cytoplasm close to the large chalazal vacuole (transverse section). The synergid cells are characterized by the presence of poorly differentiated and irregular dark-staining plastids (pl) that possess numerous small vesicles. Microtubules in transverse section (*arrowheads*) can be seen close to the synergid-central cell boundary.

(C–E) The synergid cell cytoplasm adjacent to the chalazal vacuole (longitudinal sections).

(C) The cytoplasm is rich in dictyosomes (*arrowheads*) and their associated vesicles (v, vacuole).

(D) Enlargement of a single dictyosome. Usually these organelles consist of three to six electron dense cisternae and several newly formed (ve) or forming (*arrows*) vesicles that often contain fine fibrillar material.

(E) Enlargement of the cytoplasm adjacent to the tonoplast in C. The released dictyosome vesicles (ve) can be seen apparently fusing with the tonoplast (t) of the chalazal vacuole.

Bar = 0.5 μm in **D**, **E**; 1 μm in **A**, **B**, **C**.

S.G. Mansfield

Reproduced from Mansfield et al. (1991) with permission from the National Research Council of Canada.

Ovules: The Mature Embryo Sac

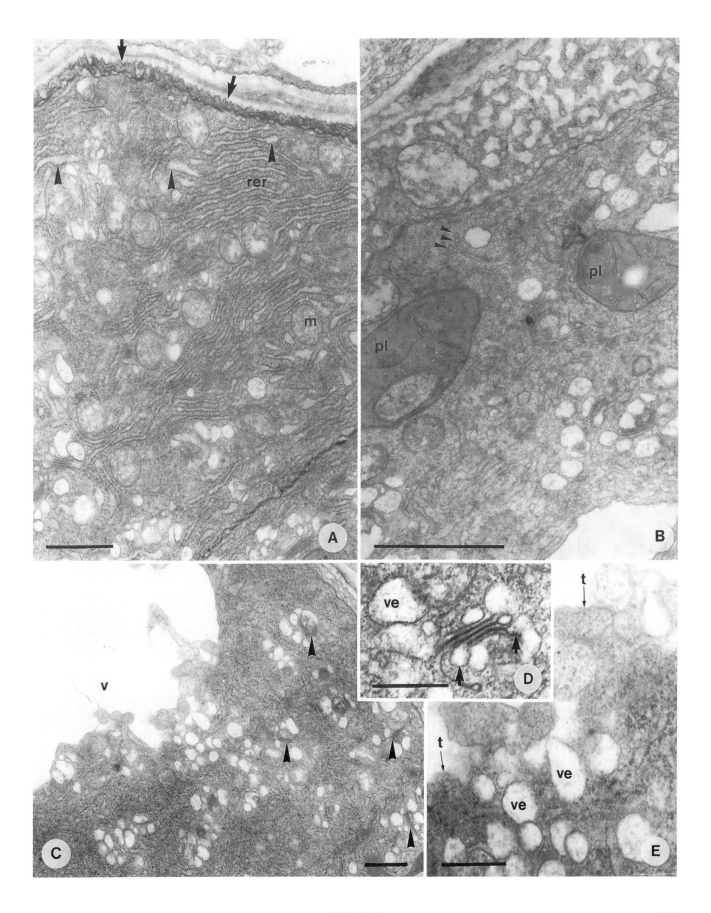

Plate 5.12

Plate 5.13
The central cell zero
hours after flowering
(HAF)

(A) Light micrograph showing a longitudinal section through the central cell (cc) prior to embryo sac maturity. The two polar nucleoli of the cell (*arrows*) can be seen here about to undergo fusion.

(B) Part of the single diploid nucleus (nu) of the central cell just prior to fertilization. Note the crenellated appearance of the nuclear membrane as compared to the regular profile of the synergid (C in Plate 5.11) and egg (C in Plate 5.10) cell nuclei. The cytoplasm immediately adjacent to the nucleus contains an abundance of well-differentiated starch (st) -containing plastids (pl) which, as seen by light microscopy (B in Plate 5.11), form an almost complete shell of PAS-positive inclusions around the nucleus. Mitochondria (m) are extremely abundant and dictyosomes (d) appear active in vesicle production. Other characteristic features of the cytoplasm are the presence of many myelin bodies (*arrowheads*) and small vacuoles (v). (PAS, periodic acid-Schiff's staining.)

The wall of the central cell is continuous except at the chalazal end of the egg, where it is absent (D in Plate 5.10), and adjacent to the synergids, where it is thinner than elsewhere, irregular, and in some areas entirely absent (not shown). The cell also shares a common wall with the antipodals, where it is slightly thicker than other regions (D in Plate 5.14). Plasmodesmata occur only in this area. Localized area of very small wall crenellations develop on the longitudinal wall of the central cell, in a region midway between the central cell nucleus and chalazal end of the synergids (A in Plate 5.12).

(C) Enlargement of the central cell nuclear membrane showing the distribution of its nuclear pores (*arrows*) (nu, nucleus).

Bar = 0.5 μm n C; 3 μm in B; 20 μm in A.

S.G. Mansfield

Reproduced from Mansfield et al. (1991) with permission from the National Research Council of Canada.

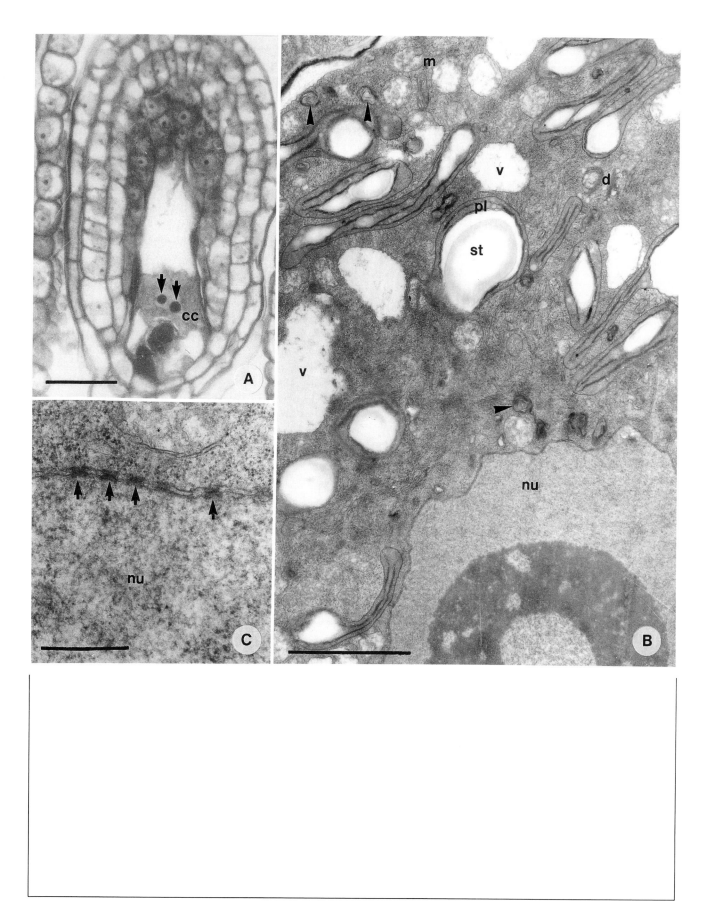

Plate 5.13

329

Plate 5.14
**The antipodal cells and
surrounding tissue zero
hours after flowering
(HAF)**

The three ephemeral antipodals are situated at the extreme chalazal pole of the embryo sac in a triangular arrangement similar to the egg apparatus.

(**A**) Low power view of the antipodal cells (ac) and surrounding chalazal nucellus (cn) tissue (longitudinal section). A narrow channel (*arrows*) is present between the nucellus tissue and the inner integument (i). This separation, together with their cuticle-like wall coverings, allows the chalazal nucellus to be crushed during the proliferation of the distal nucellar cells (approximately 2–8 hours after fertilization), without causing disruption to the integumentary tissues.

(**B**) The antipodal cells (ac). The lower two compartments are a single highly-convoluted antipodal cell. Also note how convoluted the cell wall (cw) is between the antipodal cells and the adjacent nucellar cells (cn). Plasmodesmata (*arrows*) are present between nucellar and antipodal cells and the central cell (cc).

(**C**) Enlargement of the antipodal cell cytoplasm showing the structure of the rough endoplasmic reticulum and the cells' high ribosome density. Note how the ribosomes on the cisterna appear prominent and regularly-spaced.

(**D**) View of the common cell wall between a lower antipodal cell (ac) and the central cell (cc). The concentration of plasmodesmata (*arrows*) in this particular region is very high.

(**E**) The chalazal nucellar cells adjacent to the antipodals. Nucellar cells of this region are irregular in profile and thick-walled, although all cells are interconnected by many plasmodesmata (*large arrows*). The cytoplasm is dense and contains numerous mitochondria (m) and dictyosomes (*arrowheads*), but very few vacuoles and plastids. The irregularly shaped nuclei (nu) contain many perinuclear deposits (*small arrows*). The more distal nucellar cells are regular in profile and thin-walled (**A**), with numerous plasmodesmata (not shown); these cells constitute what will eventually become (post-fertilization) the chalazal proliferating tissue.

Bar = 0.5 μm in **C**; 1 μm in **B**, **D**, **E**; 5 μm in **A**.

S.G. Mansfield

Reproduced from Mansfield et al. (1991) with permission from the National Research Council of Canada.

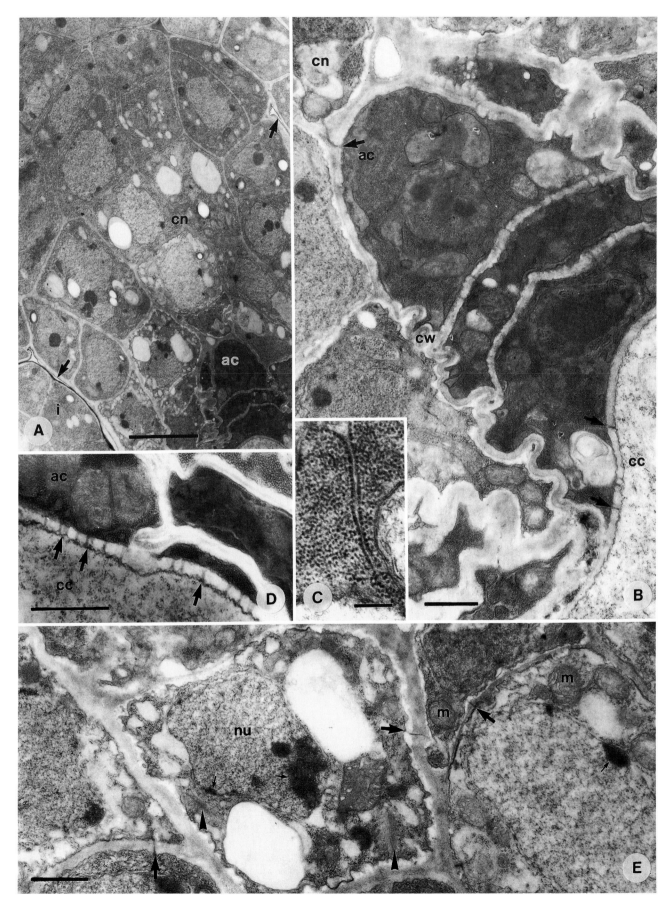

Plate 5.14

6
Pollination

Introduction

The energy-intensive processes of gametophytic development culminate in fertilization and the commencement of development of the next sporophytic generation. Successful pollination is the result of a series of events, including pollen contact, adhesion and attachment of pollen to the stigma, pollen hydration, pollen germination, pollen growth through the stigma, style, and ovary, and finally, fertilization of the ovule (Knox, 1984). Each of these steps may be regulated to insure appropriate interactions between the male and female reproductive tissues.

To insure successful propagation of the species, angiosperms have evolved a variety of ways to distribute their pollen by utilizing vectors such as wind, insects, and birds. Thus, plants must be able to distinguish compatible from incompatible pollen. In the wild, this involves not only species-specific recognition and acceptance of compatible pollen, but also, in some cases, the rejection of incompatible self-pollen (for reviews, see Haring et al., 1990; Dzelzkalns et al., 1992). Since *Arabidopsis* is self-fertile, there is no incompatible self-pollen. However, species-specific recognition events must still occur. Following the acceptance of compatible pollen, the pollen tube must be directed through the ovary to the ovules to effect fertilization. The guidance of pollen tube growth and the avoidance of polyspermy likely involve additional cellular recognition systems.

Contact, Attachment, and Adhesion

In *Arabidopsis thaliana*, the first step in the pollination process, that of getting the pollen to the stigma, is not complicated by self-incompatibility; the stamens dehisce compatible pollen directly onto the stigmatic papillae. Pollen adhesion to the stigma is likely mediated by both the dry surface of the stigma and the reticulate exine coating of the pollen grain (Knox, 1984). Soon after adhesion, the outer exine coating of the *Arabidopsis* pollen grain flows out to form a "foot" at the point of pollen–stigma contact, establishing attachment (Elleman et al., 1992). Changes also occur rapidly (within a couple of hours) in the stigmatic papillae at the point of contact, with the two layers of the stigmatic cell wall becoming visibly separated (Elleman et al., 1992). These changes in the cell walls of both the pollen grain and stigmatic papillae probably lead to hydration of the desiccated pollen grain (Heslop-Harrison, 1979; Knox, 1984; Elleman and Dickinson, 1986). These processes (adhesion, attachment, and hydration) are likely targets in the regulation of species-specific pollen recognition, and in sporophytic incompatibility (Knox, 1984).

In this regard, it is interesting to note that in *Arabidopsis*, female sterile mutants that block the penetration of growing pollen tubes through the stigma, and male sterile mutants that produce pollen capable of effecting fertilization, but which fails to be recognized by the stigmatic papillae have been isolated (Pruitt et al., 1991). The latter class of mutants can be rescued by applying wild-type pollen grains to the stigma with the mutant pollen (Pruitt et al., 1991). This is reminiscent of the mentor effect in which interspecific pollen may successfully germinate on a foreign stigma and sometimes effect fertilization if pollen from the maternal species is also applied to the stigma (Stettler and Ager, 1984).

Pollen Tube Growth and Guidance

Following hydration, *Arabidopsis* pollen grains germinate with the pollen tube usually penetrating the cuticle of the stigmatic papillae through the attachment foot (Elleman et al., 1992). Upon germination, the intine of the pollen wall becomes continuous with the growing pollen tube wall. Growth of the pollen tube occurs at the tip, where the cytoplasm and nuclei of the pollen grain accumulate. Once inside the unicellular papillae, the pollen tubes grow inside the cellulose-pectin layer of the cell wall to the base of the papillar cell, and exit into the middle lamellae of the transmitting tract tissue beneath the stigma (Webb, 1991; Elleman et al., 1992). Growth of the pollen tube through the solid style is in the intercellular spaces of the central transmitting tissue (Webb, 1991). Pollen tube growth continues in the transmitting tissue of the septum, exiting the septum between epidermal cells, and growing along its surface. Pollen tube growth along the epidermal surface of the septum is generally basipetal before growing along the funiculus and into the micropyle of an ovule (Webb, 1991).

An obvious question that arises is how do pollen tubes find their way through the ovary to an unfertilized ovule. The entire pathway taken by pollen tubes appears to be actively directed, but the molecular mechanisms involved are largely a mystery (Webb, 1991). The final stage may be directed chemotropically from substances produced in the embryo sac and secreted through the micropyle. In some species additional specialized structures such as obturators (Tilton and Horner, 1980; Hill and Lord, 1987) are thought to provide some directional guidance. Although there are no obvious specialized morphological structures in *Arabidopsis*, there may be more subtle markers. A class of male-sterile mutants has been isolated which seem to disrupt the guidance of pollen tube growth in the *Arabidopsis* ovary (Pruitt et al., 1991). These mutants produce pollen grains that germinate and grow normal pollen tubes. However, the pollen tubes fail to locate ovules in the ovary. Further insight into pollen tube guidance may be provided by classes of female-sterile mutations that cause arrest of ovule development. At least in some cases, if development is arrested early, the mutant ovules fail to attract pollen tubes, while if ovule development is arrested later, the mutant ovules are able to attract pollen tubes even though the ovules are non-functional (Pruitt et al., 1991).

Fertilization

The final step in the process is the act of double fertilization itself, resulting in the fusing of one of the sperm cells with the egg cell to form the zygote, and the fusing of the second sperm cell with the diploid central cell to produce the triploid endosperm. This process presumably follows the series of events typical for angiosperms (Esau, 1977), although it has not been directly documented in *Arabidopsis*. In the case of tenuinucellate ovules, as in *Arabidopsis*, the pollen tube reaches the embryo sac directly after passing through the micropyle. After penetration of the filiform apparatus, the pollen tube enters one of the synergid cells, and the contents of the pollen tube tip, including the vegetative nucleus and the two sperm cells, are injected into the synergid cell which soon degenerates. The mechanism by which the two sperm cells find their respective targets is largely obscure. Finally, only a single pollen tube effects fertilization of an ovule, but the mechanism by which supernumerary pollen tubes are rejected, thus avoiding polyspermy, is unknown.

J.L. Bowman

Plate 6.1
Morphology of the stigma

The *Arabidopsis* stigma is of the dry type, as defined by Heslop-Harrison (1981). As with other members of the Brassicaceae, the epidermal layer consists of papilla-like cells covered with a cuticle (see Roggen, 1972).

(**A**) Scanning electron micrograph (SEM) showing the papillae of the immature stigma, which are short and closely appressed to one another. During development, these cells elongate (possibly by tip growth, see Webb, 1991).

(**B**) SEM of a mature stigma in which the fully extended papillae are widely spaced. Each cell has a swollen base (*arrow*) which forces them apart. Note: some pollen grains (p) are present on the stigma. This pattern of development is very similar to that described for *Brassica oleracea* (Ockendon, 1972).

(**C**) Once fertilization has occurred, the stigmatic papillae begin to collapse and degenerate.

(**D**) Longitudinal section (2 μm, toluidine blue staining) showing the stigmatic papillae and part of the style during megasporogenesis. The short style which connects the stigma to the ovary is of the closed type, with a central region of transmitting tissue, resembling that reported for other crucifers (Sassen, 1974).

Bar = 50 μm.

M.C. Webb

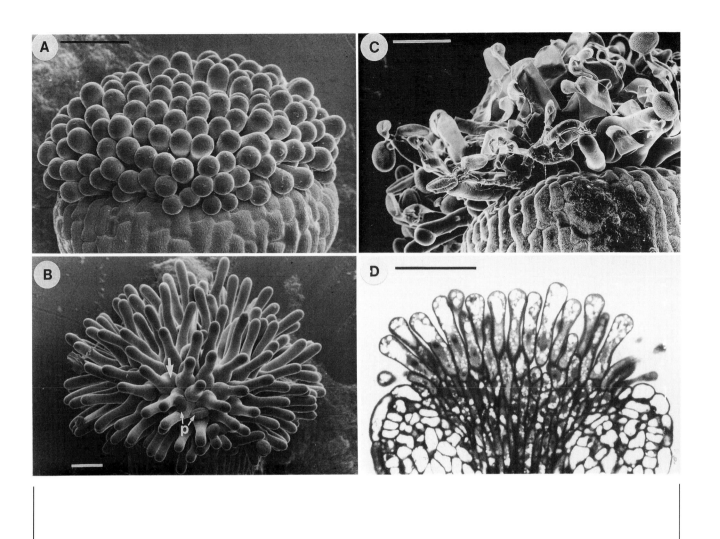

Plate 6.1

339

Plate 6.2
Pollen release and
germination

(**A**) Once released from the anthers, pollen grains interact with the papillae on the surface of the stigma and initiate pollen tube growth.

(**B**) Mature pollen has a multi-layered wall that gives rise to a highly sculptured surface which is the combined result of the action of pollen genes and maternal genes that encode the sporopollenin released from the tapetum.

S. Craig and A. Chaudhury

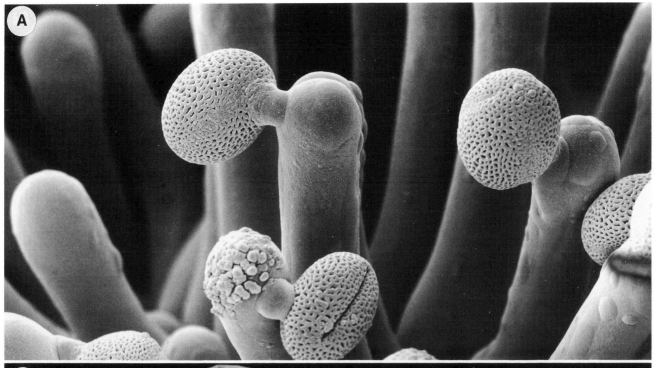

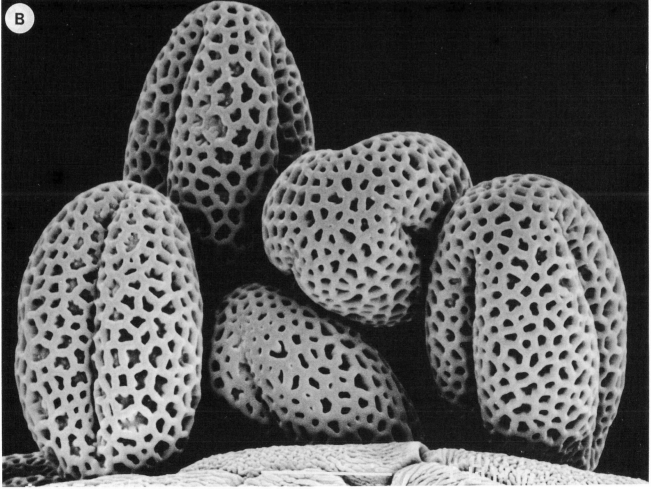

Plate 6.2

Plate 6.3
Pollen tube growth in the pistil

Shown are views of pollen germination and pollen tube growth in the stigma (st, style). Entry into the stigma appears to be achieved by the same means as reported in other members of the Brassicaceae (Kroh, 1964; Kroh and Munting, 1967; Dickinson and Lewis, 1973). The pollen tube penetrates the cuticle of the stigmatic papilla, and grows inside the cellulose–pectin layer of the cell wall (Kroh and Munting, 1967; Elleman et al., 1992). The cuticle is broken down enzymatically (Heinen and Linskens, 1961).

(**A**) Scanning electron micrograph showing pollen grains among the stigmatic papillae. One grain has germinated and penetrated a papilla cell (*arrow*).

(**B**) Longitudinal section (2 μm, toluidine blue staining) showing a pollen tube growing underneath the cuticle of the papilla (*arrows*) without disrupting the plasma membrane, growing between two distinct layers of the cell wall (Elleman et al., 1992). A second pollen tube has also penetrated and exhibits spiralling growth around the papilla (*arrowheads*), which is a common occurrence. The pollen tube grows in the central transmitting tract of the style to the ovary. The pollen tube grows between the cells of the transmitting tissue in the ovary's septum, then grows along the placenta (epidermal layer of the septum).

(**C**) and (**D**) Longitudinal sections (2 μm, toluidine blue staining) taken from a complete series, showing a pollen tube (*arrows*) emerging from the septum onto the placenta. The pollen tube travelling down the transmitting tissue (tt) has exited between two epidermal cells (**C**) and continues along the placenta (pl) (**D**). Note the distortion of one of the cells next to which the pollen tube has emerged (*arrowhead*).

Bar = 10 μm.

M.C. Webb

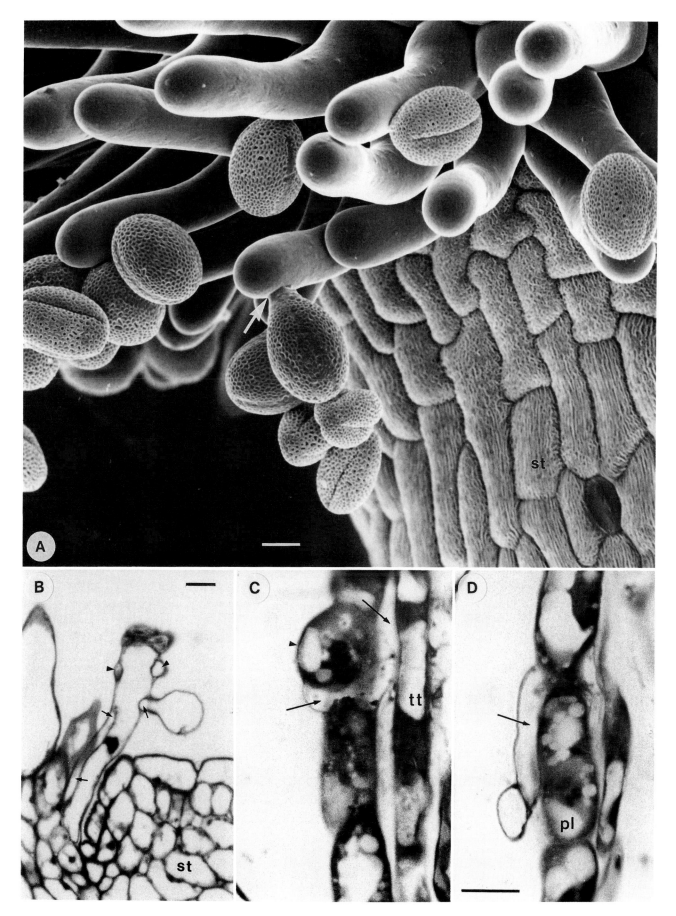

Plate 6.3

Plate 6.4
Pollen tube growth in the pistil

Decolorized aniline blue squash preparations show the pollen tube pathway within the *Arabidopsis thaliana* pistil. The morphology of the gynoecium, as well as the pathway followed by pollen tubes, closely resembles that of *Raphanus raphanistrum* (Hill and Lord, 1987) in most respects. (For details, including discussion of pollen tube guidance, see Webb, 1991).

(**A**) Pollen germinates, the pollen tubes penetrate the stigma (sg) and grow through the pistil confined to the transmitting tract of the stigma, style (st), and ovary (o), before emerging onto the placenta and each entering an ovule micropyle.

(**B**) Higher magnification of an equivalent pistil, showing the pollen tubes exiting from the septum and entering the micropyles of ovules (ov). While some micropyles are close enough to be reached directly from the placenta (*arrow*), pollen tubes usually travel up the funiculus (fu) to gain access.

Double fertilization occurs following the release of the two sperms carried by the pollen tube into a synergid cell which has degenerated. Further details of this process are discussed elsewhere (e.g., Webb, 1991; Webb and Gunning, 1993).

Bar = 20 μm in **B**; 100 μm in **A**.

M.C. Webb

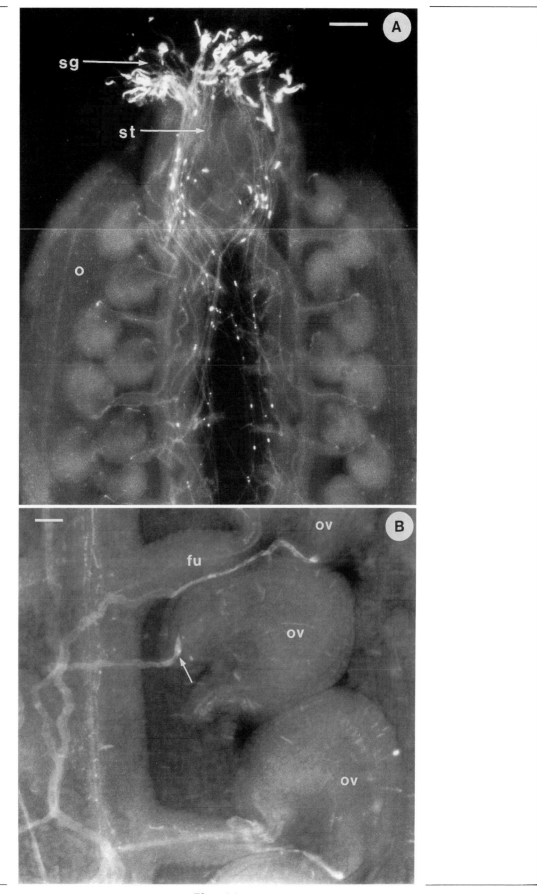

Plate 6.4

Plate 6.5
Fruit development

The fruits of *Arabidopsis* are siliques consisting of two valvately dehiscent locules separated by a false septum. Siliques elongate rapidly after pollination, reaching their maximum length (approximately 10 mm for Landsberg *erecta* and 14 mm for Columbia) two to three days after fertilization. The length of the mature silique is correlated with the number of developing seeds within (Meinke and Sussex, 1979a). By this time the sepals, petals, and stamens have senesced and fallen from the silique. Siliques remain green for more than a week following fertilization, but turn yellow, and then brown at maturity. By about three weeks following pollination, the valves of the dry siliques separate releasing the 30–60 seeds found in each. Silique development does not commence if pollination does not occur. However, silique development is not dependent on normal seed development, since plants homozygous for embryo lethal mutations can develop normal fruits (Franzmann et al., 1989).

(**A**) A pollen tube is visible (*arrowhead*) that has grown along the funiculus and entered the micropyle of an ovule.

(**B**) Close-up of **A** showing pollen tube (*large arrowhead*) along the funiculus and entering the micropyle (*small arrowhead*).

(**C**) Siliques of Landsberg *erecta* plants. A silique approximately 15 hours after fertilization is shown in the *inset* next to a fully elongated silique (about four days post-pollination).

(**D**) Close-up of the silique in the inset of **C**. The epidermis consists of files of elongate cells with interspersed stomata.

(**E**) Close-up of the fully elongated silique in **C**. Much of the elongation appears to be due to cell elongation throughout the length of the fruit. However, there is likely to be cell division occurring at least into stage 14, since *CDC2*, the p34 protein kinase required for entry into mitosis, is highly expressed in ovary walls of stage 13–14 flowers (Martinez et al., 1992).

(**F**) Abscission zone of the fully elongated silique in **C**. Bulbous cells are present at the abscission zones of the sepals, petals, and stamens, which senesce and abscise within a few days following fertilization. Collapsed nectaries (*arrowheads*) are also visible.

Bar = 10 μm in **B**; 20 μm in **A**; 50 μm in **D**, **E**, **F**; 500 μm in **C** and inset of **C**.

J.L. Bowman

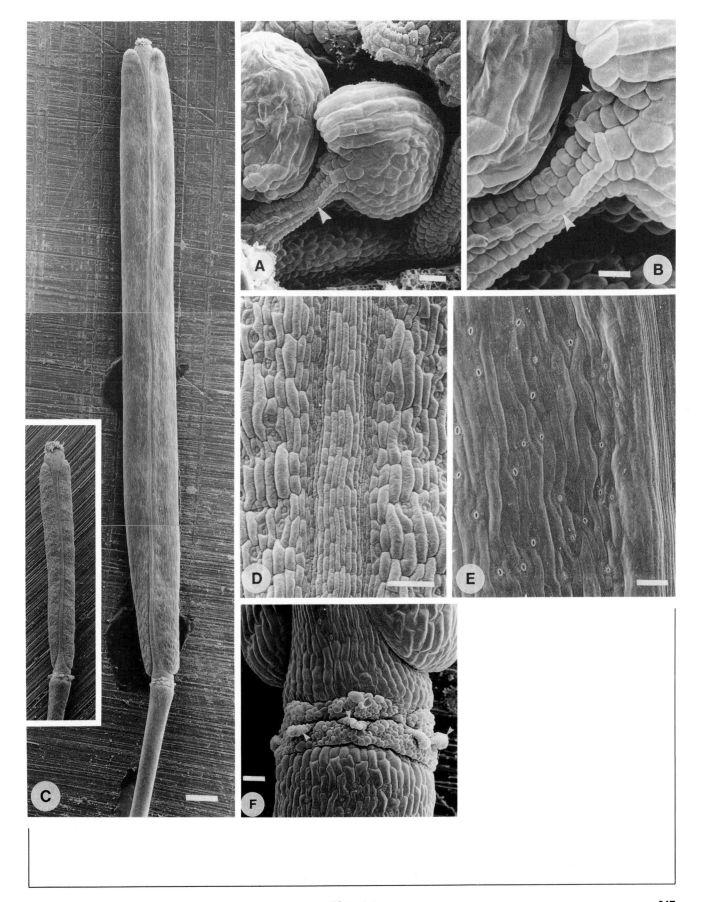

Plate 6.5

7
Embryogenesis

Introduction

During embryogenesis the basic body plan of the mature plant is generated (Jürgens et al., 1991; Mayer et al., 1991). This involves the production of both shoot and root meristems, cotyledons, radicle, and hypocotyl, resulting in the establishment of the apical–basal axis of the plant and formation of the radial pattern elements, epidermis, ground tissue, and vasculature. These pattern elements are established early in embryogenesis and are stably maintained throughout the lifetime of the plant.

Early Stages of Embryogenesis

Development of *Arabidopsis* embryos closely follows that of *Capsella bursa-pastoris*, another member of the Brassicaceae, which is often used as a model of embryo development in dicotyledonous plants (Maheshwari, 1950; Schulz and Jensen, 1968a; Schulz and Jensen, 1968b; Esau, 1977). Numerous descriptions of early embryogenesis in *Arabidopsis* are available (Vandendries, 1909; Reinholz, 1959; Misra, 1962; Müller, 1963; Gerlach-Cruse, 1969; Yakovlev and Alimova, 1976; Meinke and Sussex, 1979a; Huang, 1986; Mansfield et al., 1991; Mansfield and Briarty, 1991; Webb and Gunning, 1991; Mayer et al., 1991; Jürgens et al., 1991; Guignard et al., 1991; Mansfield and Briarty, 1992), including detailed analyses of the ultrastructure of embryonic cells at several stages of embryogenesis (Mansfield et al., 1991; Mansfield and Briarty, 1991; Mansfield and Briarty, 1992). Stages of embryogenesis have been defined according to the number of cells derived from the apical cell (see below) or the shapes of the developing embryo: zygote, two-terminal cell, quadrant, octant, dermatogen, globular, triangular, heart, torpedo, walking-stick, upturned-U. These stages and the progression of cell differentiation within the developing embryo are outlined in Figures 1 and 2.

Zygote to Single Terminal Cell Stages

Early embryo differentiation in *Arabidopsis* follows the *Capsella* variation of the Onagrad type, characterized by a longitudinal division of the apical cell of a transversely divided zygote. Following double fertilization, the polarized zygote divides transversely, producing a small apical or terminal (chalazal) cell and an elongate basal (micropylar) cell. The smaller apical cell will give rise to the embryo, except for the root. The larger basal cell will give rise to the radicle as well as the extraembryonic suspensor through which nutrients are supplied to the developing embryo from the mother plant. The polarized nature of the zygote is evident at the ultrastructural level prior to the first division. Thus, the apical–basal axis of the developing embryo is established in part due to the polarized nature of the zygote. However, there may not be an absolute requirement for a polarized zygote since, in many species, somatic embryos can be induced to develop from somatic cells maintained in a defined culture or from differentiated structures of mature plants (for review see Williams and Maheswaran, 1986).

Two-Terminal Cell to Octant Stages

Prior to the first division of the terminal cell, the basal cell may undergo one or two transverse divisions, or may remain undivided. The terminal cell undergoes a longitudinal division to produce the two-terminal cell embryo, each cell being equal in size (Figure 2). Each of the cells of the two-terminal cell embryo

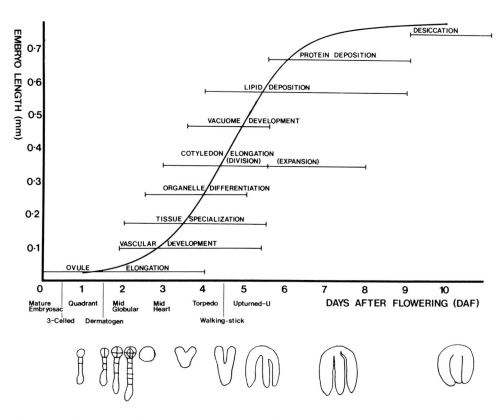

Figure 1. Summary of embryogenesis of *Arabidopsis*. Stages of embryo development and cellular differentiation within the developing embryo related to days after flowering. Flowering is defined as the time the length of the medial (long) stamens exceeds that of the gynoecium. It is essentially the time of pollination [stage 14, as defined by Müller (1961) and Smyth et al. (1990)]. Plants were grown under continuous lighting at 25°C and 70% humidity, and embryos examined were taken from the middle of the silique of the third to seventh flowers on the main flowering stem (Mansfield et al., 1991).

then undergo longitudinal divisions resulting in a quadrant embryo. These divisions of the apical cell are essentially cleavage divisions, with little increase in size of the embryo. This is reflected in the change in size of the embryo itself, outlined in Table 1. Each of the four apical cells divides transversely, producing the octant embryo. During this time the basal cell has completed two transverse divisions, forming the suspensor between the basal cell and the apical cells.

Dermatogen to Early Globular Stages

Each of the eight cells derived from the apical cell of the octant embryo divides periclinally to produce a 16-cell embryo (dermatogen or early globular stage) with eight inner and eight outer, or protodermal cells (Figure 2). The protodermal cells of the globular embryo divide anticlinally nonsimultaneously while the eight inner cells divide longitudinally. The dermatogen stage is noteworthy as it marks the first step towards radial pattern elements. The protodermal cells will give rise to the epidermis, while the ground tissue and procambium are derived from the inner cells. Development of the suspensor, consist-

Embryogenesis: Introduction

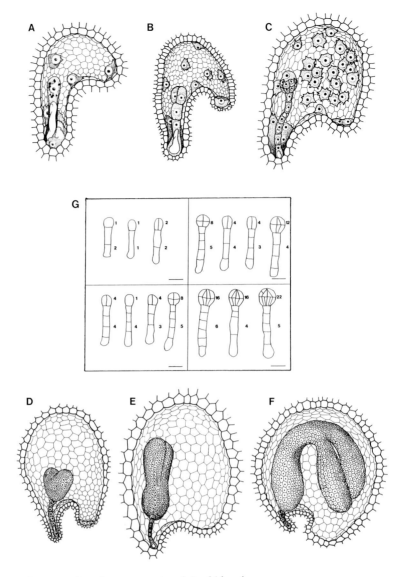

Figure 2. Stages of embryogenesis of *Arabidopsis*.

(**A**) Zygote at 4 hours after flowering (HAF). Zero HAF is the beginning of stage 14, as defined by Müller (1961) and Smyth et al. (1990), when the length of the medial (long) stamens exceeds that of the gynoecium.

(**B**) Two-terminal cell embryo (18 HAF).

(**C**) Dermatogen embryo (36 HAF).

(**D**) Heart stage embryo (72 HAF).

(**E**) Torpedo stage embryo (96 HAF).

(**F**) Upturned-U stage embryo (120 HAF).

(**G**) Temporal variation of embryo development within single siliques at four different stages of fruit development. Each box shows a sample of three or four embryos from a single silique, with the number of cells in the embryo-proper and in the suspensor of each embryo indicated.

A–F, drawings by S. Erni (Kunstgewerbeschule der Stadt Zürich). Reproduced with permission from Prof. A.R. Kranz, ed. *Arabidopsis Information Service*.

Table 1. Change in embryo-proper length[a] during early embryogenesis.

Embryo stage	HAF	Length (μm)
Early zygote	nd	
Late zygote	4	45.0 (2.5)
Single-terminal cell	12	21.4 (0.2)
Two-terminal cell	18	21.6 (0.3)
Quadrant	24	21.7 (0.4)
Octant	30	24.5 (0.4)
Dermatogen	36	30.2 (0.4)
Early globular	42	37.1 (0.4)
Midglobular	48	45.1 (0.4)
Late globular	60	52.2 (0.3)
Midheart	72	104.9 (0.8)
Late heart	84	174.9 (2.4)
Torpedo	96	253.1 (1.4)

[a] Except for zygote values (mean of four measurements), values represent mean of 100–150 measurements, with standard errors in parentheses.

ing of 7–9 highly vacuolate cells including the basal cell and the hypophyseal cell, is complete by the early globular stage. The cell division patterns that generate the suspensor are highly variable (Figure 2). The hypophyseal cell, or incipient root primordium, will give rise to the root meristem and root cap via a series of longitudinal and transverse divisions.

The development of embryos within a single silique is not synchronous, with embryos exhibiting a wide range of developmental stages. A sample of the extent of variation observed is presented in Figure 2.

Midglobular to Late Globular Stages

The outer embryonic cells continue to develop by nonsimultaneous anticlinal division producing 16 (midglobular stage) and then 32 cell protoderms. Following longitudinal divisions (midglobular stage), the daughters of the original inner 8 cells then divide transversely. By late globular stage, the apical cell of a transversely divided embryo has given rise to a 64-cell embryo. At this time, although the size of the embryo has increased marginally (Table 7.1), the average cell size is a fraction of that of the zygote or even the single terminal cell. Throughout embryogenesis, the cells of the embryo-proper and the suspensor are interconnected via plasmodesmata, but there is no symplastic communication between the embryo and its surrounding tissues, either the free-nuclear or cellularized endosperm, or the integuments.

By the late globular stage, most of the pattern elements, both apical–basal and radial, of the embryo can be visualized. The procambium primordium, consisting of 8 narrow cells derived from the inner cells, lies directly above the root primordium (hypophyseal cell). In addition, regionally specific increased cell division in the protoderm, which has has been distinct since the dermatogen stage embryo, marks the site of the future cotyledons in late globular stage embryos.

It has been hypothesized that no more than 100 cells comprise the embryo (late globular to triangle stage embryo) by the time the basic pattern of the embryo is established (Jürgens et al., 1991). Given that the cells of the embryo are interconnected via plasmodesmata (Mansfield and Briarty, 1991) and considering the small size of the embryo at this stage, it is feasible that stable concentration gradients of diffusable morphogens could theoretically be estab-

Embryogenesis: Introduction

lished in the embryo at least through this stage of development (Jürgens et al., 1991).

Triangular to Torpedo Stages

Cell divisions in the region of the future cotyledons and cell elongation in the region of the procambium result in a short-lived triangular embryo. The cell divisions producing the cotyledon primordia are rapid, leading to the heart stage embryo (Figure 2). During the heart stage, the cotyledon primordia enlarge quickly and cell divisions and cell elongation in the hypocotyl and radicle (root primordium) regions cause the embryo to assume an elongate shape, the torpedo stage (Figure 2; Table 1). In addition, provascular tissue starts to differentiate at the base of the cotyledons.

Continued elongation causes the torpedo-shaped embryo to force its way into the cellularized endosperm. During this stage, the central cells of the hypocotyl and radicle primordium differentiate, forming the primary vascular tissue. Surrounding the vascular tissue are three layers of large parenchymous cells. Differentiation of tissues begins to accelerate with plastid differentiation leading to the "greening" of embryos.

Origin of the Shoot and Root Meristems

The root and shoot meristems, both of which have their origins in embryogenesis, are ultimately responsible for the postembryonic architecture of the plant. The two meristems differ markedly in the timing of their origin as well as their cellular organization. The root meristem is formed relatively early in embryogenesis (by the mid-heart stage) and is comprised of three tiers of initials which give rise to well-defined cell lineages (Mayer et al., 1993; Dolan et al., 1993). In contrast, the shoot meristem arises later in embryogenesis (during the torpedo stage) and displays a tunica-corpus organization with no strictly defined pattern of cell lineage (Furner and Pumphrey, 1992; Irish and Sussex, 1992; Barton and Poethig, 1993).

The radicle, or embryonic root, is derived from the hypophysis (uppermost cell of the suspensor) and the lowest tier of cells of the globular embryo (Dolan et al., 1993). The hypophysis undergoes an asymmetric division with the upper lenticelar cell giving rise to the quiescent central cells of the root meristem and the larger lower cell giving rise to the columella root cap (Mayer et al., 1993; Dolan et al., 1993). The remainder of the root meristem, including the three tiers of initials, and the lateral root cap are derived from the embryo proper.

The shoot meristem is derived from a subset of cells of the upper half of the globular stage embryo. During the transition from the heart stage to the torpedo stage embryo, three distinct layers of cells overlying a core of provascular cells become recognizable in the upper hemisphere of the developing embryo (Barton and Poethig, 1993). These layers are the precursors of the two tunica layers (L1 and L2) and the corpus (L3) of the shoot meristem. The upper tunica layer is a derivative of the protoderm whereas the lower tunica layer and the corpus are derived from the inner embryonic cells, or hypodermal cells. Already during the torpedo stage embryo, the cell division patterns of the shoot meristem are characteristic of a tunica-corpus organization (Barton and Poethig, 1993). The shoot meristem is easily identifiable in later stage embryos as storage granules do not accumulate in the cells of the shoot meristem.

Late Stages of Embryogenesis

Walking Stick to Upturned-U Stages

Later stages of embryogenesis are characterized by cell differentiation and an increase in embryo size (Mansfield and Briarty, 1991; Mansfield and Briarty, 1992). The torpedo stage embryo elongates into the "walking stick" embryo before the growing cotyledons curve over, forming the "upturned-U" embryo (Figure 2). The increase in cotyledon size up to the upturned-U stage embryo is due primarily to cell division.

However, most subsequent growth is due to cell expansion. Deposition of protein and lipid reserves in the cotyledons takes place over several days, with lipid deposition starting at the torpedo stage and protein deposition at the upturned-U stage. During these stages, expansion of the embryo forces it into the fully cellularized endosperm. Cellularization of the initially free-nuclear endosperm begins at the late globular embryo stage, the process starting in the micropylar chamber and proceeding centripetally towards the embryo and peripherally in the direction of the chalazal chamber Mansfield and Briarty, 1990a; Mansfield and Briarty, 1990b). Cellularization is essentially complete by the torpedo stage.

The Mature Seed

Continued cell expansion throughout the embryo causes it to fill most of the embryo sac, crushing the endosperm formed earlier during embryogenesis and presumably absorbing it (Mansfield and Briarty, 1992). By maturity, the cotyledons are tightly appressed to the radicle, leaving little open space in the embryo sac, with only a single persistent layer of endosperm remaining. With the onset of desiccation, the rate of reserve deposition is curtailed significantly, and there is a slight decrease in embryo size. The seed coat, or testa, is derived from the two integuments. The outer epidermis of the outer integument become filled with a layered mucilaginous material. This outer cell wall collapses and the mucilage swells when it comes in contact with water covering the surface of imbibed seeds.

In most species of the Brassicaceae, the cells of the inner epidermis of the outer integument, the palisade layer, develop lignified thickenings on the radial and tangential walls, making it the strongest layer structurally (Esau, 1977). The inner epidermis of the inner integument becomes the pigment layer. In the case of *Arabidopsis*, anthocyanins accumulate in the pigment layer, giving the mature seeds a brown color. The embryo-proper (cotyledons) is yellowish at maturity, the chlorophyll formed earlier in embryogenesis being broken down before maturation (Koornneef, 1981).

The time from fertilization through desiccation of the seed varies with genotype and environmental conditions, taking approximately two weeks for Landsberg *erecta* plants grown at 25°C under continuous light and at 70% relative humidity. The mature desiccated embryo enters a state of dormancy that may persist several years before favorable conditions induce germination and subsequent postembryonic growth.

Mutations Affecting Embryogenesis

Literally hundreds of mutations have been isolated that disrupt embryogenesis in *Arabidopsis* (for reviews see Jürgens et al., 1991; Meinke, 1991a; Meinke, 1991b). The spectrum of isolated mutants can be broken down into three broad categories: (1) embryonic lethal mutants in which embryogenesis is arrested; (2) abnormal seedling mutants with altered pigmentation; and (3) abnormal seedling mutants with altered morphology. Based on allele frequencies in a large scale mutagenesis that yielded approximately 25,000 embryo-lethal mutants, it has been estimated that there are about 4,000 essential genes required for embryogenesis in *Arabidopsis* (Jürgens et al., 1991).

Embryonic Lethal Mutations

A large heterogeneous collection of embryonic lethal mutants has been selected on the basis of aborted embryonic development (Müller, 1963; Meinke and Sussex, 1979b; Meinke, 1985; Patton et al., 1991; Errampalli et al., 1991; Meinke, 1991a). The stage at which development is arrested varies greatly between mutants, with the duration of the lethal phases varying as well. In some mutants, developmental arrest occurs earlier than the globular stage, while in other mutants arrest does not occur until after germination. Mutants may exhibit a well-defined distinct lethal phase, or the lethal phase may extend over several stages of embryogenesis. The embryo-lethal mutations display a broad range of phenotypes in terms of changes in color, size, and morphology, and can be categorized based on these alterations. Many of these embryo-lethal mutants have been examined with respect to their general morphology, response in tissue culture (Baus et al., 1986; Franzmann et al., 1989), and ultrastructure (Patton and Meinke, 1990) in order to facilitate the classification of corresponding genes as having developmental or more general housekeeping roles.

The *in vitro* culture of several embryo lethals has led to some clues to the function of the mutated genes (Baus et al., 1986; Franzmann et al., 1989). Most commonly, mature plants cannot be obtained from the embryo-lethal lines, suggesting that the mutant gene is required for both embryogenesis and later stages of the life cycle. However, this is not always the case. For example, mature *embryo-defective24* plants with 100% fully aborted seeds can be obtained via tissue culture, indicating that the wild-type *EMBRYO-DEFECTIVE24* gene product is required only during embryogenesis (Franzmann et al., 1989). *In vitro* culture techniques can also be used to detect auxotrophs, as in the case of *biotin1*, a biotin auxotroph (Schneider et al., 1989).

Some common themes can be detected in many of the embryo-lethals arrested in early stages of embryogenesis (Meinke, 1991a). For example, in many mutants arrested at the globular stage, the suspensor is also abnormal, growing to a much larger size than in wild-type embryos (50–200 cells in the mutants versus 6–8 cells in wild type), suggesting that the growth of the suspensor may be inhibited by the developing embryo (Marsden and Meinke, 1985). In addition, globular or preglobular arrested mutants often continue to grow in size, although normal morphogenesis has ceased. Ultrastructural studies have shown that when normal morphogenesis is arrested, cell differentiation, such as accumulation of storage proteins, is disrupted as well (Patton and Meinke, 1990). Many embryo-lethal mutants exhibit nonrandom distribution of mutant seeds in siliques of heterozygous plants, suggesting that the corresponding genes are expressed prior to fertilization (i.e., gametophytic expression) as well as having an essential function during embryogenesis (Meinke, 1982; Meinke et al., 1991a).

Seedling Mutants with Altered Pigmentation

A broad spectrum of abnormal pigmentation mutants, within which distinct subclasses can be recognized, have been isolated (Müller, 1963; Jürgens et al.,

1991). The defects in pigmentation range from pure white to yellowish-green to purplish-green. Mutants with altered seedling pigmentation represent a heterogeneous group in phenotype as well as probable cause. In most cases the primary defect is not known, but for some, specific biochemical lesions have been identified.

Several hundred seedling-lethal *albino* mutants have been isolated (van der Veen and Blankestijn-de Vries, 1973; Fischerová, 1975; Relichová, 1976; Jürgens et al., 1991). The pure white seedlings generally fail to produce true leaves, dying without any increase in size beyond the seedling stage. Some of these are conditional lethals, in that a wild-type phenotype can be restored with the addition of exogenous chemicals. For example *thiamine requiring1* (*th1*), *thiamine requiring2* (*th2*), *thiazole requiring* (*tz*), and *pyrimidime requiring* (*py*) mutants can be rescued with thiamine or, in some cases, locus-specific thiamine precursors (Langridge, 1955; Langridge, 1958; Feenstra, 1964; Rédei, 1965b; Li and Rédei, 1969b).

Mutations in at least two loci cause seedlings to have variegated green and white leaves, and may result in lethality, depending on growth conditions. The recessive mutation *immutans* results in variegated leaves with accumulation of inorganic phosphorous in the white sectors whose extent is increased with light intensity (Rédei, 1967; Röbbelen, 1968). Variegation is also observed in plants homozygous for the recessive nuclear mutation *chloroplast mutator* (*chm*) (Rédei, 1973a; Rédei, 1973b). The *chm* plants display abnormal plastid morphology (Rédei, 1973a; Rédei, 1973b) and rearrangements in their mitochondrial genome (Martinez-Zapater et al., 1992).

Yellowish-green mutants comprise the second major class of seedling pigmentation mutants (Röbbelen, 1957b; Hirono and Rédei, 1965; Fischerová, 1975; Relichová, 1976; Yakubova et al., 1980; Koornneef et al., 1983; Jürgens et al., 1991). Mutants may be entirely yellowish-green, as in the *chlorina* mutants, or they may exhibit altered pigmentation localized to certain organs or at specific developmental stages, as in *yellow inflorescence*. Yellowish-green mutants can be lethal, fully viable, or viable only in specific environmental conditions. The best characterized mutants of this class are the pigment deficiency mutants *chlorina1* and *chlorina42*.

Plants homozygous for the *chlorina1* mutation lack both chlorophyll *b* (Rédei and Hirono, 1964) and light harvesting chlorophyll binding protein (Murray and Kohorn, 1991), and yet have relatively normal chloroplast structure (Murray and Kohorn, 1991). The *chlorina42* mutation causes a pale green phenotype and lethality in the absence of exogenous sucrose (Fischerová, 1975). *CHLORINA42* encodes a light-regulated chloroplast protein of unknown function (Koncz et al., 1990).

The *fusca* mutations comprise the third class of pigmentation mutants (Müller, 1963; Jürgens et al., 1991; Feldmann, 1991; Errampalli et al., 1991). The *fusca* mutants are characterized by abnormal anthocyanin accumulation late in embryogenesis, causing *fusca* seeds to appear dark brown and *fusca* seedlings to appear purplish-green. Mutations in at least 10 loci result in the fusca phenotype. Although the exact basis of lethality in *fusca* mutants is unclear, the primary defect is not related to accumulation of anthocyanins, since lack of anthocyanins does not suppress their lethality (Jürgens et al., 1991; Errampalli et al., 1991).

Seedling Mutants with Altered Morphology

Selection of mutants on the basis of seedling lethality rather than embryonic lethality has led to the isolation of mutants which exhibit altered seedling mor-

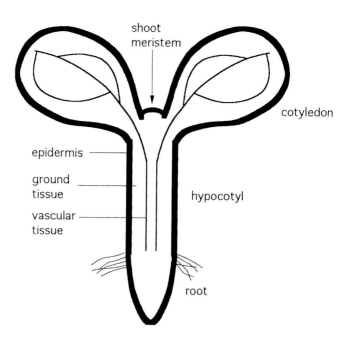

Figure 3. Diagram depicting the basic pattern elements of the *Arabidopsis* seedling. Along the apical-basal axis, four elements can be recognized: the shoot meristem, the cotyledons, the hypocotyl, and the root including the root meristem. The radial pattern consists of three tissues: the epidermis, the ground tissue, and the vascular elements. Adapted from Mayer et al. (1991).

phology, yet complete embryogenesis and germinate normally (Mayer et al., 1991; Jürgens et al., 1991). Mutants isolated in this manner display either nonspecific morphological alterations (the vast majority fall into this class) or specific pattern defects (a small number, approximately 40 loci) of seedling morphology (Jürgens et al., 1991). The mutants that affect pattern formation exhibit detects in the basic body plan of the seedling, either in the apical–basal axis or in the radial pattern of tissues.

Apical–Basal Pattern Mutants. Four basic pattern elements can be distinguished along the apical–basal axis. This axis consists of the root including the root meristem, the hypocotyl, the cotyledons, and the epicotyl including the shoot apical meristem (Figure 3). Vascular tissue is continuous from the root through the hypocotyl, and branches into each of the cotyledons. Mutations affecting the apical–basal patterning of the embryo result in either the deletion of one or more of the pattern elements, or in one case, cause a deletion with a concommitant duplication of a different pattern element (Mayer et al., 1991; Jürgens et al., 1991). The four primary classes of pattern deletion mutants are apical, basal, central, and terminal (both apical and basal). For each of the mutants described, several alleles of varying severity of phenotype have been isolated.

In the basal deletion class of mutants is *monopteros* (Mayer et al., 1991; Berleth and Jürgens, 1993). These seedlings are missing both hypocotyl and root, and thus consist of only the apical portion of the seedling, including the apical shoot meristem and the cotyledons. The cotyledons are well differentiated although they may be altered in terms of number and fusion. The internal structure, in terms of radial pattern elements, corresponds to the external phenotype, with vascular strands forming in the cotyledons. The characteristic divisions of the hypophyseal cell that give rise to the root in wild-type embryos are lacking in *monopteros* embryos. Alterations in the development of *monopteros* embryos can be traced to the octant-stage embryo (Berleth and Jürgens, 1993).

The central deletion class is represented by *fackel* mutants (Mayer et al., 1991). In *fackel* seedlings the cotyledons appear to be directly connected to the root, with continuous vascular strands between the root and cotyledons despite the loss of hypocotyl. Again, the cotyledons of *fackel* mutants may be altered in shape and number. Morphological differences are detectable as early as the heart stage, with *fackel* embryos being broader than wild-type embryos.

The third class of deletion mutants, known as apical deletions, is represented by *gurke* mutants, in which the shoot apical meristem and cotyledons are affected (Mayer et al., 1991). The internal phenotype corresponds to the external phenotype, with vascular strands terminating apically without any evidence of branching towards the positions the cotyledons would normally occupy. The apical surfaces of heart stage *gurke* embryos are round, with no evidence of cotyledon primordia, while the basal regions appear to be normal.

The terminal class of deletions is represented by the extensively characterized *gnom* mutants, in which both the root and shoot meristems are deleted, the root fails to form, and the cotyledons are reduced to varying degrees (Mayer et al., 1991; Jürgens et al., 1991; Mayer et al., 1993). In the extreme case, *gnom* seedlings are ball-shaped, with no evidence of apical–basal polarity. However, the majority of *gnom* seedlings, even putative complete loss-of-function alleles, display some degree of apical–basal polarity. Despite this, all the radial pattern elements can be found in *gnom* mutants, although in most cases the vascular tissue does not form distinct strands.

Analysis of developing *gnom* embryos reveals that the initial transverse cleavage of the zygote gives rise to apical and basal cells of almost equal size, in contrast to wild type where the apical cell is substantially smaller than the basal (Mayer et al., 1993). Subsequent cell division patterns in embryos through the dermatogen stage are also altered and variable. Later in embryogenesis, *gnom* mutants lack the characteristic asymmetric cell divisions of the hypophyseal cell that give rise to the root in wild-type embryos. In addition, *gnom* mutants fail to produce roots in tissue culture, forming only callus-like growth. Cotyledon primordia do not form normally in the apical region of the heart stage *gnom* embryos, resulting in cone-shaped, oblong, or ball-shaped embryos (Mayer et al., 1993). Thus, the *GNOM* gene product appears to promote asymmetric cell division in the zygote, and perhaps during other stages of embryogenesis (Mayer et al., 1993).

Another class of apical–basal pattern mutants, represented by *doppelwurzel* mutants, is not a simple deletion, but rather an apical deletion with a concomitant mirror-image duplication of the basal elements in the place of the missing apical elements (Jügens et al., 1991). Thus, *doppelwurzel* seedlings consist of two basal elements, including hypocotyl and some cotyledon tissue, but no apical shoot meristem.

Radial Pattern Mutants. Three basic tissues comprise the radial pattern of the seedling: epidermis, ground tissue, and vascular tissue (Figure 3). The establishment of the epidermis is disrupted in two mutants (Mayer et al., 1991). In *knolle* embryos, the formation of the epidermis at the dermatogen stage of embryogenesis is altered, with outer cells never becoming morphologically distinguishable from the inner cells (Mayer et al., 1991). Mature *knolle* embryos are irregularly shaped and do not have a well formed layer of epidermal cells. In *keule* embryos, although the epidermal precursors appear to arise normally, they follow an abnormal developmental pathway (Mayer et al., 1991). Mature *keule* have bloated and irregularly arranged epidermal cells, giving the embryos a rough surface and an irregular shape.

Other Pattern Defects. Several other pattern defects have been observed in embryo-pattern mutants (Mayer et al., 1991; Jürgens et al., 1991; Meinke, 1992). The number of cotyledons can be increased in *häuptling* seedlings (Jürgens et al., 1991). Cotyledons may be transformed into shoot-like structures, as in *toro* mutants (Jürgens et al., 1991), or leaf-like organs, as in *leafy cotyledon* mutants (Meinke, 1992). In the case of *leafy cotyledon* embryos, which are desiccation intolerant and occasionally viviparous, the mutant cotyledons have stellate trichomes and vasculature reminiscent of post-embryonic leaves, and do not accumulate embryo-specific protein bodies.

Finally, mutant seedlings with an altered shape, but with all pattern elements present, have been characterized (Mayer et al., 1991). Both cell and organ shape throughout the entire seedling is altered in short, stout *fass* mutants. In contrast, *knopf* and *mickey* mutations result in shape alterations in specific regions of the seedling. Both epidermal and vascular tissues are altered in *knopf* embryos, although the tissues are initiated normally. The vascular tissue of *knopf* seedlings fails to differentiate normally and their epidermal cells are densely packed. The cotyledons of *mickey* mutants are large and thickened, and the vasculature of the hypocotyl is not well developed. The developmental defects in *mickey* mutants are not detectable until after the heart stage.

J.L. Bowman and S.G. Mansfield

Plate 7.1

The zygote between two and four hours after flowering (HAF)

Embryo differentiation in *Arabidopsis thaliana* follows the classical *Capsella* variation of the Onagrad type, characterized by a longitudinal division of the terminal cell of a transversely divided zygote. Longitudinal sections shown.

(**A**) and (**B**) Light micrographs of the zygote (z) just after fertilization. Although the cell remains highly polarized (c.f. the egg cell, Plate 5.10) with a chalazally-sited nucleus, many small vacuoles (arrowheads) replace the single, large, chalazal vacuole of the egg cell. Both zygotes have already started to elongate. **B** is at a slightly later stage than **A**.

(**C**) Light micrograph of the fully elongated zygote (z) 4 HAF. The terminal region of the cell contains the single, large nucleus and a large proportion of the cytoplasm. An extremely large micropylar vacuole (v) dominates the cell. Note the accumulation of PAS-positive starch inclusions around the nucleus (*arrowheads*) (PAS, periodic acid-Schiff's staining).

(**D**) The elongated zygote (z) showing its relationship with the surrounding tissues: the integuments (i) and free-nuclear endosperm (en). Several cytoplasmic strands (*arrows*) extend through the large micropylar vacuole (v).

(**E**) The zygote tip 4 HAF. The cytoplasm contains many mitochondria (*arrowheads*), a number of poorly-differentiated starch-containing plastids (pl), and small vacuoles (v). Ribosome density has increased compared to the egg-cell stage. The cell wall is incomplete or absent in this region of the zygote. A superficial layer of free nuclear endosperm cytoplasm (en) completely surrounds the cell.

Bar = 5 μm in **D**, **E**; 10 μm in **A**, **B**, **C**.

S.G. Mansfield

Reproduced from Mansfield and Briarty (1991) with permission from the National Research Council of Canada.

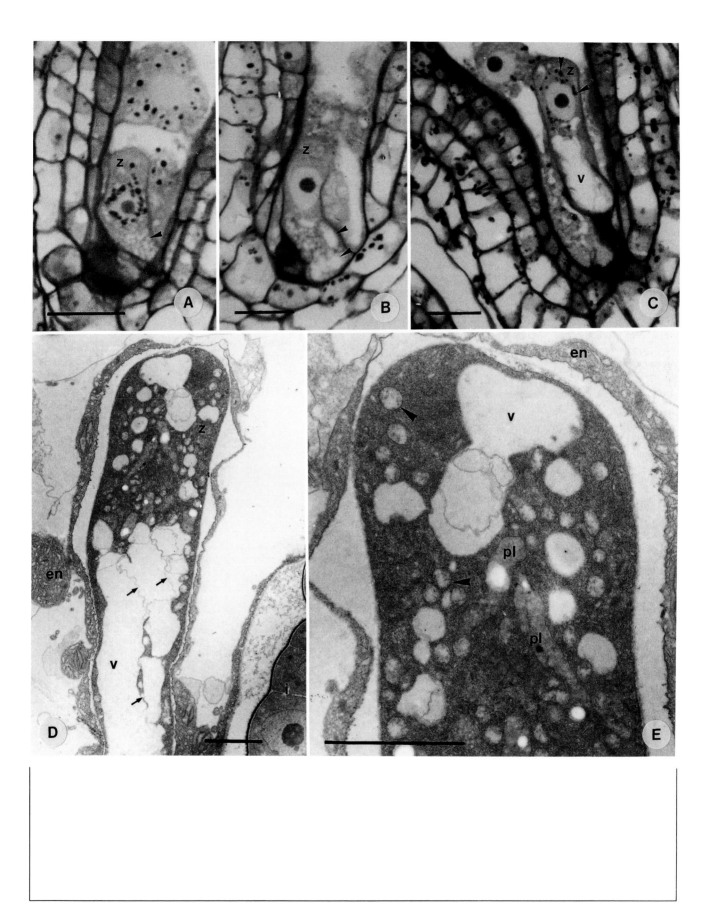

Plate 7.1

Plate 7.2
The zygote and surrounding tissues four hours after flowering (HAF)

Longitudinal sections shown.

(**A**) Low-power micrograph showing the appearance of the zygote base (z) with its extremely thickened and crenellated cell wall (*arrowhead*). No plasmodesmatal connections exist with the surrounding endosperm (en).

(**B**) Enlargement of the extreme chalazal cytoplasm of the zygote. Many dictyosomes (d) and regularly-shaped mitochondria (m) are present.

(**C**) Micropylar chamber of the expanding embryo sac, showing the degenerating synergid (ds) and filiform apparatus (fa), endosperm cytoplasm (en) and nucleus (nu), and the base of the elongated zygote (z). The cytoplasm of the degenerating synergid consists of a mass of electron-dense material containing barely discernible profiles of mitochondria and endoplasmic reticulum. In contrast, the adjacent cytoplasm of the early endosperm appears metabolically active. Mitochondria (*arrows*) and differentiated plastids (pl) are abundant. Embryo sac wall projections (*arrowheads*) develop and proliferate throughout the micropylar chamber.

(**D**) Enlargement of the embryo sac wall projections (wp) in the micropylar chamber. The occurrence of many mitochondria (m), dictyosomes (d), and short strands of endoplasmic reticulum (*arrowheads*) adjacent to the wall projections is a significant feature of this region.

Bar = 1 μm in **B**, **D**; 3 μm in **C**; 5 μm in **A**.

S.G. Mansfield

Reproduced from Mansfield and Briarty (1991) with permission from the National Research Council of Canada.

Embryogenesis: Early Embryo Development

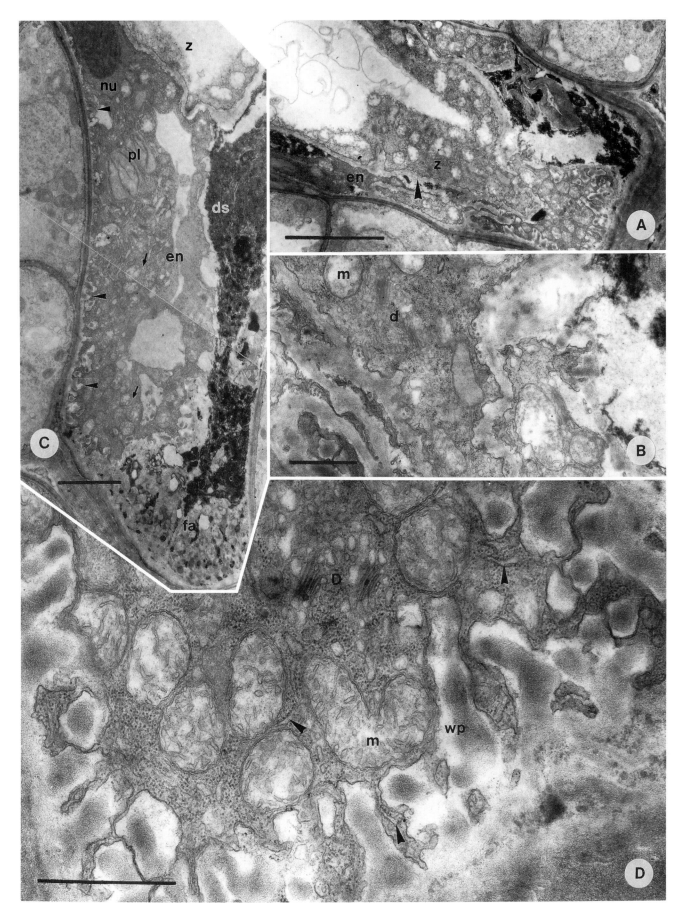

Plate 7.2

Plate 7.3

Single- and two-terminal-cell embryos 12 and 18 hours after flowering (HAF)

Longitudinal sections shown.

(**A**) Light micrograph of a single-terminal-cell embryo 12 HAF. The lower cell has undergone one division, forming two highly vacuolate cells; a basal cell (bc) and a suspensor cell (sc). Prior to the first division of the terminal cell, the basal cell of a two-celled embryo may undergo two divisions, one division or remain undivided.

(**B**) The two-terminal-cell embryo 18HAF as seen by light microscopy. Note that the embryo-proper is of similar size to that of the single-terminal-cell stage. The two cells are highly vacuolate.

(**C**) Enlargement of the two terminal cells. The longitudinal cell wall (*arrows*) of the embryo-proper is thin and irregular at this early stage; formation occurs from the apex to the cell base. Both terminal cells possess a fully formed outer cell wall. Each cell is usually dominated by the presence of a single large vacuole (v) together with several minor structures. Mitochondria (*arrowheads*) are very abundant in both cells. A thin layer of endosperm (en) cytoplasm envelops the embryo-proper.

(**D**) Enlargement of the nucleus and adjacent cytoplasm of the embryo-proper cells showing the structure of mitochondria (*arrows*), undifferentiated plastids (pl), and rough endoplasmic reticulum (*arrowheads*).

(**E**) The extreme micropylar cytoplasm of the basal cell of a two-terminal-cell embryo. The cell wall in this region is extremely thickened, especially the common wall between the basal cell and embryo sac (cw), and small wall projections are present (*arrowheads*). No plasmodesmata occur in this common cell wall or in the endosperm–embryo boundary. The bulk of the basal cell cytoplasm is situated at the chalazal end surrounding the nucleus. In the micropylar cytoplasm (not shown), plastid abundance is lower compared to the nuclear-associated cytoplasm, but mitochondrial (m) abundance and dictyosome (d) activity are greater. Ribosome density is high throughout the cell.

Bar = 1 μm in **E**; 3 μm in **C,D**; 20 μm in **A,B**.

S.G. Mansfield

Reproduced from Mansfield and Briarty (1991) with permission from the National Research Council of Canada.

Embryogenesis: Early Embryo Development

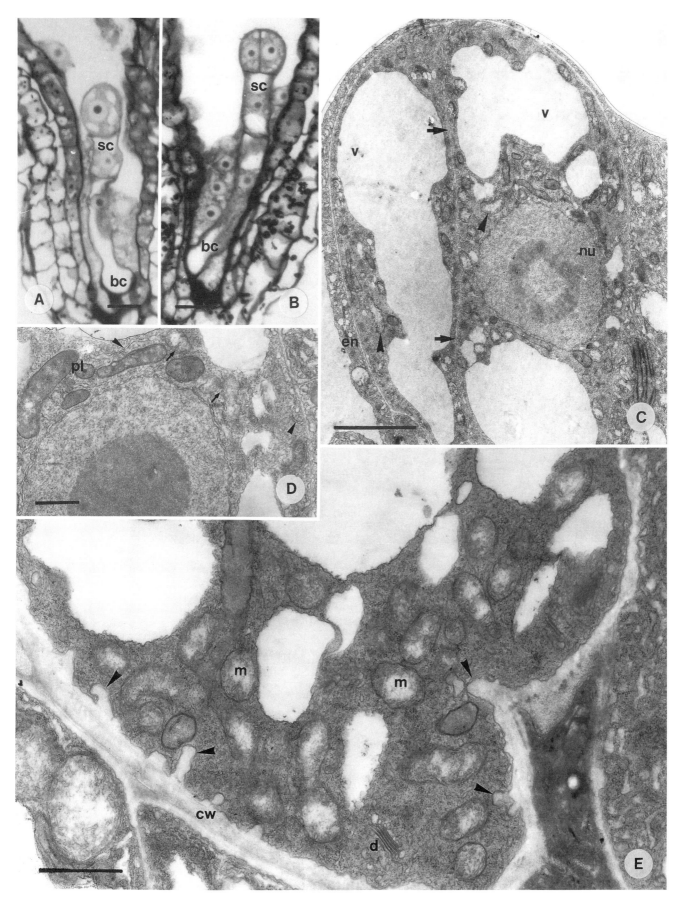

Plate 7.3

Plate 7.4
Quadrant to Early-globular embryos 24–42 hours after flowering (HAF).

Longitudinal sections shown.

(**A**) The quadrant embryo 24 HAF. Longitudinal divisions in the two terminal cells have produced a four-celled embryo-proper (e). Total embryo-proper cell volume remains the same up to this point in development, and each cell is highly vacuolate (*arrowheads*) with a prominent nucleus and nucleolus. The suspensor (su) now consists of three cells including the basal cell (bc).

(**B**) The octant embryo 30 HAF. The embryo-proper (e) consists of eight cells formed by each of the quadrant embryo cells dividing transversely. Each cell is of similar size, but the embryo-proper diameter is greater than in earlier stages (single-terminal to quadrant embryos; su, suspensor).

(**C**) The dermatogen embryo 42 HAF. Each of the octant cells has undergone a periclinal division. This, the formation of a protoderm, is the first recognizable tissue differentiation. There is developmental variation during the formation of this stage, in that division within the right or left cell quartet was observed to occur simultaneously, but division never occurred simultaneously in the upper or lower cell quartet (e, embryo-proper; su, suspensor).

(**D**) The early-globular embryo 42 HAF. The protoderm and inner cells of the embryo-proper (e) divide by anticlinal and longitudinal divisions, respectively, although these divisions do not occur simultaneously. All cells are interconnected by many plasmodesmata, but none are observed connecting the embryo-proper with the endosperm. Division within the suspensor (su) is complete by this stage and, including the hypophyseal (h) and basal cells, consists of between seven and nine (eight in this example) highly vacuolate cells. It is noticeable that the embryo-proper cells possess smaller but more regular vacuoles (*arrowheads*) compared to earlier stages (c.f. **A**). Compared to the embryo-proper cells, the hypophyseal cell is conspicuously vacuolate.

Bars = 20 μm.

S.G. Mansfield

Reproduced from Mansfield and Briarty (1991) with permission from the National Research Council of Canada.

Embryogenesis: Early Embryo Development

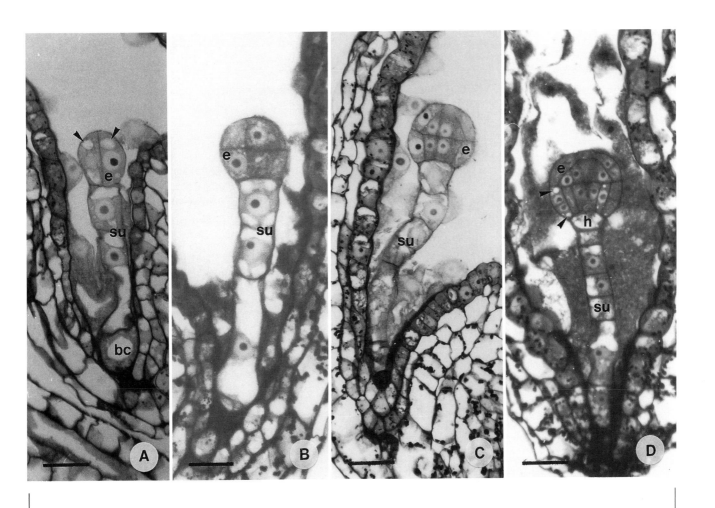

Plate 7.4

369

Plate 7.5

**The suspensor 36–42
hours after flowering
(HAF)**

(**A**) Longitudinal section of the lower half of a middle suspensor cell from a dermatogen embryo 42 hours after flowering. The cell has a single large vacuole (v) and, in most regions, a comparatively thin layer of cytoplasm which contains a large prominent nucleus (nu) located midway along the cell. Note the presence of poorly differentiated elongated plastids (pl), and fewer mitochondria (*arrowheads*) and dictyosomes (not shown) compared to the embryo-proper cells. Plasmodesmata (*arrow*) are present in the thin end-walls of the cell.

(**B**) Light micrograph (transverse section) of a single ovule 36 HAF showing the relationship between the suspensor (su), surrounded by a free-nuclear endosperm (en), and integuments (i). By this stage the endosperm envelops most of the suspensor length, providing mechanical support for the embryo.

Bar = 2 μm in **A**, 20 μm in **B**.

S.G. Mansfield

Reproduced from Mansfield and Briarty (1991) with permission from the National Research Council of Canada.

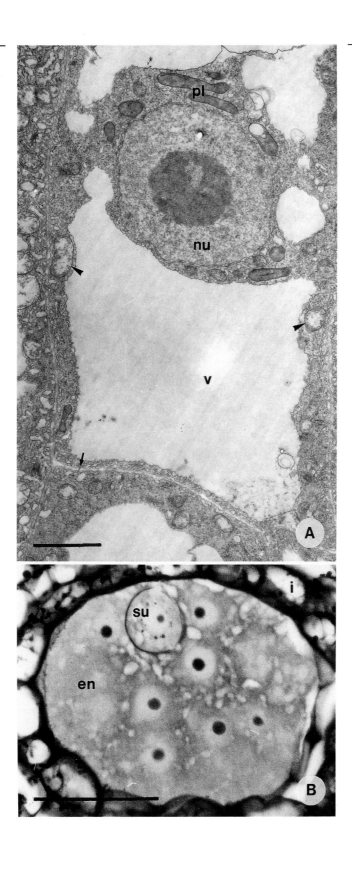

Plate 7.5

371

Plate 7.6
Mid- to late-globular
embryos 48–60 hours
after flowering (HAF)

The outer embryonic cells of the embryo-proper continue to develop by non-simultaneous anticlinal divisions, eventually forming a 16- (mid-globular) and 32-cell (late-globular) protoderm, and the original 8 inner cells divide longitudinally and then transversely (dermatogen stage), producing a 64-cell embryo-proper (late-globular).

(A) and (B) Light microscope section (A) and a chemically-cleared preparation (B) of the late-globular embryo (e) 60 HAF. The suspensor (su), as seen in B, is very elongated by this stage, having already completed its divisions, and holds the embryo-proper in a more favorable position within the embryo sac (em). The micropylar chamber of the embryo sac contains a large volume of free-nuclear endosperm (en) which surrounds virtually the whole of the suspensor. Often, free-nuclear endosperm "cells" adhere to the embryo-proper surface (*arrowhead* in A). A single mitotic figure (*arrow* in A) can be seen in a protodermal cell.

(C) Oblique section of the embryo-proper of the late-globular embryo 60 HAF. Since the embryo-proper cells are continually dividing, cell walls often appear irregular (*arrow*), although plasmodesmata are present between all cells (*small arrowheads*). Large accumulations of mitochondria (m) occur throughout the embryo. Nuclei (nu) are extremely prominent, and micronucleoli and perinuclear deposits (*large arrowheads*) occur in the the nucleoplasm by this stage. A free-nuclear endosperm "cell" (en) adheres to the embryo-proper. Note the presence of several differentiated plastids and the low ribosome density of the endosperm cytoplasm compared to the embryo-proper cells.

(D) Enlargement of the cell wall and adjacent cytoplasm of the embryo-proper inner cells 60 HAF. The cell wall has an irregular profile and the middle lamella (*arrow*) is very distinct. Microtubules in transverse section (*arrowheads*) can be seen in close proximity to the cell wall. The network of endoplasmic reticulum by this stage consists of many single, elongated profiles encrusted with ribosomes. Similarly, the Golgi system is well-developed; dictyosomes (d) are active in vesicle production. Plastids (not shown) remain relatively undifferentiated, appearing variable in size and shape, with a simple bithylakoid system if present at all, and are devoid of starch.

(E) One of many "giant" mitochondria observed in the late-globular embryo 60 HAF. These atypical organelles may appear cup-shaped, elongated, or multiply-branched in profile, and contain many cristae (*arrows*) and a matrix interspersed with granular material and fibrillar mitochondrial DNA (*arrowheads*).

(F) Part of two cells from the inner tissues of a late-globular embryo 60 HAF. Ribosomes accumulate in small circular conspicuous groups (*arrowheads*) throughout the cells, giving them a "blotchy" appearance.

Bar = 0.5 μm in D; 1 μm in E, F; 5 μm in C; 20 μm in A, B.

S.G. Mansfield

Reproduced from Mansfield and Briarty (1991) with permission from the National Research Council of Canada.

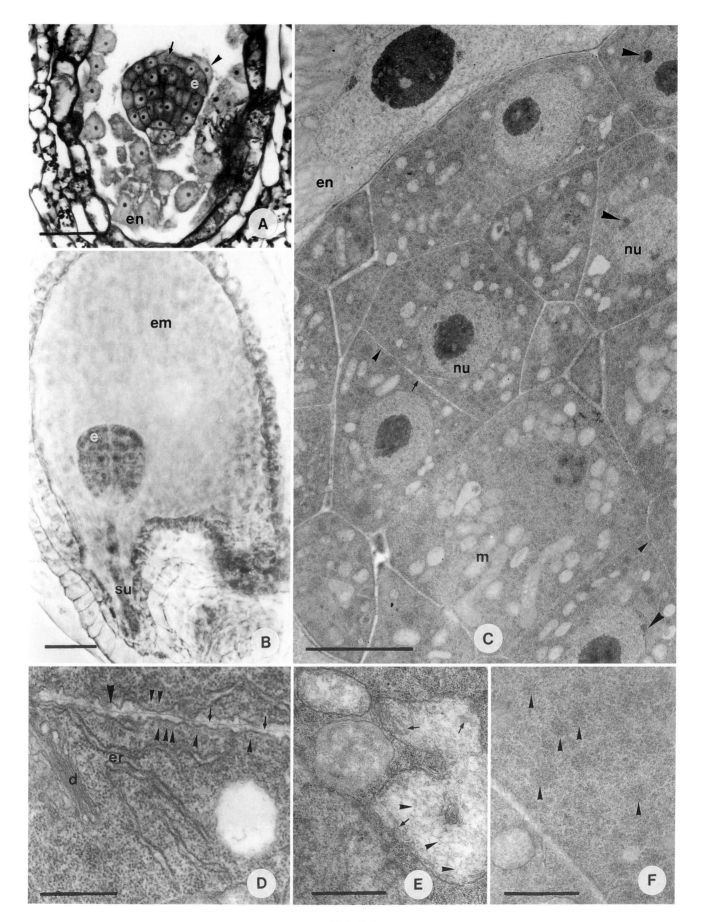

Plate 7.6

Plate 7.7
Heart and torpedo stage
embryos 66–96 hours
after flowering (HAF)

The Heart Stage Embryo 66–84 HAF

By the late-globular stage 60 HAF, protodermal divisions increase in frequency in the region of the future cotyledons. As the embryo passes from globular to heart stage, it assumes a triangular profile (not shown). Differences between cells of the embryo become evident by the mid-heart stage 72–78 HAF. Cells of the procambium and shoot apex are more vacuolate than those of the protoderm, but ribosome density is comparable in all regions.

(A) Light micrograph of an early-heart embryo 66 HAF (longitudinal section). Cotyledon initials (co) are present and the central cells are beginning to elongate and divide to form the provascular tissue (pr). The embryo is enclosed by a network of cellularized endosperm (ce), and the uppermost cell of the suspensor has started to differentiate to form the hypophysis (h).

(B) Chemically-cleared preparation of the mid-heart embryo 72 HAF, showing the relationship with the embryo sac (em). The cotyledon initials (co) can be seen quite clearly by this stage. The embryo-proper is maintained in a favorable position by the suspensor and surrounding cellularized endosperm.

The Torpedo Stage Embryo 96 HAF

(C) Light micrograph of the torpedo embryo (longitudinal section). The cotyledons (co) are elongate by this stage, having forced their way into the fully cellularized endosperm (ce). The embryo central cells of the radicle (r) and hypocotyl (hy) have differentiated, forming the provascular tissue (pr). Immediately adjacent to the provascular cells, particularly in the hypocotyl, the parenchyma cells are more vacuolate.

(D) Enlargement of parenchyma cells in the cotyledon base. Numerous mitochondria and plastids (pl) with rudimentary grana (living embryos show "greening" at this stage) and numerous electron dense bodies (ob) are present in most cells. Nuclei (nu) are large and prominent and often contain nucleolar organizers (*arrow*) and micronucleoli (not shown). Plasmodesmata (*arrowheads*) are present between all cells.

(E) Enlargement of the cytoplasm in a cotyledon parenchyma cell showing the structure of the rough endoplasmic reticulum (*small arrowhead*), mitochondria (m), and Golgi system (d). Cell walls (*large arrowhead*) have a more regular profile by this stage. Vacuoles (v) at this stage are small and regular.

(F) Enlargement of an electron-dense body in a cotyledon cell of a torpedo embryo (greatest concentration occurs in the epidermal cells). These organelles have an amorphous matrix and, in this example, a fine dark boundary (*arrowheads*) that resembles a half-unit membrane.

Bar = 0.5 μm in **E**, **F**; 1 μm in **D**; 30 μm in **A**, **B**, **C**.

S.G. Mansfield

Reproduced from Mansfield and Briarty (1991) with permission from the National Research Council of Canada.

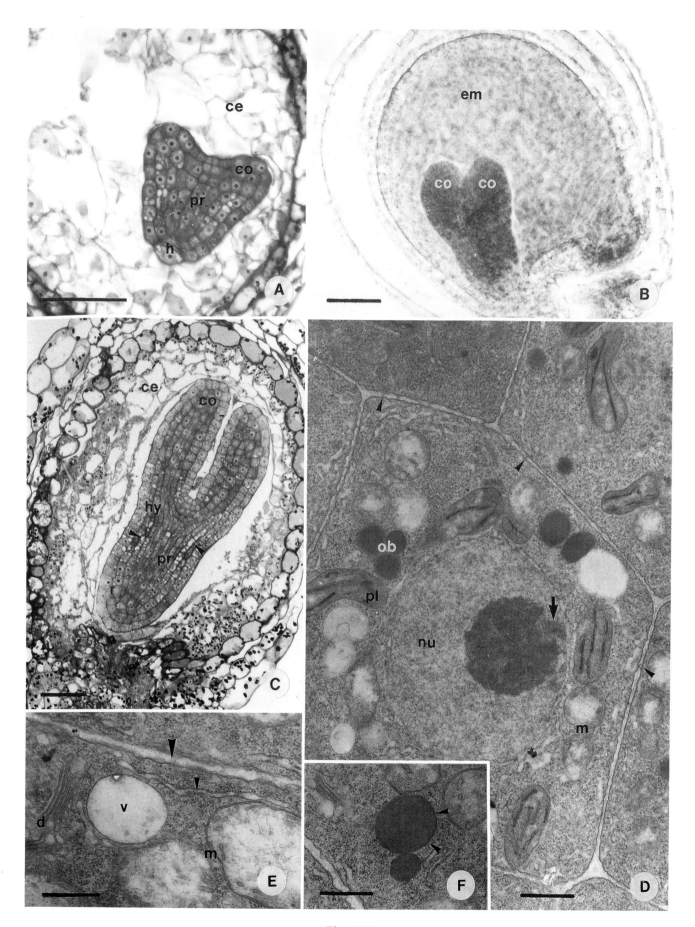

Plate 7.7

Plate 7.8
Cotyledon cell
development

Between 96 and 120 hours after fertilization (HAF) the cotyledons elongate very rapidly, forming the characteristic walking-stick stage embryo by 108 HAF, and the upturned-U stage by 120 HAF (not shown). Cell division in the cotyledons has ceased by 144 HAF but cell expansion continues during the subsequent phases of reserve deposition 144–216 HAF (see Table 1). During this time, the storage protein content increases rapidly (Figure 1).

(**A–C**) Light micrographs of the developing embryo.
(**A**) 168 HAF.
(**B**) 192 HAF.
(**C**) 240 HAF.

Small cavities are present between the cotyledons (c), radicle (r), and peripheral endosperm (en) during early development (*arrowheads* in **A** and **B**). The procambium (pr) is well differentiated by 168 HAF (**A**) extending throughout the embryo and branching within the cotyledons. Embryo size decreases slightly with the onset of seed maturation and desiccation (**C**) (h hypocotyl; i, inner integument).

Bar = 100 μm.

Table 1. Change in cotyledon cell number and cell volume during reserve deposition.

HAF	No. of cells per cotyledon[a]	Cotyledon cell volume[b] (μm^3)
120	3663 (227)	—
144	5385 (285)	816 (361)
168	5266 (271)	951 (490)
192	5490 (254)	1064 (509)
216	5397 (307)	1132 (548)

[a] Values are means of 10 counts, with standard errors in parentheses, and exclude vascular and epidermal cells.
[b] Values are means of 10 cotyledons, rounded to the nearest μm^3; 100 cells per cotyledon were measured. Standard errors are in parentheses.

S.G. Mansfield

Reproduced from Mansfield and Briarty (1992) with permission from the National Research Council of Canada.

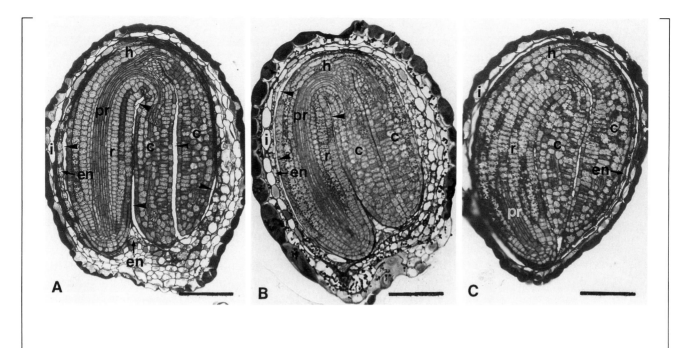

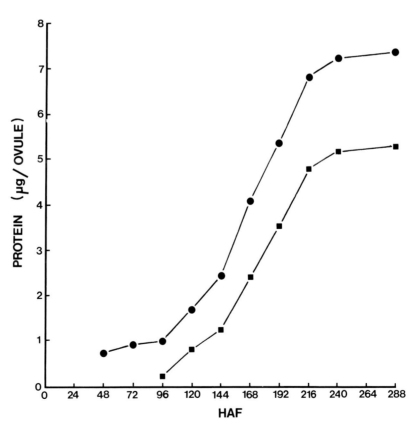

Figure 1. Storage (■) and total (●) protein content of developing ovaries. Values represent means of three samples of 100–150 ovules.

Plate 7.9
Subcellular structure of the cotyledons

The first structural indications of storage protein accumulation within the vacuome appear between 120 and 144 hours after fertilization (HAF). Storage lipid deposition begins approximately 48 hours prior to this.

(**A**) Enlargement of cotyledon storage parenchyma cells 168 HAF. Chloroplasts (pl) have a well-differentiated but less organized granal system than observed at 120 HAF. Vacuoles (*large arrowheads*) are almost full of protein; matrix structure is often different in adjacent cells. Nuclei (nu) are regular in profile and contain peripheral chromatin deposits (*small arrowheads*) (lb, lipid body).

(**B**) Part of two plastids (pl) in a cotyledon storage parenchyma cell 192 HAF. The large starch body (st) distorts the granal system (gr); in later development (192+ HAF) the starch inclusions decrease is size significantly (**H**). Note the long strands of endoplasmic reticulum (*arrowheads*) close to the plastid envelopes (pb, protein body).

(**C**) Prominent rough endoplasmic reticulum cisternae (*arrowheads*) adjacent to reserve body membranes 192 HAF; connections between the endoplasmic reticulum and protein bodies (pb) are not observed (lb, lipid body).

(**D**) Characteristic vacuoles (v) in the cotyledon parenchyma cells just prior to protein deposition 120 HAF.

(**E**) An early protein body (pb) of the inner storage parenchyma cells 144 HAF. This example contains a continuous peripheral layer of protein (pr) and a number of small globoids (*small arrowheads*).

(**F**) Matrix structure of protein bodies (*arrowheads*) shows distinct variation within cells of the inner storage parenchyma 168 HAF (lb, lipid body; nu, nucleus).

(**G**) Matrix structure of epidermal cell (ep) protein bodies (pb) is different than elsewhere in the cotyledon (144 HAF). Protein material accumulates in condensed masses at one point on the tonoplast, and the stage of development of these bodies is advanced compared to those of the inner storage parenchyma (*). Here, the vacuoles contain small deposits of protein on the tonoplast (*arrows*) and around globoids (*arrowheads*), which act as nuclei for deposition of material (lb, lipid body; nu, nucleus).

(**H**) Storage parenchyma cells of the cotyledon 240 HAF. Protein bodies (pb) are completely full of material except for small irregular spaces which are probably areas from which globoids have been lost during tissue processing (*large arrowheads*). Lipid bodies (lb) are now compressed into a homogeneous mass, and where they impinge on the membrane of protein bodies, a scalloped effect is produced (*large arrows*). Plasmodesmatal connections are present between all cells (*small arrowheads*). Plastids (pl) now contain very small starch bodies (*small arrow*).

Bar = 0.5 μm in **B**, **C**; 1.0 μm in **E**; 2.0 μm in **F**, **G**; 3.0 μm in **A**; 5.0 μm in **D**, **H**.

S.G. Mansfield

Reproduced from Mansfield and Briarty (1992) with permission from the National Research Council of Canada.

(*Text continued on p. 381*)

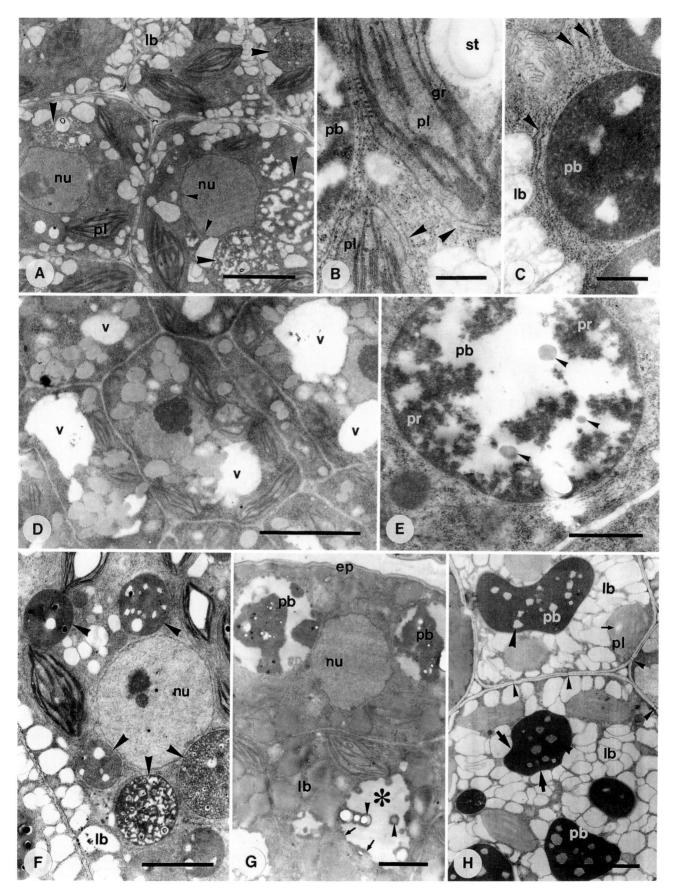

Plate 7.9

Plate 7.9
Subcellular structure of the cotyledons
(*continued*)

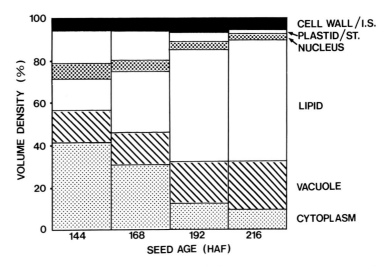

Figure 1. Volume density of major cotyledon tissue compartments (cell wall-intercellular space (I.S.), vacuole, lipid body, plastid-starch (ST.), nucleus, cytoplasm) during development, measured on 4,200× electron micrographs. Cytoplasm includes the subcompartments mitochondria, endoplasmic reticulum, dictyosomes, and ground cytoplasm.

Plate 7.10
**Storage protein transport
and the structure of the
lipid body membrane**

(A) Golgi vesicles (forming and released; *arrows*) in the cotyledon parenchyma cells 144 hours after fertilization (HAF) contain dense-staining material which is similar to the peripheral deposits of protein (pr) seen in adjacent vacuoles (pb). Vesicles are never actually observed fusing with the protein body membrane. Similarly, although profiles of endoplasmic reticulum are often seen in close proximity to developing protein bodies (*arrowheads*; also see **B** and **C** on Plate 7.9), no direct connections are observed (g, golgi).

(B) Large spherical bodies (maximum profile diameter of 0.45 μm) of unknown origin, with a dense-staining structured matrix (*arrows*) are observed in the storage cells by 168 HAF. These inclusions may participate in storage protein transport.

(C) Nascent lipid bodies (lb) of a developing parenchyma cell 96 HAF appear partly encircled by a cisterna of rough endoplasmic reticulum (*arrows*), which form cup-shaped arcs at a distance of about 30 nm from the lipid surface. Often a membrane (*arrowheads*) is detectable at the lipid–cytoplasm interface of these bodies which has a thickness of approximately 3 nm. No direct connections with these cisternae are present, however.

(D) Lipid bodies at the periphery of a storage cell 192 HAF. The association with the endoplasmic reticulum, seen in early development, is lost as the lipid bodies become densely packed at the cell periphery. Here, the membranes (*small arrowheads*) of neighboring bodies do not fuse, forming a bipartite structure with an approximate thickness of 6 nm (*large arrowheads*).

Bar = 0.5 μm in **A**, **B**, **D**; 1.0 μm in **C**.

Table 1. Absolute values (volume or area per cell), and secondary parameters of cytoplasmic subcompartments during reserve deposition measured on 14,000× electron micrographs.

	Volume, area, or width at selected HAF[a] times			
	144	168	192	216
ER[b] volume (μm^3)	27.2	38.0	16.5	13.0
ER surface area (μm^3)	757	1413	637	514
ER strand thickness (μm^3)	0.036	0.027	0.028	0.025
Dictyosome volume (μm^3)	11.8	13.6	7.7	5.0
Mitochondria volume (μm^3)	32.9	30.2	17.2	12.0
Ground cytoplasm Volume (μm^3)	265.1	207.2	81.6	65.0

[a] HAF, hours after flowering.
[b] ER, endoplasmic reticulum.

S.G. Mansfield

Reproduced from Mansfield and Briarty (1992) with permission from the National Research Council of Canada.

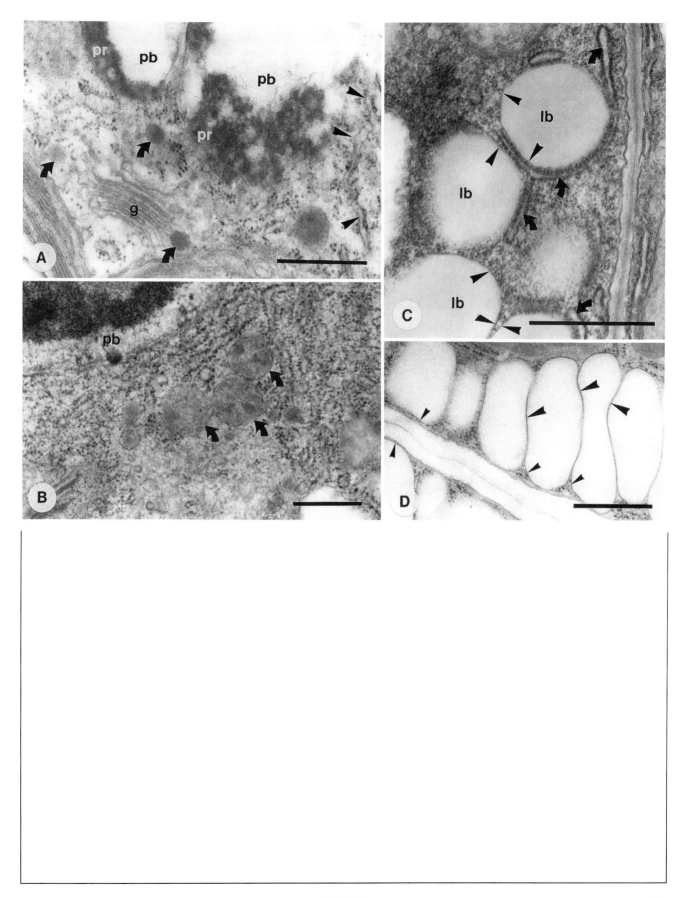

Plate 7.10

Endosperm Development

Embryogenesis in the angiosperms commences with a double fertilization. One male gamete fuses with the egg cell, giving rise to the zygote, while the second gamete fuses with the diploid central cell, forming a triploid cell known as the primary endosperm nucleus. It is this product that eventually forms the coenocytic endosperm tissue.

Apart from the classically assigned function of nourishing the embryo (Brink and Cooper, 1947), the endosperm may play other roles including hormonal regulation of embryo growth, maintenance of a high osmotic potential around the embryo, providing mechanical support during early embryo growth, or acting as a store for reserves after compartmentalization by cell wall formation.

The development of the free-nuclear (coenocytic) and cellular endosperm has been described in *Arabidopsis* by Mansfield and Briarty (1990a; 1990b) and Marsden and Meinke (1985). In *Arabidopsis*, its close relative *Capsella bursa-pastoris* (Schulz and Jensen, 1974), and other crucifers there are some common features of endosperm development. First, the two daughter nuclei formed from the first division of the primary endosperm nucleus migrate to opposite poles of the embryo sac. Second, the numerous nuclei eventually formed always accumulate on the periphery of the embryo sac. Third, significant structural and developmental differences exist between the endosperm of the micropylar and chalazal chambers. In *Arabidopsis* this is so pronounced that the endosperm could be defined as helobial.

Endosperm cellularization in *Arabidopsis* begins during cotyledon initiation and is complete throughout the embryo sac as storage deposition begins in the embryo. Prior to this, during cotyledon elongation and expansion, much of the endosperm is crushed and absorbed by the embryo. A single layer of endosperm containing storage reserves persists during seed desiccation. The process of cellularization occurs by means of freely-growing cell walls and formation from wall projections (Mansfield and Briarty, 1990b).

S.G. Mansfield

Plate 7.11
Endosperm development

Development of the Free-Nuclear Endosperm:
The Two- to Four-Nucleate Endosperm

(**A**) Two daughter nuclei (*arrowheads*) formed from the division (2–4 hours after fertilization, HAF) of the primary endosperm nucleus (cpt, chalazal proliferating tissue; i, integument).

(**B**) An embryo sac 4 HAF containing 4 endosperm nuclei (*arrowheads*), 2 on the longitudinal walls of the embryo sac, a single nucleus that has migrated to the chalazal chamber, and a 4th (not visible in this section) which is adjacent to the zygote (z).

In the nuclei that migrate to the extreme chalazal and micropylar poles of the embryo sac, division is temporarily arrested, but their associated cytoplasm proliferates. The second division series, producing the 4-nucleate endosperm, is synchronous, but later, with rapid enlargement of the endosperm, divisions become asynchronous.

(**C**) Two endosperm nuclei on the longitudinal walls of an embryo sac (em) containing six nuclei (4 HAF).

(**D**) The micropylar nucleus (nu) and associated cytoplasm adjacent to the zygote (z) base (4 HAF) (i, integument).

(**E**) Enlargement of the nucleus (nu) and cytoplasm of a single free-nuclear endosperm "cell" (4 HAF). The cytoplasm is very similar to the central cell cytoplasm in the mature embryo sac in many respects: a high density of spherical to oval chloroplasts (pl) and mitochondria (m), high ribosome density, and abundant endoplasmic reticulum. However, the small vacuoles present in the nuclear-associated cytoplasm of the central cell disappear, as do the large starch bodies of the plastids.

Bar = 2 μm in **E**; 3 μm in **D**; 10 μm in **C**; 20 μm in **A**, **B**.

S.G. Mansfield

Reproduced from Mansfield and Briarty (1990a) with permission from Prof. A.R. Kranz, ed. *Arabidopsis Information Service*.

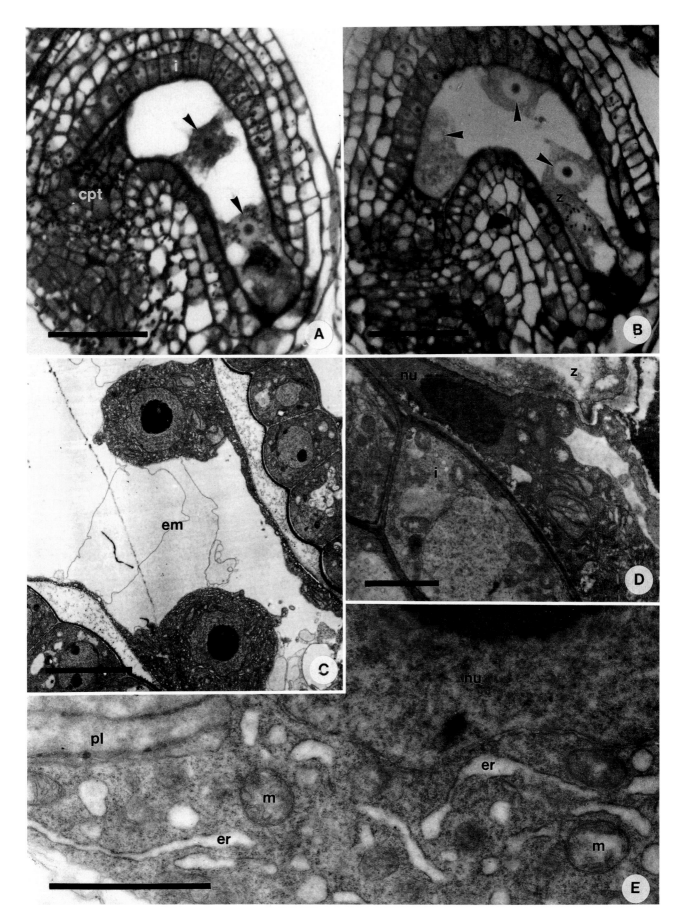

Plate 7.11

Plate 7.12
Development of the free-nuclear endosperm: The micropylar endosperm

A large central vacuole occupies the major portion of the curved embryo sac between the endosperm nuclei, and there is also a thin peripheral layer of endosperm lining the whole embryo sac and connecting each of the nuclei.

(**A**) Quadrant embryo endosperm 24 hours after fertilization (HAF). Several nuclei (*arrowheads*; total number of nuclei is 14) are present in the embryo sac, and endosperm cytoplasm proliferates around the suspensor (su) of the embryo-proper (e) (cpt, chalazal proliferating tissue; i, integument).

(**B**) Enlargement of a single endosperm "cell" situated on a longitudinal wall of the embryo sac. Several small vacuoles (v) are present in the cytoplasm adjacent to the integument (i) cell wall. Plastids (pl) are well-differentiated and mitochondria (m) are very abundant.

(**C**) Late-globular embryo endosperm 60 HAF. By this stage the free-nuclear endosperm attains maximum development with the completion of nuclear divisions prior to cellularization. Endosperm nuclei (*arrowheads*) extend around the whole embryo sac periphery. The embryo proper (e) and suspensor (su) are submerged in endosperm tissue (en).

(**D**) Enlargement of two endosperm "cells" from a late-globular embryo endosperm 60 HAF. Compared to earlier stages of development the "cells" appear irregular in profile. The surface area of endosperm tissue adhering to the embryo sac wall is reduced by means of pulling away of the plasmalemma (i, integument; nu, nucleus).

(**E**) Enlargement of the cytoplasm in **D**. Plastid (pl) shape and internal structure is different compared to that observed in the quadrant embryo endosperm (**B**). Osmiophilic bodies (*arrowheads*) are present in the plastid matrix.

Bar = 0.5 μm; in **E**; 2 μm in **B**; 5 μm in **D**; 20 μm in **A**, **C**.

S.G. Mansfield

Reproduced from Mansfield and Briarty (1990a) with permission from Prof. A.R. Kranz, ed. *Arabidopsis Information Service*.

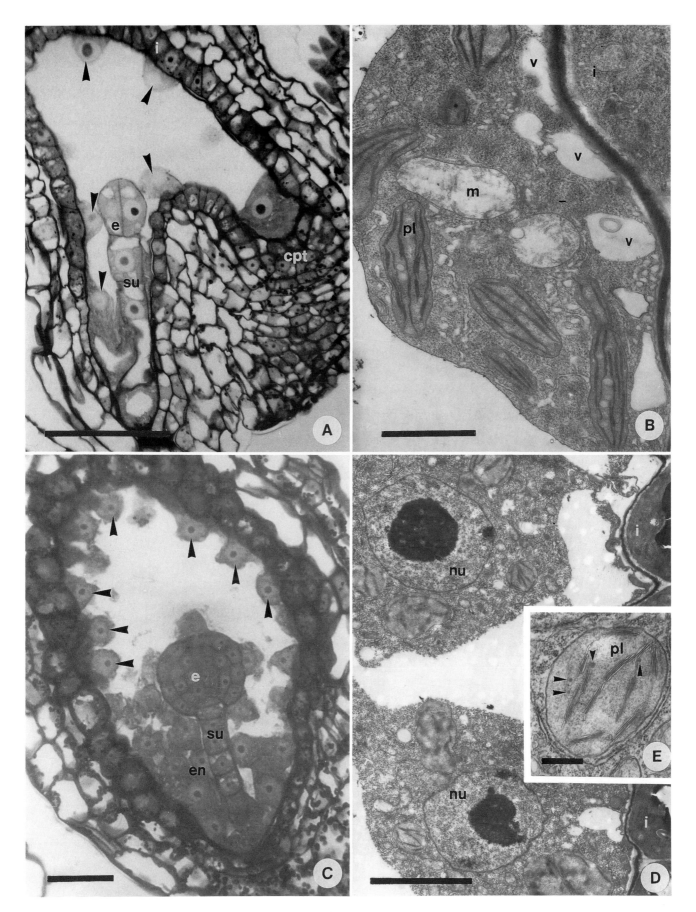

Plate 7.12

The daughter endosperm nucleus that migrates to the chalazal pole of the embryo sac remains undivided for approximately 18–24 hours after fertilization (HAF), although its cytoplasm continues to proliferate, increasing in volume. As the embryo sac reaches the quadrant stage (24 HAF) and the micropylar endosperm consists of peripherally situated nuclei, the chalazal nucleus divides.

(**A**) Binucleate mass of tissue (*arrowhead*) of the chalazal endosperm 24 HAF. Note that the nuclear diameter is considerably greater than the micropylar endosperm nuclei (e, embryo-proper).

(**B**) Wall projections (wp) in the chalazal cytoplasm (mid-globular embryo endosperm 48 HAF) adjacent to the degenerated antipodals and nucellus (not shown). Present by this stage are numerous mitochondria (m), dictyosomes (d) active in vesicle production, an abundance of rough endoplasmic reticulum, and many irregular vacuoles (v) of variable size.

(**C**) Heart embryo endosperm 72 HAF. By this stage the chalazal endosperm consists of between 8 and 12 prominent nuclei. Numerous plastids (pl) accumulate around each nucleus (nu), and the cytoplasm is extremely vacuolate (v).

(**D**) Maturing chalazal endosperm prior to cellularization 96 HAF. Dictyosomes (d) and rough endoplasmic reticulum are very abundant; cisternae are arranged in parallel stacks (v, vacuole).

(**E**) Crystal-like inclusion (cr) observed occasionally in the maturing chalazal endosperm 84 HAF.

Bar = 0.5 μm in **D**; 1 μm in **B**, **E**; 3 μm in **C**; 20 μm in **A**.

S.G. Mansfield

Reproduced from Mansfield and Briarty (1990a) with permission from Prof. A.R. Kranz, ed. *Arabidopsis Information Service*.

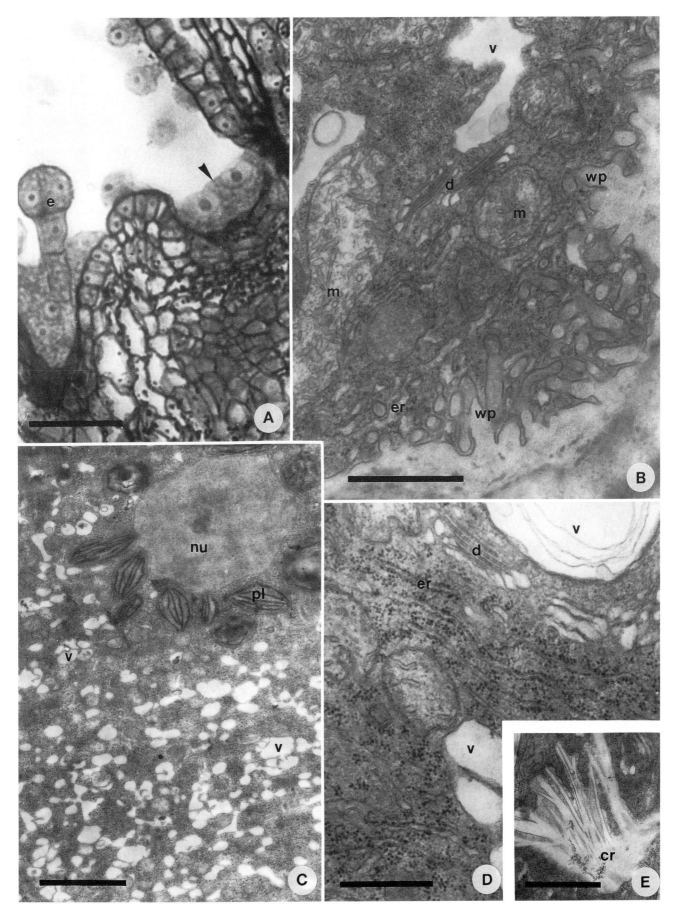

Plate 7.13

Plate 7.14
Pattern of endosperm cellularization in the embryo sac

The endosperm remains in a free-nuclear state for approximately 60 hours after fertilization (HAF). It then begins to cellularize rapidly. This process is almost complete by 84 HAF, although it continues up to 120 HAF. Cell wall formation begins at cotyledon initiation during the late-globular/early-heart embryo stage of embryogenesis (**B**). The walls are initiated at the edges of the micropylar region of the embryo sac, growing centripetally towards the embryo, and extending peripherally in the direction of the chalazal chamber over a period of approximately 24 hours (**C, D, E**). Endosperm in the chalazal region remains free-nuclear at 96 HAF (**E**), but is fully cellularized by 120 HAF (**F**), forming a fused network with the micropylar tissue (e, embryo-proper).

Embryo stage	HAF	Free-nuclear endosperm	
(A) Mid-globular	48	Cellular peripheral endosperm	☐
(B) Late-globular	60	Second cellular endosperm layer	◩
(C) Mid-heart	72	Maturation layers of cellular endosperm	▦
(D) Late-heart	84	Maturation layers of cellular endosperm	▦ ◪
(E) Torpedo	96	Cellular chalazal endosperm	⊠
(F) Upturned-U	120	Degenerating cellular endosperm (chalazal region)	▧

Bar = 50 μm.

S.G. Mansfield

Reproduced from Mansfield and Briarty (1990b) with permission from Prof. A.R. Kranz, ed. *Arabidopsis Information Service*.

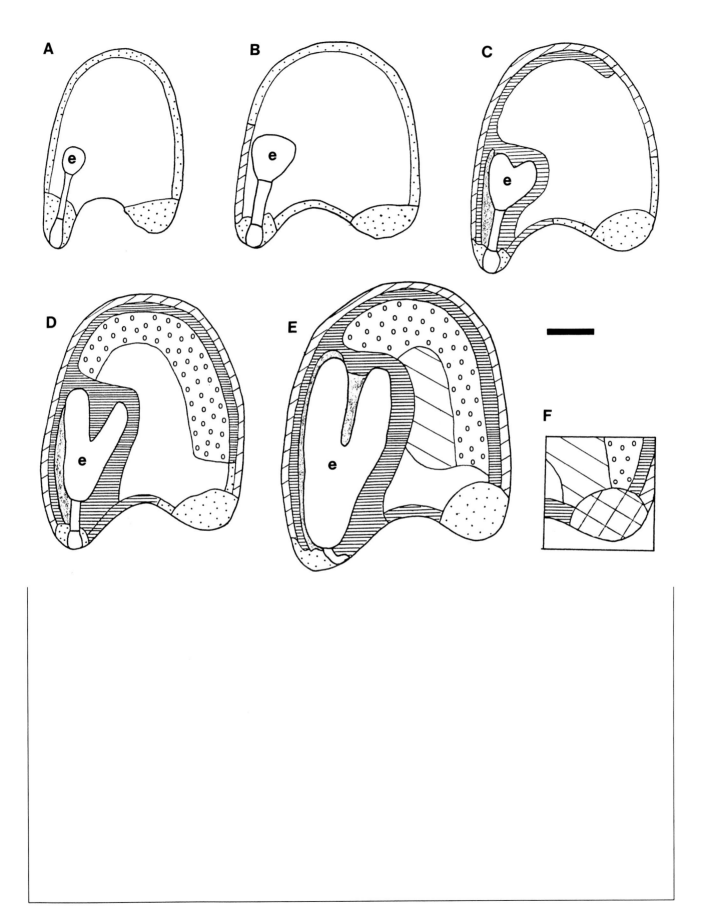

Plate 7.14

Plate 7.15
Mechanism of endosperm cellularization in the micropylar chamber

(**A**) A single ovule 68 hours after fertilization (HAF). Cellularized endosperm (ce) surrounds the embryo-proper (e) and suspensor (su). The leading edges of the forming cell walls are indicated with *arrowheads*. Tissue further towards the chalazal endosperm (chen) is still in a free-nuclear state (i, integument).

(**B**) Cellularizing micropylar endosperm 72 HAF. A network of highly convoluted cell walls (*arrowheads*) containing many plasmodesmata (*arrows*) has developed in the cytoplasm adjacent to the embryo sac wall (v, vacuole).

(**C**) Cellular endosperm 96 HAF. Several layers of highly vacuolate cellular endosperm (ce) have now formed. Note that the chalazal mass of proliferating endosperm (chen) remains in a free-nuclear state even by this stage (ce, cellularized endosperm).

(**D**) and (**E**) Enlargement of the tips of freely growing endosperm cell walls (cw) 72 HAF. The growing tips are usually thickened and sometimes appear bulb-shaped or hooked. Associated with these structures are masses of well ordered vesicles (ve) and microtubules (*arrowheads* in **E**), dictyosomes (d), and mitochondria (m). Further from the tip (towards the embryo sac wall), a normal, continuous wall containing plasmodesmata (*arrows* in **D**) is present. Adjacent wall tips are often observed in close proximity but these are not observed fusing to form a cell. However, cells may be formed by branching of the growing point. In **E** the two arrays of vesicles extending from the growing tip may represent early stages of branch formation.

Bar = 1 μm in **D**, **E**; 2 μm in **B**; 20 μm in **C**; 40 μm in **A**.

S.G. Mansfield

Reproduced from Mansfield and Briarty (1990b) with permission from Prof. A.R. Kranz, ed. *Arabidopsis Information Service*.

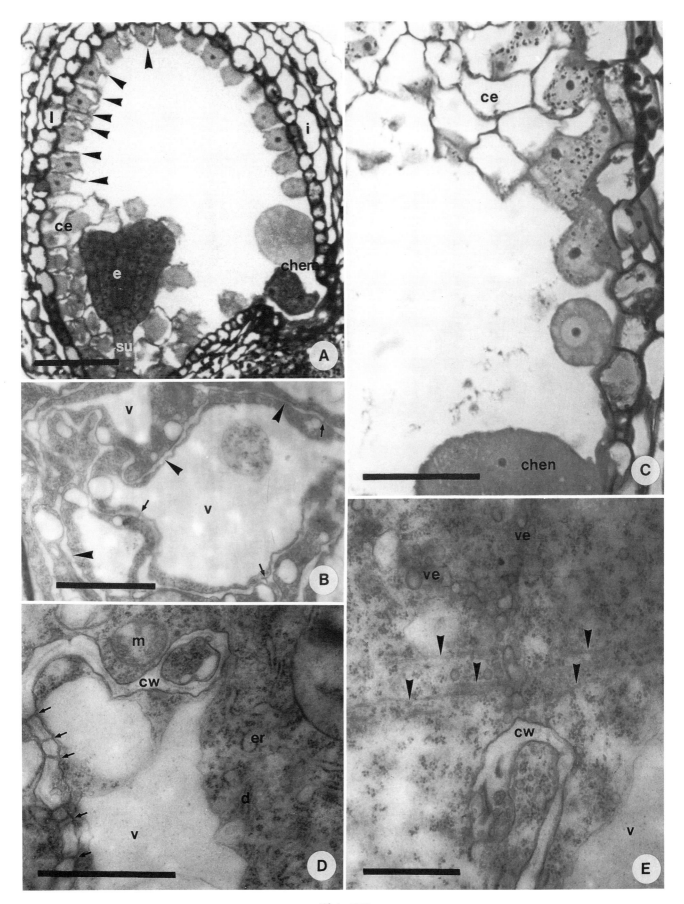

Plate 7.15

Plate 7.16
Endosperm
cellularization in the
chalazal chamber

(**A**) The chalazal endosperm cytoplasm prior to cellularization 108 hours after fertilization (HAF). Masses of endoplasmic reticulum and dictyosomes (d) are ubiquitous. The crystalline inclusions observed earlier in the development of the chalazal endosperm are also present at this stage (not shown).

(**B**) Cellularizing chalazal endosperm 120 HAF. Wall projections (wp) at the chalazal base provide a template for the growth of cell walls. Associated with the wall projections are many dictyosomes (d) and mitochondria (m).

(**C**) Enlargement of the cytoplasm and cell walls (cw) of the chalazal endosperm 120 HAF. Several nodule-like structures (*arrows*) are present on the main walls. Plasmodesmata exist between all cells (*arrowheads*) (m, mitochondrion; v, vacuole).

Bar = 0.5 μm in **A**, **B**; 1 μm in **C**.

S.G. Mansfield

Reproduced from Mansfield and Briarty (1990b) with permission from Prof. A.R. Kranz, ed. *Arabidopsis Information Service*.

Embryogenesis: Endosperm Development

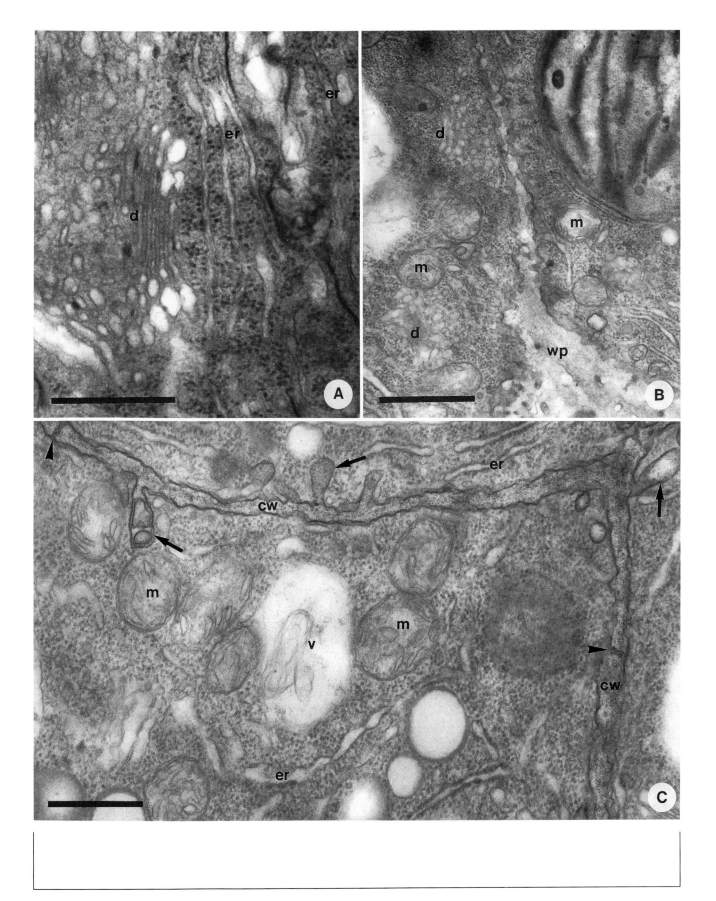

Plate 7.16

Plate 7.17
Mutations affecting seed
morphology

Mutations at the *transparent testa glabra* (*ttg*) locus result in the absence of trichomes as well as the presence of yellow seeds as opposed to wild-type brown. The normal brown color of seeds is due to accumulation of anthocyanins in the testa. The seed color in *ttg* mutants is a result of having a transparent seed coat (testa), with the yellow color attributable to the normal color of the cotyledons of the ripe seed. Mutants at several other loci, *TRANSPARENT TESTA* (*tt1*, *tt2*, *tt3*, *tt4*, and *tt5*), also have seeds that are yellow due to a transparent testa (Koornneef, 1981). These mutants lack detectable anthocyanin in their seeds, although two (*tt1* and *tt2*) have anthocyanin in their leaves. Mutants at five additional loci (*tt6*, *tt7*, *tt8*, *tt9*, and *tt10*) with yellowish seeds have been identified (Koornneef, et al., 1982a; Koornneef, 1990a). As in *ttg* mutants, mutations in at least three other loci (*GLABRA*; *GL1*, *GL2*, and *GL3*) also result in a lack of trichomes, but each of these mutants has normal seed coloring.

Two of these mutants, *ttg* and *gl2*, are also characterized by the absence of mucilage on the surface of the seeds, and by an altered seed coat morphology. The *ttg* mutants also exhibit reduced seed dormancy. Mutations at the *APETALA2* locus also result in an abnormal seed morphology, with both the seed coat morphology and overall shape of *apetala2* seeds altered. In addition, *apetala2* seeds are larger than wild-type seeds. Altered seed coat morphology and color are maternally inherited traits for each of the described mutants. This is summarized in Table 1 (p. 401).

Table 1 is based on data from Koornneef (1990a). Many (except *tt8*, *tt9*, and *tt10*) of these mutants havew been mapped (Koornneef, 1990b).

Several of the genes have been assigned biochemical or regulatory functions in the anthocyanin pathway. The *TT3* gene has been shown to encode dihydroflavonol 4-reductase (DFR) (Shirley et al., 1992), the *TT5* gene encodes chalcone flavanone isomerase (CHI) (Shirley et al., 1992), and the *TT7* gene encodes flavanoid 3'-hydroxylase (Koornneef et al., 1982c). The *TT4* gene likely encodes chalcone synthase (CHS) since *tt4* plants lack detectable flavanoid compounds, and the chalcone synthase gene maps to a position similar to that of the *TT4* gene (Chang et al., 1988; Koornneef, 1990a; Shirley et al., 1992). The *TTG* gene may be the *Arabidopsis* homologue of the *R* locus of maize, since the *R* gene (which encodes a protein with a helix-loop-helix DNA binding and dimerization motif; Ludwig et al., 1989) can complement the *ttg* mutant phenotype in transgenic *Arabidopsis* plants (Lloyd et al., 1992).

Scanning electron micrographs of the seed surface of wild-type and mutant *Arabidopsis* seeds.

(A) Wild-type seed.

(B) Detail of wild-type seed coat showing cell-like patterns with a central elevation. The outer cell layer of the seed coat contains layers of mucilage. Upon imbibition, the outer cell wall of the seed coat collapses, resulting in the excretion of the stored mucilage.

(C) *ttg* seed.

(D) Detail of *ttg* seed where the central elevation seems to be collapsed, presumably due to a failure to form mucilage.

(E) *gl2* seed.

(F) Detail of *gl2* seed showing abnormal morphology of cell-like structures.

Bar = 10 μm in **B**, **D**, **F**, **I**; 100 μm in **A**, **B**, **C**, **G**; 300 μm in **H**.

(*Text continued on p. 401*)

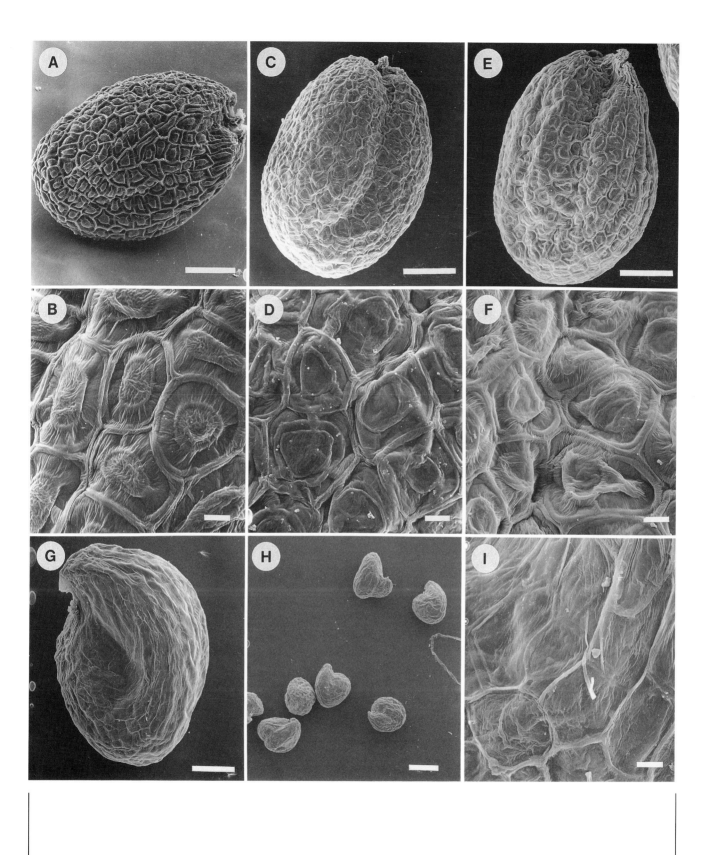

Plate 7.17

399

Plate 7.17
Mutations affecting seed morphology
(continued)

(**G**) *ap2-1* seed showing abnormal seed size and morphology.
(**H**) *ap2-1* seeds exhibiting altered seed shape.
(**I**) Detail of *ap2-1* seed showing lack of cell-like morphology of the seed coat.

Table 1 is based on data from Koornneef (1981) and Koornneef (1990a). Many (except *tt8*, *tt9*, and *tt10*) of these mutants have been mapped (Koornneef, 1990b).

Table 1. Phenotype of *tt* and *gl* mutants compared to wild type.

Genotype	Seed coat pigment	Anthocyanin in leaves	Trichomes	Mucilage
Wild type	+	+	+	+
gl1, gl3	+	+	−	+
gl2	+	+	−	−
ttg	−	−	−	−
tt1, tt2, tt8	−	+	+	+
tt3, tt4, tt5	−	−	+	+
tt6, tt7	+/−	+/−	+	+
tt9, tt10	+/−	+	+	+

J.L. Bowman and M. Koornneef

8
Pathogens

Introduction

Plants live in a world that is at times hostile. As sessile organisms, they have no recourse to environmental insults but to "hunker down" and survive whatever hits they receive. Fascinating adaptive strategies have evolved by which plants perceive and respond to abiotic environmental changes, such as heat, cold, drought, flood, and wind. Plants are in constant contact with fungal spores floating on the wind, bacteria carried by ground water and rain or living in the soil, nematodes foraging through the underground, insects and herbivores scouring the landscape for a meal and an appropriate place to deposit their larvae, and often, hitchhiking viruses. Against this harsh backdrop, it is not surprising that plants have evolved varied and complex mechanisms for recognition and response to pathogens.

The first layer of defense is literally armor. Many potential pathogens simply cannot gain access to those plant tissues necessary for them to begin a successful life cycle. Among the plant's arsenal of armor are thick wax and cuticle layers coating epidermal tissues, complex cell wall structures, and preformed anti-microbial compounds stored as activatable reserves in cell walls and vesicles. If pathogens do find a particular plant environment to their nutritional liking, or have evolved ways to skirt the preformed barriers and begin their life cycle (fungal germination, nematode attachment, bacterial division, viral particle uncoating), then the plant must have other systems to ensure its survival.

Acting as sentries beyond the preformed defense mechanisms are active systems which have evolved to recognize some molecular structure of the pathogen as "non-self," and to then set into action a series of molecular and cellular responses aimed at stopping further pathogen ingress. Finally, it was recognized decades ago that a successful resistance response can often lead to the establishment of "systemic acquired resistance" in tissues distant from the initial point of infection. The systemic immunity generated in this case can be maintained for periods of up to several months, and is characterized by resistance to not only the original challenge pathogen, but to a broad range of subsequently infecting pathogens.

Active defense mechanisms have received an immense amount of scientific attention, both because of the inherent interest in understanding how an organism distinguishes friend from foe, and because the extent plant and crop loss due to pathogen predation is agronomically important. We are, however, still largely ignorant about the signal perception and transduction mechanisms that lead to either the locally or systemically resistant phenotype. We also have very little firm knowledge of how recognition of pathogens has evolved in plants, and how disease resistance is deployed in natural plant communities.

Plant breeders recognized, fast on the heels of the rediscovery of Mendel's rules of inheritance, that plant resistance to particular isolates of many pathogens can be inherited as a dominant or co-dominant trait at a single locus. Pioneering work by H. H. Flor in North Dakota in the 1940s and 1950s formalized these observations into a genetic model known subsequently as the "gene for gene" hypothesis. In its modern essence, this model states that resistance determined by single genes is mediated via perception of either the direct or indirect product of a corresponding dominant or co-dominant avirulence gene function in the pathogen. This model explains observations for essentially all classes of plant–pathogen interactions where genetic analysis of both partners can be performed.

Moreover, plant breeders manipulate hundreds of "resistance specificities" (resistance of a particular plant genotype against a particular pathogen isolate) in the major crop species. From this wealth of material, some fascinating genetic paradigms have emerged. First, plant resistance gene loci can exhibit significant allelic diversity. Second, there can be multiple specificities encoded at one locus in what is thought to be a multigenic (or at least a multispecific) cluster. Third, some complex resistance loci are capable of specifying resistance to more than one species of pathogen. Fourth, some resistance alleles are inherently unstable during meiosis, possibly suggesting recombinational mechanisms for generation of cognitive diversity. And fifth, analysis of highly inbred "wild species" has shown that plant populations can carry broad assemblages of resistance specificities, even in very localized communities. This point again begs the question of how cognitive diversity is generated and maintained.

What does this all have to do with *Arabidopsis*? In the last five years, several of the major paradigms described above have been demonstrated using *Arabidopsis* in combination with either natural pathogen isolates or pathogens isolated from related crucifer species. The utility of *Arabidopsis* as a genetically tractable model has attracted the attention of researchers interested in all major pathogen types. Already, several loci conditioning resistance to defined avirulence genes from phytopathogenic Pseudomonads have been described. As well, several phytopathogenic fungi infect *Arabidopsis* naturally, and these systems are being genetically analyzed. They appear to be as complex as those in crop plants. Agronomically important nematode species also colonize *Arabidopsis* roots, and viral pathogens abound. Lastly, systemic acquired resistance is inducible in *Arabidopsis*, and its course appears identical to that described in classic species such as tobacco and cucumber.

We can therefore expect that genetic definition and molecular isolation of signal cognition, signal transduction, and resistance "effector function" genes from *Arabidopsis* will have a major impact on molecular understanding of disease resistance. We also can expect that identification of such loci, and the genes they encode, will enhance our understanding of normal development. This will come as we learn to appreciate the developmental and cell-type specific expression of resistance against particular pathogens. It will also derive from a molecular characterization of resistance genes. Are they a "dedicated" system, or do they carry out primary functions related to normal growth and development usurped in certain circumstances by a secondary pathogen recognition function?

The same question can be addressed with respect to signal transduction and effector function genes. In this case, we can imagine that "normal" biochemical functions will indeed have been recruited to play key roles in halting pathogen growth. This is already supported by volumes of literature in other species, but requires clear genetic support not provided to date. Finally, and potentially of the most interest, is the fact that *Arabidopsis* has evolved outside the long-armed reach of agronomic intervention. Its interactions with pathogens is also essentially unaffected by human cultivation habits. Thus, we should have ample opportunity to understand the molecular basis of resistance as it has evolved and dispersed within *Arabidopsis* and its wild and cultivated relatives, and how this evolution impinges on the pathogen communities colonizing them. Application of this knowledge to biorational control of pathogens and designer resistance is not a far-fetched goal of researchers in this area.

In the following sections, colonization of *Arabidopsis*, and typical resistance responses are pictured. This is only a sampling of this very intriguing and exciting aspect of *Arabidopsis* research. Many reviews cover plant–pathogen interaction in general, and two concentrate on the *Arabidopsis* systems (Dangl, 1993a; Dangl, 1993b).

J. Dangl

Plate 8.1

Arabidopsis **interactions with phytopathogenic** *Pseudomonas*: **The hypersensitive response**

Plant–bacterial interactions of the "gene-for-gene" sort are classified into two types. During incompatible interactions, a particular plant genotype is resistant to a particular pathogen isolate, which in this instance is avirulent. During compatible interactions, a particular plant genotype is susceptible to infection by a virulent pathogen isolate. The accompanying plates illustrate these interactions. Recent articles on this and related systems using *Arabidopsis* include Debener et al. (1991), Dong et al. (1991), Whalen et al. (1991), and Dangl et al. (1992).

(**A**) Incompatible interaction. A resistant *Arabidopsis* ecotype was infiltrated with a relatively low density *Pseudomonas syringae* isolate. Three days post-inoculation, no bacteria are visible, but many cells in the area of inoculation exhibit collapsed cell membranes (*arrow*). This is indicative of a typical hypersensitive response. Cells just outside the inoculated zone (*arrowhead*) show no sign of membrane collapse.

(**B**) Compatible interaction after inoculation of the same bacterial strain onto leaves of a susceptible *Arabidopsis* ecotype. In this case, after 3 days, many bacteria (*arrow*) are visible in intercellular spaces, and cell membranes remain intact (*arrowhead*). Note that **B** is enlarged relative to **A**.

Bar = 20 μm.

J. Dangl, H. Liedgens, and T. Debener

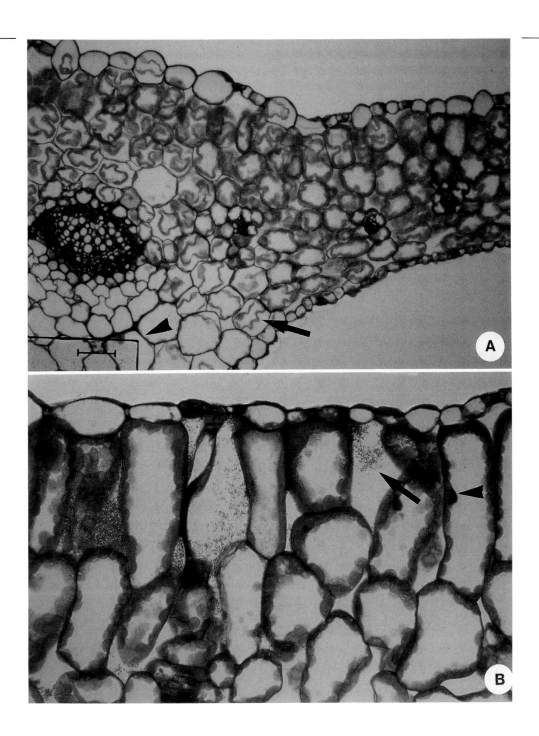

Plate 8.1

409

Plate 8.2

Arabidopsis **interactions with phytopathogenic** *Pseudomonas*: **The hypersensitive response**

This plate is analogous to Plate 8.1, except that a 1,000-fold higher initial bacterial density was inoculated.

(**A**) Incompatible interaction. The high bacterial density leads to massive and rapid local cell collapse (hypersensitive response) after only 12–16 hours (*arrow*). Note that there is no spread of collapsing tissue beyond the inoculation site, and that no bacteria are apparent.

(**B**) Compatible interaction. Under these inoculation conditions during a compatible interaction, many bacteria are visible in intercellular spaces (*arrow*) but very little tissue damage is apparent.

Bar = 20 µm.

J. Dangl, H. Liedgens, and T. Debener

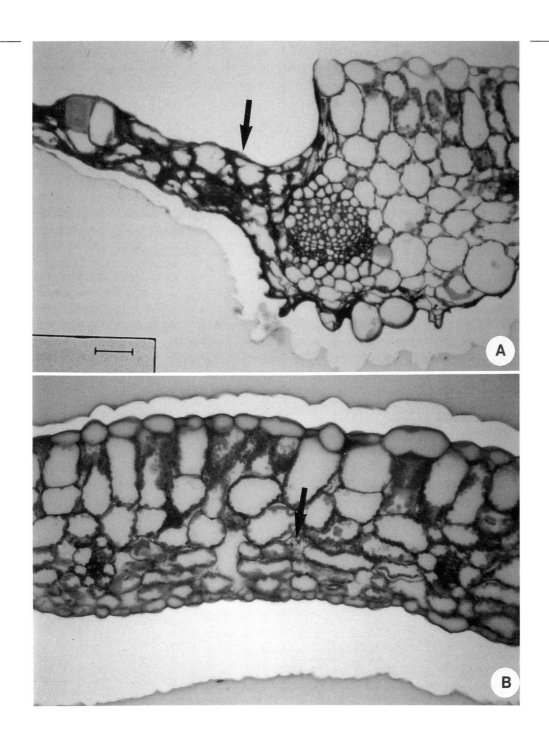

Plate 8.2

411

Plate 8.3
Arabidopsis **interactions**
with phytopathogenic
Pseudomonas:
Compatible interactions

The timing and severity of disease can vary between different compatible interactions. Electron micrographs show infection of the same *Arabidopsis* ecotype (Niederzenz) by two different virulent bacterial strains.

(**A**) This particular bacterial strain grows throughout the intercellular spaces, but causes little damage 3 days postinoculation. This is the same interaction as illustrated in **B** of Plate 8.1.

(**B**) In contrast, a more aggressive pathogenic strain causes greater plant cell damage at the same point, corresponding to more severe symptoms with the bacteria invading the broken cells.

Bar = 10 μm.

J. Dangl, H. Liedgens, and T. Debener

Reproduced from Debener et al. (1991) with permission from Blackwell Scientific Publications.

Pathogens: *Pseudomonas*

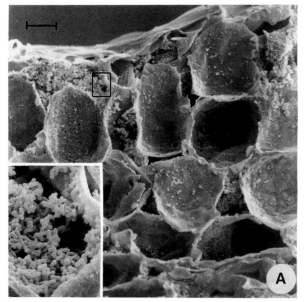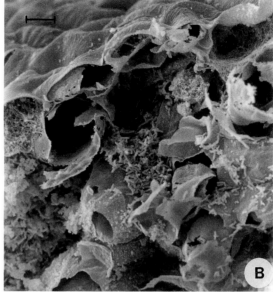

Plate 8.3

Plate 8.4
Downy mildew of
Arabidopsis thaliana

Downy mildew, caused by the Oomycete *Peronospora parasitica*, occurs on cultivated brassicas and other crucifers throughout the world. Downy mildew of *Arabidopsis* was reported early this century (Gäumann, 1918), but the first comprehensive description of this host–pathogen interaction was published more than 70 years later (Koch and Slusarenko, 1990). The fungus is an obligate biotroph and during the initial stages of the infection very little apparent damage can be seen in the host. However, in heavy infection, substantial necrosis can occur after sporulation, and in cool, wet years disease outbreaks can be serious. Very often, however, climatic factors limit the development and spread of the disease so that losses among cultivated brassicas remain limited.

(**A**) Infection of the aerial parts of the plant begins from wind-dispersed or water-carried conidia. A germinated conidium (c) with an appressorium (a) can be seen on this leaf surface. The appressoria form over or close to the periclinal junction between two adjacent epidermal cell walls (1,450×).

(**B**) A penetration hypha grows down between the epidermal cells, and the fungal cytoplasm is translocated into the fungal structures inside the plant. Only the empty spore remains on the surface. As shown here, the fungus grows intercellularly through the leaf. Branched, coenocytic intercellular hyphae (ih) expand in shape to fit the intercellular spaces between the mesophyll cells. This gives a characteristically irregular "string of sausages" appearance to the mycelium. The hyphae put out feeding structures called haustoria (h) into the plant cells. Although the haustoria form within the cell wall boundary, the host cell's plasmalemma is invaginated but not penetrated by the fungus (220×).

(**C**) After a period of intercellular growth lasting from 3–12 days, depending upon the particular combination of *Arabidopsis* host and fungal isolate, sexual and asexual spores are formed. An oogonium (o) in the mesophyll is shown being fertilized by an antheridium (an) through a fertilization tube (ft). The antheridium is in a paragynous attitude relative to the oogonium, which is typical for *Peronospora parasitica* (1,450×).

(**D**) An oogonium (o) and antheridium (an) adjacent to a mature oospore (osp) which exhibits very clearly the typical layered structure of the mature zygote. The oospore is separated by the oospore wall (osw) from the periplasm (pe) which in turn is bounded by the oogonial wall (ow) (1,450×).

(**E**) A conidiophore initial (ci) beginning to grow out of an open stoma from the substomatal cavity (2,030×).

(**F**) After emerging from the stoma, the conidiophore may branch initially, giving rise later to several conidiophores.

B. Mauch-Mani and A.J. Slusarenko

Technical assistance from Urs Jauch is gratefully acknowledged. **A** reproduced from Koch and Slusarenko (1990) with permission from the American Society of Plant Physiologists.

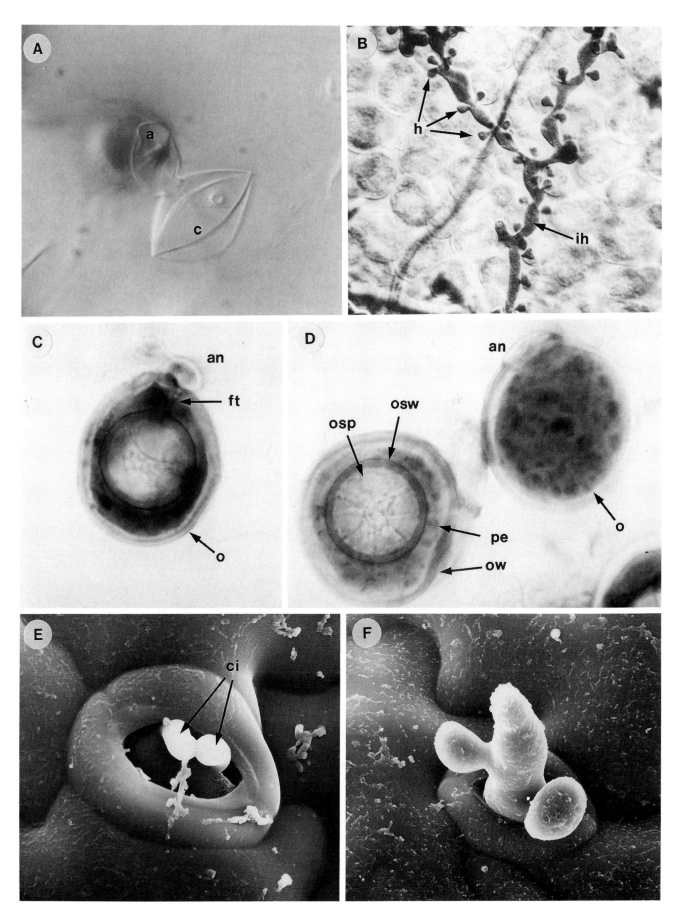

Plate 8.4

Plate 8.5
Downy mildew of
Arabidopsis thaliana

(**A**) and (**B**) Clusters of conidiophore initials and young conidiophores growing out of stomata on the adaxial epidermis of an *Arabidopsis* leaf. Such profuse branching is encouraged by very high relative humidity. **A**, 388×; **B**, 545×.

(**C**) Elongation of the conidiophore initials out of the stomatal pore continues until they begin to branch dichotomously and cut off single conidia at the tips (1,555×).

(**D**) and (**E**) Mature conidiophores bearing numerous conidia. **D**, 280×; **E**, 233×.

(**F**) The distal part of a dichotomously-branching conidiophore is shown bearing single, spherical conidia at the tip of each branch. The conidia detach very easily from the tips of the conidiophore, especially when the humidity drops and the conidiophore twists as it dries out (2,030×).

(**G**) Profuse asexual sporulation has occurred, and aerial conidiophores (*arrowed*) can be seen on the leaf laminae and petioles of *Arabidopsis thaliana* infected with *Peronospora parasitica*. Note that the conidiophores are approximately the same size as the trichomes, and when sporulation is light, it is easy to overlook, especially in the field.

B. Mauch-Mani and A.J. Slusarenko

Technical assistance from Urs Jauch is gratefully acknowledged. **G** reproduced from Koch and Slusarenko (1990) with permission from the American Society of Plant Physiologists.

Pathogens: *Peronspora*

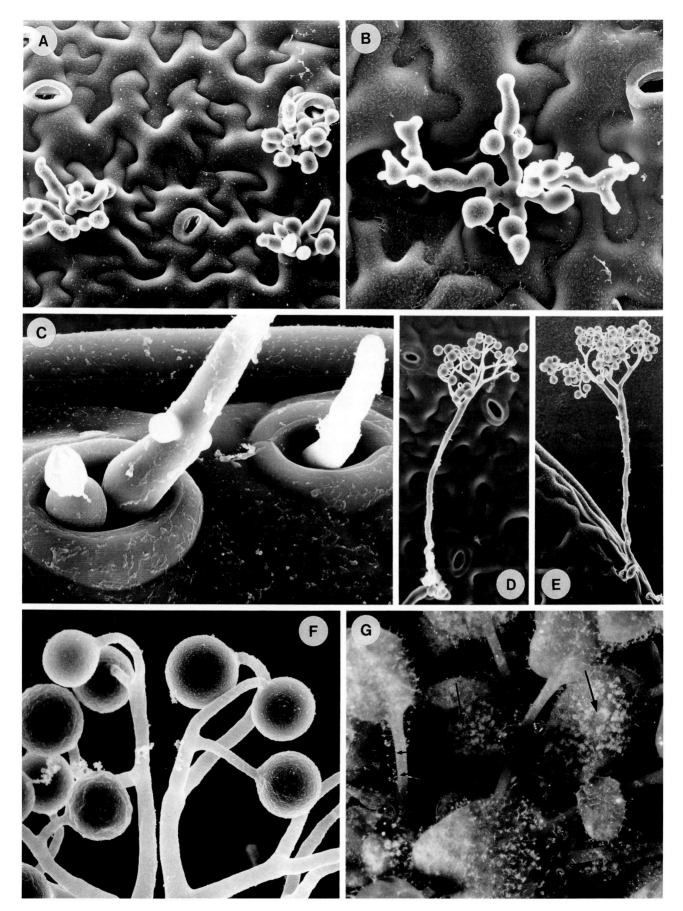

Plate 8.5

Plate 8.6
Sedentary endoparasitic nematodes in *Arabidopsis thaliana*: *Meloidogyne incognita*

Several nematode species are pathogens of *Arabidopsis* (Sijmons et al., 1991). The most intriguing ones are sedentary endoparasites which invade the roots and induce permanent feeding sites. Economically, root-knot nematodes (*Meloidogyne* spp.) and cyst nematodes (*Heterodera* spp. and *Globodera* spp.) are the most important. Their feeding sites consist of cells within the vascular cylinder, which become completely reorganized in a specific way. These hypertrophied and metabolically highly active transfer cells (Jones, 1981) serve as food sources throughout the life cycle of the nematodes.

In *Arabidopsis*, the thin and translucent roots allow microscopic *in vivo* observations on the behavior of the nematodes and on plant responses to their activity. Complementing these observations with histological investigations, the processes of interaction between sedentary nematodes and *Arabidopsis* can be examined in detail.

Wyss (1992) described the parasitic behavior of *Meloidogyne incognita* from root penetration until the induction of the giant cells.

(**A–D**) The infective juveniles (J2) penetrate the roots in the elongation zone behind the root tip. They either perforate epidermal cells (**A** and **B**) or enter the space between two cells (**C** and **D**) thus reaching the cortical layer in which they immediately orient themselves towards the root tip (s, stylet).

(**E**) After a period of migration within the intercellular space which is widened by stylet (s) thrusts and, probably, enzymes, they finally reach the meristem of the root tip.

(**F**) Once at the meristem, they turn around and enter the developing vascular cylinder (s, stylet).

(**G**) Along the stele, the nematodes move toward the root base. Forward movement comes to an end when the juveniles enter the differentiation zone. Although, in this stage juveniles still show the same behavioral pattern as during intercellular migration, thrustings of the stylet(s) alternate with short phases of median bulb activity. This indicates withdrawal of material from a number of cells around the anterior end of juveniles. During stylet insertion into cells, secretions from gland cells are probably injected (x, xylem).

(**H**) Within a few hours, the cells respond with synchronous nuclear divisions without cytokinesis, becoming transformed into multi-nuclear "plasmodial" giant cells (n, *arrows* indicate nuclei). In regular turns, the juveniles insert the stylet (s) into the cells and withdraw food with the aid of their pumping organ, the median bulb. As a response to feeding, the cells become greatly enlarged and thus expand the stelar region. As the cells become filled with cytoplasm, the central vacuole disintegrates into smaller units (x, xylem).

At later stages, the giant cells form extensive cell wall ingrowths, primarily around xylem vessels. The juveniles slowly become saccate, molt three times in succession, and turn into globular-shaped females. Nematode development is accompanied by proliferation of surrounding root tissue, which usually embeds the entire female. At maturity, the females release a gelatinous material (egg sac) into which eggs are laid. This egg sac protrudes from the gall.

Bars = 10 μm.

F.M.W. Grundler, U. Wyss, and W. Golinowski

H reproduced from Sijmons et al. (1991) with permission from Blackwell Scientific Publications.

Pathogens: Nematodes

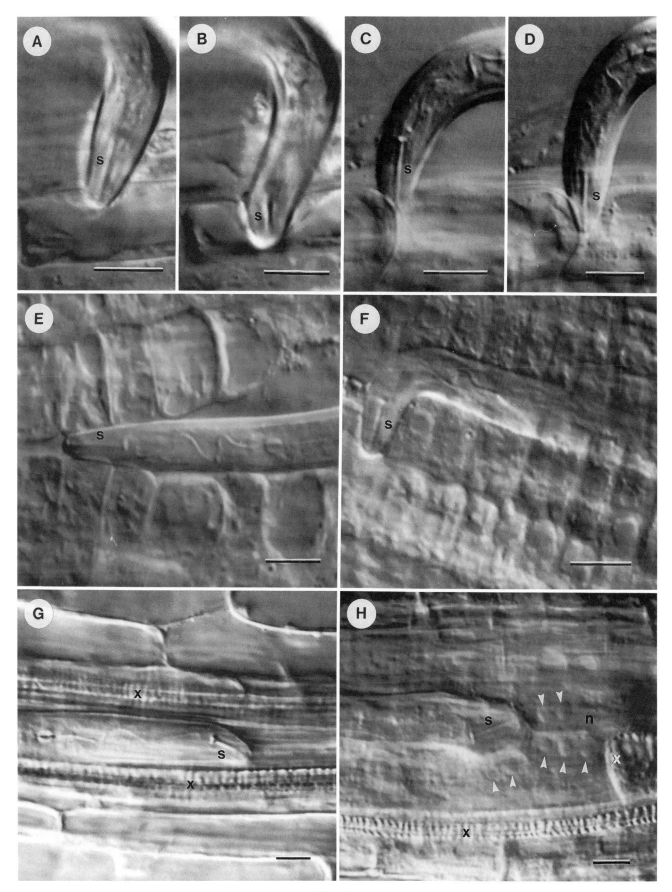

Plate 8.6

Plate 8.7
Sedentary endoparasitic
nematodes in
Arabidopsis thaliana:
Heterodera schachtii
(Part 1)

The behavior of *H. schachtii* and its development inside roots of *Arabidopsis* and other cruciferous plants has been described and analyzed by Wyss (1992). After penetration into the epidermis, the infective juveniles (J2) migrate intracellularly more or less directly towards the vascular cylinder. During migration, cell walls are perforated with coordinated stylet thrustings. Thus, a path of necrotic cells through the root tissue is formed.

(A) Within the central cylinder, a single cell of the stelar parenchymatic tissue is selected as the initial feeding site. This cell is carefully pierced and the stylet (s) stays protruded for several hours within the cell. During this period, the nematodes do not feed, and distinct cell responses are not discernable.

Feeding behavior consists of three distinct phases, during which: (1) the stylet is inserted for food withdrawal, then (2) is redrawn, and (3) is finally inserted again for the release of secretions. Under the influence of nematode secretions, which are injected into the cell, dramatic physiological and cytological alterations are triggered. The selected cell undergoes a phase of redifferentiation. First indications of this process are the acceleration of cytoplasmic streaming and the enlargement of the nucleus. Further redifferentiation occurs by fragmentation of the central vacuole, giving rise to many small vacuoles dispersed within the cytoplasm.

(B) and (C) At the same time, parts of the cell walls of adjacent parenchymatic cells successively dissolve so that the protoplasts of the cells form a syncytium (sy) (s, stylet).

(D) The cell walls of the syncytium (sy) undergo striking structural changes, especially around xylem vessels, where a network of ingrowths (*arrows*) is formed (v, vacuole; se, sieve element).

(E) The syncytium (sy) permanently expands along the central cylinder during the development of the nematode and, in the case of females (f), can be as long as 10 mm. As a response to the expansion and hypertropism of the syncytium, numerous cells are produced within the vascular cylinder that ensheath the hypertrophying syncytial cells (Golinowski and Grundler, 1992).

(F) In contrast to root-knot nematodes, female cyst nematodes feed during all parasitic stages. Only during molting is feeding interrupted. During the third stage, juveniles grow to such an extent that they emerge from the roots with only the head remaining at the feeding site inside the root. At maturity the females (f) have a lemon shape and are about 0.5–1 mm in length. At the posterior end, gelatinous material is released, which contains sex pheromones. Male juveniles finish feeding when they start molting to the fourth stage. After one additional molt, the vermiform adult males emerge and migrate towards a female, being attracted by its pheromones. After mating, the female produces up to 500 eggs, which remain within the female body. After death, the female cuticle turns brown and becomes hardened. Most of the juveniles hatch when they are stimulated by exudates of host plant roots.

Bar = 2 μm in **D**; 10 μm in **A**, **B**; 30 μm in **C**; 300 μm in **E**; 1,000 μm in **F**.

F.M.W. Grundler, U. Wyss, and W. Golinowski

B reproduced from Sijmons et al. (1991) with permission from Blackwell Scientific Publications.

Pathogens: Nematodes

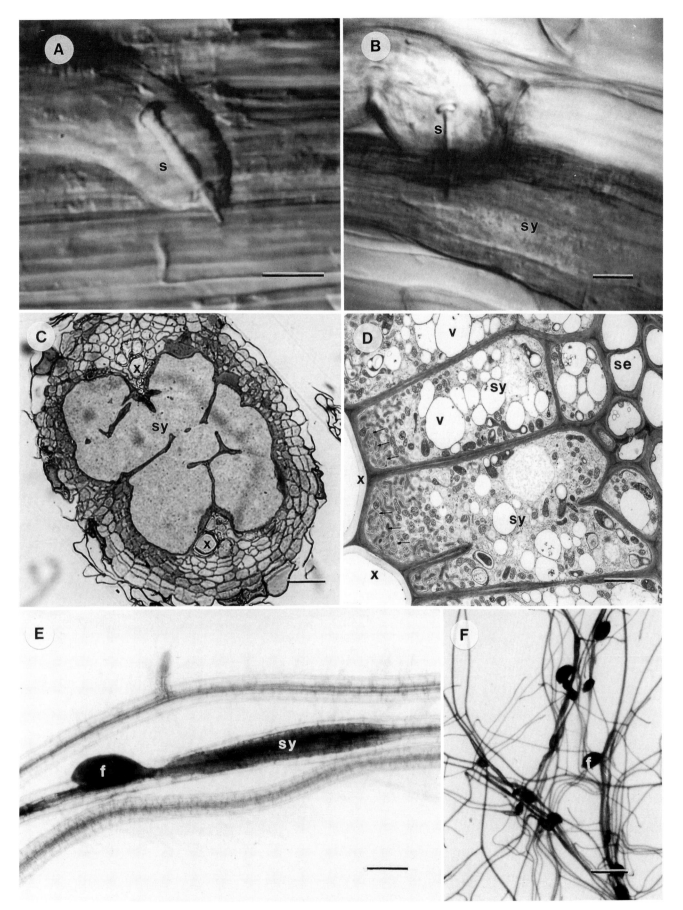

Plate 8.7

Plate 8.8
Sedentary endoparasitic nematodes in *Arabidopsis thaliana*: *Heterodera schachtii* (Part 2)

Under certain circumstances, *H. schachtii* J2 juveniles are able to invade the hypocotyl and leaves of *Arabidopsis*. In the hypocotyl, the syncytium is induced in the pith parenchyma tissue. From this central location the syncytium expands towards the procambial tissue of the stem during the rosette phase of vegetative growth.

(**A**) Under healthy conditions this procambial tissue produces thick-walled sclerenchymatic parenchyma (sp) cells (pa, paranchyma; pr, procambium; se, sieve element).

(**B**) Under nematode influence, the procambial cells differentiate into a syncytium (sy), while the abutting phloem tissue proliferates and differentiates a high number of sieve elements (se). *Arrows* indicate network of ingrowths from the cell walls of the syncytia (pa, parenchyma; sp, sclerenchymatic parenchyma; v, vacuole; p, plastid; m, mitochondrion).

F.M.W. Grundler, U. Wyss, and W. Golinowski

The authors thank Maria Gagyi and Miroslaw Sobczak for their technical assistance.

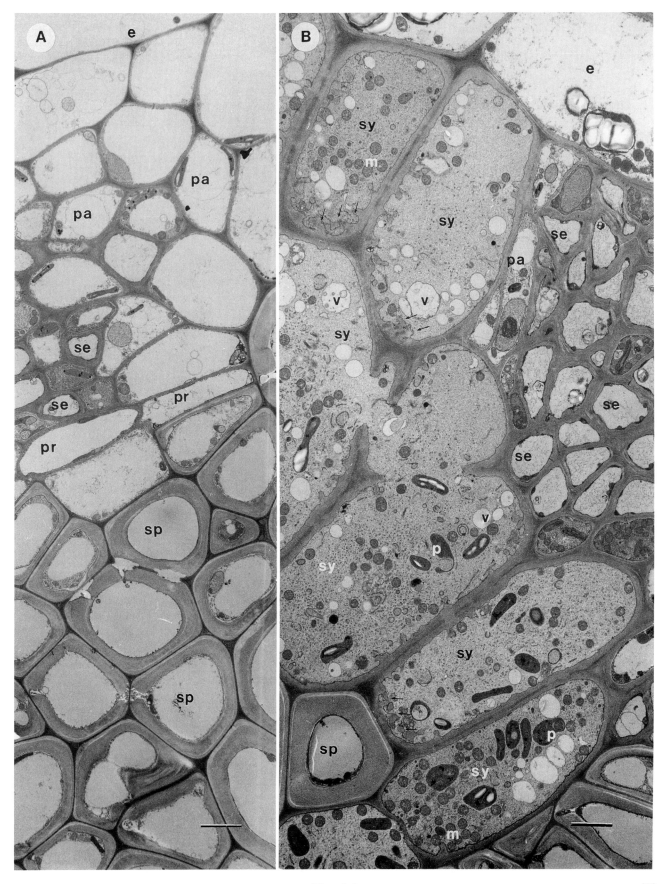

Plate 8.8

423

References

Abramov, V.I., Fedorenko, O.M., Shevchenko, V.A. Genetic consequences of radioactive contamination of *Arabidopsis* populations. Arab. Inf. Serv. 27: 19–20; 1990.

Albertsen, M.C., Palmer, R.G. A comparative light- and electron-microscopic study of microsporogenesis in male sterile (ms₁) and male fertile soybeans (*Glycine max* (L.) Merr.). Am. J. Bot. 66: 253–265; 1979.

Albertsen, M.C., Phillips, R.L. Developmental cytology of 13 genetic male sterile loci in maize. Can. J. Genet. Cytol. 23: 195–208; 1981.

Alexander, I. Entwicklungsstudien an Blüten von Cruciferen und Paperveraceen. Planta 40: 125–144; 1952.

Alvarez, J., Guli, C.A., Yu, X.-H., Smyth, D.R. *terminal flower*: a gene affecting inflorescence development in *Arabidopsis thaliana*. Plant J. 2: 103–116; 1992.

Araki, T., Komeda, Y. Analysis of the role of the late-flowering locus, *GI*, in the flowering of *Arabidopsis thaliana*. Plant J. 3: 231–239; 1993.

Arber, A. Studies in floral morphology. I. On some structural features of the cruciferous flower. New Phytol. 30, 11–46; 1931a.

Arber, A. Studies in floral morphology. II. On some normal and abnormal crucifers: with a discussion on teratology and atavism. New Phytol. 30: 172–203; 1931b.

Arondel, V., Lemieux, B., Hwang, I., Gibson, S., Goodman, H.M., Somerville, C.R. Map-based cloning of a gene controlling omega-3 fatty acid desaturation in *Arabidopsis*. Science 258: 1353–1355; 1992.

Bagnall, D.J. Control of flowering in *Arabidopsis thaliana* by light, vernalisation, and gibberellins. Aust. J. Plant Physiol. 19: 401–409; 1992.

Bagnall, D.J. Light quality and vernalization interact in controlling late flowering in *Arabidopsis thaliana* eco-types and mutants. Ann. Bot. 71: 75–83; 1993.

Barendse, G.W.M., Kepczynski, J., Karssen, C.M., Koornneef, M. The role of endogenous gibberellins during fruit and seed development: studies on gibberellin deficient genotypes of *Arabidopsis thaliana*. Physiol. Plant. 67: 315–319; 1986.

Barton, M.K., Poethig, R.S. Formation of the shoot apical meristem in *Arabidopsis thaliana*: an analysis of development in the wild type and in the *shoot meristemless* mutant. Development 119: 823–831; 1993.

Baskin, J.M., Baskin, C.C. Ecological life cycle and physiological ecology of seed germination of *Arabidopsis thaliana*. Can. J. Bot. 50: 353–360; 1972.

Baskin, T.I., Betzner, A.S., Hoggart, R., Cork, A., Williamson, R.E. Root morphology mutants in *Arabidopsis thaliana*. Aust. J. Plant Physiol. 19: 427–438; 1992a.

Baskin, T.I., Busby, C.H., Fowke, L.C., Sammut, M., Gubler, F. Improvements in immunostaining samples embedded in methacrylate: localization of microtubules and other antigens throughout developing organs in plants of diverse taxa. Planta 187: 405–413; 1992b.

Bateson, W. Materials for the study of variation. Macmillan, London; 1894.

Bauer, W.D. Infection of legumes by Rhizobia. Annu. Rev. Plant Physiol. 32: 407–449; 1981.

Baus, A.D., Franzmann, L., Meinke, D.W. Growth in vitro of arrested embryos from lethal mutants of *Arabidopsis thaliana*. Theor. Appl. Genet. 72: 577–586; 1986.

Beadle, G.W. Genes in maize for pollen sterility. Genetics 17: 413–431; 1932.

Bell, A.D. Plant form: an illustrated guide to flowering plant morphology. Oxford University Press; 1991.

Bell, C.J., Maher, E.P. Mutants of *Arabidopsis thaliana* with abnormal gravitropic responses. Mol. Gen. Genet. 220: 289–293; 1990.

Berger, B. The taxonomic confusion within *Arabidopsis* and allied genera. In: G. Röbbelen, ed. *Arabidopsis* Research, Report of an International Symposium, Arab. Inf. Serv, Göttingen; 1965, pp. 19–25.

Berleth, T., Jürgens, G. The role of the *monopteros* gene in organizing the basal body region of the *Arabidopsis* embryo. Development 118: 575–587; 1993.

Bernier, G. Evolution of the apical meristem of *Sinapis alba* L. (long day plant) in long days, in short days, and during the transfer from short days to long days. Caryologia 15: 303–325; 1962.

Bernier, G. Etude histophysiologique et histochimique de l'evolution du méristeme apical de *Sinapis alba* L. cultivé en milieu conditionné et en diverses durées de jour favorable ou défavorable a la mise a fleurs. Mem. Acad. R. Belg. Cl. Sci. 16: 1–150; 1964.

Bernier, G., Bronchart, R., Kinet, J.-M. Nucleic acid synthesis and mitoic activity in the apical meristem of *Sinapis alba* during floral induction. In: G. Bernier, ed. Cellular and Molecular Aspects of Floral Induction. Longman, London; 1970, pp. 51–79.

Besnard-Wibaut, C. Modification des syntheses d'acides nucléiques dans l'apex de l'*Arabidopsis thaliana* lors du passage a l'état reproducteur. In: G. Bernier, ed. Cellular and Molecular Aspects of Floral Induction. Longman, London; 1970, pp. 37–50.

Besnard-Wibaut, C. Histoautoradiographic analysis of the thermoinductive processes in the shoot apex of *Arabidopsis thaliana* L. Heynh, vernalized at different stages of development. Plant Cell Physiol. 18: 949–962; 1977.

Blackmore, S., McConchie, C.A., Knox, R.B. Phylogenetic analysis of the male ontogentic program in aquatic and terrestrial monocotyledons. Cladistics 3: 333–347; 1987.

Bleecker, A.B., Estelle, M.A., Somerville, C., Kende, H. Insensitivity to ethylene conferred by a dominant mutation in *Arabidopsis thaliana*. Science 241: 1086–1089; 1988.

Bowman, J.L., Alvarez, J., Weigel, D., Meyerowitz, E.M., Smyth, D.R. Control of flower development in *Arabidopsis thaliana* by *APETALA1* and interacting genes. Development 119: 721–743; 1993.

Bowman, J.L., Drews, G.N., Meyerowitz, E.M. Expression of the *Arabidopsis* floral homeotic gene *AGAMOUS* is restricted to specific cell types late in flower development. Plant Cell 3: 749–758; 1991a.

Bowman, J.L., Meyerowitz, E.M. Genetic control of pattern formation during flower development in *Arabidopsis*. In: G.I. Jenkins and W. Schuch, eds. Molecular Biology of Plant Development, Symposium of the Society of Experimental Biology, XLV. The Company of Biologists Ltd., Cambridge; 1991, pp. 89–127.

Bowman, J.L., Sakai, H., Jack, T., Weigel, D., Mayer, U., Meyerowitz, E.M. *SUPERMAN*, a regulator of floral homeotic genes in *Arabidopsis*.

Development 114: 599–616; 1992.

Bowman, J.L., Smyth, D.R., Meyerowitz, E.M. Genes directing flower development in *Arabidopsis*. Plant Cell 1: 37–52; 1989.

Bowman, J.L., Smyth, D.R., Meyerowitz, E.M. Genetic interactions among floral homeotic genes in *Arabidopsis*. Development 112: 1–20; 1991b.

Bowman, J.L., Yanofsky, M.F., Meyerowitz, E.M. *Arabidopsis thaliana*: A review. Oxford Surv. Plant Molec. Cell Biol. 5: 57–87; 1988.

Braaksma, F.J., Feenstra, W.J. Isolation and characterization of nitrate-deficient mutants of *Arabidopsis thaliana*. Theor. Appl. Genet. 64: 83–90; 1982.

Braaksma, O.-F.J., Feenstra, W.J. Isolation and characterization of chlorate-resistant mutants of *Arabidopsis thaliana*. Mut. Res. 19: 175–185; 1973.

Braun, A. Ueber ein gefülltes und durchwachsenes Exemplar von *Arabis thaliana*. Sitzungs. Gesell. Naturforsch. Freunde z. Berlin 75; 1873.

Brink, R.A., Cooper, D.C. The endosperm in seed development. Bot. Rev. 13: 423–541; 1947.

Brown, J.A.M., Klein, W.H. Photomorphogenesis in *Arabidopsis thaliana* (L.) Heynh. Plant Physiol. 47: 393–399; 1971.

Brown, J.A.M., Miksche, J.P., Smith, H.H. An analysis of ^{3}H-thymidine distribution throughout the vegetative meristem of *Arabidopsis thaliana* (L.) Heynh. Radiat. Bot. 4: 107–113; 1964.

Browse, J., McCourt, P., Somerville, C.R. A mutant of *Arabidopsis* lacking a chloroplast-specific lipid. Science 227: 763–765; 1985.

Buggert, F. and Röbbelen, G. Some observations on McKelvie's *axillaris* mutant. Arab. Inf. Serv. 7: 33–34; 1970.

Bünning, E. Über die differenzierungsvorgange in der Cruciferenwurzel. Planta 39: 126–153; 1951.

Burn, J.E., Bagnall, D.J., Metzger, J.D., Dennis, E.S., Peacock, W.J. DNA methylation, vernalization, and the initiation of flowering. Proc. Natl. Acad. Sci. USA 90: 287–291; 1993.

Carlquist, S. Toward acceptable evolutionary interpretations of floral anatomy. Phytomorphology 19: 332–362; 1969.

Caspar, T., Huber, S.C., Somerville, C. Alterations in growth, photosyntheis, and respiration in a starchless mutant of *Arabidopsis thaliana* (L.) deficient in chloroplast phosphoglucomutase activity. Plant Physiol. 79: 11–17; 1985.

Caspar, T., Pickard, B.G. Gravitropism in a strachless mutant of *Arabidopsis*. Planta 177: 185–197; 1989.

Chang, C., Bowman, J.L., DeJohn, A.W., Lander, E.S., Meyerowitz, E.M. Restriction fragment length polymorphism linkage map for *Arabidopsis thaliana*. Proc. Natl. Acad. Sci. USA 85: 6856–6860; 1988.

Chaudhury, A.M., Craig, S., Bloemer, K.C., Farrell, L., Dennis, E.S. Genetic control of male fertility in higher plants. Aust. J. Plant Physiol. 19: 419–426; 1992.

Cheng, C., Dewdney, J., Nam, H., en Boer, B.G.W., Goodman, H.M. A new locus (NIA1) in *Arabidopsis thaliana* encoding nitrate reductase. EMBO J. 7: 3309–3314; 1988.

Chory, J. Light signals in leaf and chloroplast development: photoreceptors and downstream responses in search of a signal transduction pathway. New Biol. 3: 538–548; 1991.

Chory, J. A genetic model for light-regulated seedling development in *Arabidopsis*. Development 115: 337–354; 1992.

Chory, J., Nagpal, P, Peto, C.A. Phenotypic and genetic analysis of *det2*, a new mutant that affects light-regulated seedling development in *Arabidopsis*. Plant Cell 3: 445–459; 1991.

Chory, J., Peto, C.A. Mutations in the *DET1* gene affect cell-type-specific expression of light-regulated genes and chloroplast development in *Arabidopsis*. Proc. Natl. Acad. Sci. USA 87: 8776–8780; 1990.

Chory, J., Peto, C.A., Ashbaugh, M., Saganich, R., Pratt, L., Ausabel, F. Different roles for phytochrome in etiolated and green plants deduced from characterization of *Arabidopsis thaliana* mutants. Plant Cell 1: 867–880; 1989a.

Chory, J., Peto, C., Feinbaum, R., Pratt, L., Ausabel, F. *Arabidopsis thaliana* mutant that develops as a light-grown plant in the absence of light. Cell 58: 991–999; 1989b.

Christ, C., Tye, B.-K., Functional domains of the yeast transcription/replication factor MCM1. Genes Dev. 5: 751–763; 1991.

Clarkson, D.T. Factors affecting mineral nutrient acquisition by plants. Annu. Rev. Plant Physiol. 36: 77–115; 1985.

Clowes, F.A.L. The difference between open and closed meristems. Ann. Bot. 48: 761–767; 1981.

Coen, E.S., Meyerowitz, E.M. The war of the whorls: genetic interactions controlling flower development. Nature 353: 31–37; 1991.

Conrad, D. Über eine Röntgenmutante von *Arabidopsis thaliana* (L.) Heynh. mit verändertem Blütenbau und Blütenstand. Biol. Zentralbl. 90: 137–144; 1971.

Cormack, R.G.H. Development of root hairs in angiosperms. II. Bot. Rev. 28: 446–464; 1962.

Crawford, N.M. Smith, M., Bellissimo, D., Davis, R.W. Sequence and nitrate regulation of the *Arabidopsis thaliana* mRNA encoding nitrate reductase, a metalloflavoprotein with three functional domains. Proc. Natl. Acad. Sci. USA 85: 5006–5010; 1988.

Crawford, N.M., Wilkinson, J.Q., LaBrie, S.T. Control of nitrate reduction in plants. Aust. J. Plant Physiol. 19: 377–385; 1992.

Crone, W. Growth patterns in floral organogenesis. Ph.D. Thesis, University of California, Riverside, CA; 1992.

Cronquist, A. An integrated system of classification of flowering plants. Columbia University Press, New York; 1981.

Cutter, E.G. Recent experimental studies of the shoot apex and shoot morphogenesis. Bot. Rev. 31: 7–113; 1965.

Dangl, J.L. Applications of *Arabidopsis thaliana* to outstanding issues in plant–pathogen interactions. Int. Rev. Cytol. 144: 53–93; 1993a.

Dangl, J.L. The emergence of *Arabidopsis thaliana* as a model for plant–pathogen interactions. Adv. Plant Pathol. 10: 127–155; 1993b.

Dangl, J.L., Ritter, C., Humphrey, M.J., Woods, J.W., Mur, L.A.J., Goss, S., Mansfield, J.W., Vivian, A. Functional homologs of the *Arabidopsis RPM1* disease resistance gene in bean and pea. Plant Cell 4: 1359–1369; 1992.

Davis, A.R. Nectaries—secretory structures important to beekeepers. In: J. Rhodes, ed. Proceedings of the Second Australian and International Bee Congress. 1988, pp. 175–177.

Davis, A.R. Physiological and structural aspects of floral nectar secretion. Ph.D.

Thesis, Research School of Biological Sciences, The Australian National University, Canberra, Australia; 1992.

Davis, A.R., Gunning, B.E.S. The anatomy and secretion of the floral nectary of *Arabidopsis thaliana var.* Columbia. Abstract Robertson Symposium: *Arabidopsis thaliana* and the molecular basis of plant Biology; Research School of Biological Sciences, The Australian National University, Canberra, Australia; 1991.

Davis, A.R., Gunning, B.E.S. The modified stomata of the floral nectary of *Vicia faba* L. 1. Development, anatomy, and ultrastructure. Protoplasma 166 (3/4): 134–152; 1992.

Dawson, J., Wilson, Z.A., Aarts, M.G.M., Braithwaite, A.F., Briarty, L.G., Mulligan, B.J. Microspore and pollen development in six male sterile mutants of *Arabidopsis thaliana*. Can. J. Bot.: 629–638; 1993.

Debener, T., Lehnackers, H., Arnold, M., Dangl, J.L. Identification and molecular mapping of a single *Arabidopsis* locus conferring resistance against a phytopathogenic *Pseudomonas* isolate. Plant J. 1: 289–302; 1991.

De Candolle, A.-P. Mémoire sur la famille des Crucifères. Mém. Mus. Hist. Nat. 7: 169–252; 1821.

Deng, X.-W., Caspar, T., Quail, P.H. *cop1*: a regulatory locus involved in light-controlled development and gene expression in *Arabidopsis*. Genes Dev. 5: 1172–1182; 1991.

Deng, X.-W., Matsui, M., Wei, N., Wagner, D., Chu, A.M., Feldmann, K.A., Quail, P.H. *COP1*, an *Arabidopsis* photomorphogenic regulatory gene, encodes a novel protein with both a Zn-binding motif and a domain homologous to the β-subunit of trimeric G-proteins. Cell 71: 791–801; 1992.

Deng, X.-W., Quail, P.H. Genetic and phenotypic characterization of *cop1* mutants of *Arabidopsis thaliana*. Plant J. 2: 83–95; 1992.

Dharmawardhana, D.P., Ellis, B.E., Carlson, J.E. Characterization of vascular lignification in *Arabidopsis thaliana*. Can. J. Bot. 70: 2238–2244; 1992.

Dickinson, H.G., Lewis, D. Cytoplasmic and ultrastructural differences between intraspecific compatible and incompatible pollinations in *Raphanus*. Proc. R. Soc. Lond. B. 183: 21–38; 1973.

Dolan, L., Janmaat, K., Willemsen, V., Linstead, P., Poethig, S., Roberts, K.,

Scheres, B. Cellular organisation of the *Arabidopsis thaliana* root. Development 119: 71–84; 1993.

Dong, X., Mindrinos, M., Davis, K.R., Ausubel, F.M. Induction of *Arabidopsis* defense genes by virulent and avirulent *Pseudomonas syringae* strains, and by a cloned avirulence gene. Plant Cell 3: 61–72; 1991.

Drews, G.N., Bowman, J.L., Meyerowitz, E.M. Negative regulation of the *Arabidopsis* homeotic gene *AGAMOUS* by the *APETALA2* product. Cell 65: 991–1002; 1991.

Dzelzkalns, V.A., Nasrallah, J.B., Nasrallah, M.E. Cell–cell communication in plants: self-incompatibility in flower development. Dev. Biol. 153: 70–82; 1992.

Effmertová, E. The behavior of "summer annual," "mixed," and "winter annual" natural populations as compared with early and late races in field conditions. Arab. Inf. Serv. 4: 8–9; 1967.

Elleman, C.J., Dickinson, H.G. Pollen stigma interactions in *Brassica*. IV. Structural reorganisation in the pollen grains during hydration. J. Cell Sci. 80: 141–157; 1986.

Elleman, C.J., Franklin-Tong, V., Dickinson, H.G. Pollination in species with dry stigmas: the nature of the early stigmatic response and the pathway taken by pollen tubes. New Phytol. 121: 413–424; 1992.

Endress, P.K. Evolution and floral diversity: the phylogenetic surroundings of *Arabidopsis* and *Antirrhinum*. Int. J. Plant Sci. 153: S106–S122; 1992.

Engell, K. Embryology of barley. II. Synergids and egg cell, zygote, and embryo development. In: M. Cresti, P. Gori, E. Pacini, eds. Sexual Reproduction in Higher Plants. Springer-Verlag, Berlin; 1988, pp. 383–388.

Errampalli, D., Patton, D., Castle, L., Mickelson, L., Hansen, K., Schnall, J., Feldmann, K., Meinke, D. Embryonic lethals and T-DNA insertional mutagenesis in *Arabidopsis*. Plant Cell 3: 149–157; 1991.

Esau, K. Anatomy of seed plants. John Wiley and Sons, New York; 1977.

Eskins, K. Light-quality effects on *Arabidopsis* development. Red, blue, and far-red regulation of flowering and morphology. Physiol. Plant. 86: 439–444; 1992.

Estelle, M. The plant hormone auxin: insight in sight. BioEssays 14: 439–444; 1992.

Estelle, M.A., Somerville, C.R. The mutants of *Arabidopsis*. Trends Genet. 2: 89–93; 1986.

Estelle, M.A., Somerville, C. Auxin-resistant mutants of *Arabidopsis thaliana* with an altered morphology. Mol. Gen. Genet. 206: 200–206; 1987.

Feenstra, W.J. Isolation of nutritional mutants in *Arabidopsis thaliana*. Genetica 35: 259–269; 1964.

Feenstra, W.J. Contiguity of linkage groups I and IV as revealed by linkage relationship of two newly isolated markers *dis-1* and *dis-2*. Arab. Inf. Serv. 15: 35–38; 1978.

Feldmann, K.A. T-DNA insertional mutagenesis in *Arabidopsis*: mutational spectrum. Plant J. 1: 71–82; 1991.

Finkelstein, R.R., Somerville, C.R. Three classes of abscisic acid (ABA)-insensitive mutations of *Arabidopsis* define genes that control overlapping subsets of ABA responses. Plant Physiol. 94: 1172–1179; 1990.

Fischerová, H. Linkage relationships of recessive chlorophyll mutations in *Arabidopsis thaliana* (L.) Heynh. Biol. Plant. 17: 182–188; 1975.

Franzmann, L., Patton, D.A., Meinke, D.W. In vitro morphogenesis of arrested embryos from lethal mutants of *Arabidopsis thaliana*. Theor. Appl. Genet. 77: 609–616; 1989.

Furner, I.J., Pumphrey, J.E. Cell fate in the shoot apical meristem of *Arabidopsis thaliana*. Development 115: 755–764; 1992.

Galbraith, D.W., Harkins, K.R., Knapp, S. Systemic endopolyploidy in *Arabidopsis thaliana*. Plant Physiol. 96: 985–989; 1991.

Gäumann, E. Ueber die Formen der *Peronospora parasitica* (Pers.) Fries. Belh. Bot. Zentralbl. 35: 1. Abt., 395–533; 1918.

Gerlach Cruse, D. Embryo- und endospermentwicklung nach einer röntgenbestrahlung der fruchknoten von *Arabidopsis thaliana* (L.) Heynh. Radiat. Bot. 9: 433–442; 1969.

Giraudat, J., Hauge, B.M., Valon, C., Smalle, J., Parcy, F. Goodman, H.M. Isolation of the *Arabidopsis ABI3* gene by positional cloning. Plant Cell 4: 1251–1261; 1992.

Goethe, J.W. von. Versuch die Metamorphose der Pflanzen zu erklären. Gotha: C.W. Ettinger.; 1790 (*transl.*

Arber, A. Goethe's botany. Chronica Bot. 10: 63–126; 1946).

Golinowski, W., Grundler, F.M.W. The structure of feeding sites (syncytia) of *Heterodera schachtii* in roots, stems, and leaves of *Arabidopsis thaliana*. Abstracts of the 21st International Nematology Symposium, 1992 April 11–17, Albufeira, Portugal. European Society of Nematologists: 27; 1992.

Goto, N., Katoh, N., Kranz, A.R. Morphogenesis of floral organs in *Arabidopsis*: predominant carpel formation of the *pin-formed* mutant. Jpn. J. Genet. 66: 551–567; 1991a.

Goto, N., Kumagai, T., Koornneef, M. Flowering responses to light-breaks in photomorphogenic mutants of *Arabidopsis*, a long-day plant. Physiol. Plant. 83: 209–215; 1991b.

Goto, N., Starke, M., Kranz, A.R. Effect of gibberellins on the flower development of the *pin-formed* mutant of *Arabidopsis thaliana*. Arab. Inf. Serv. 23: 66–71; 1987.

Gottschalk, W., Kaul, M.L.H. The genetic control of microsporogenesis in higher plants. Nucleus 17: 133–166; 1974.

Griffing, B., Scholl, R.L. Qualitative and quantitative genetic studies of *Arabidopsis thaliana*. Genetics 129: 605–609; 1991.

Grinikh, L.I., Shevchenko, V.V. Number of initial cells forming the reproductive tissue in vegetating *Arabidopsis thaliana* plants. Genetika 12: 690–692; 1976.

Grinikh, L.I., Shevchenko, V.V., Grigoreva, G.A., Draginskaya, L.Y. Study of chimerism in the reproductive tissue of *Arabidopsis thaliana* plants following irradiation of seeds. Genetika 10: 18–28; 1974.

Guignard, J.-L., Agier, C., Bury, M. Développement de l'embryon zygotique chez l'*Arabidopsis thaliana* Schur. Bull Soc. bot. Fr. 138, Lettres bot.: 149–154; 1991.

Gunning, B.E.S., Hardham, A.R. Microtubules. Annu. Rev. Plant. Physiol. 33: 651–698; 1982.

Guttenberg, H. von. Studien über die entwicklung des wurzelvegetationsounktes des dikotyledonen. Planta 35: 360–396; 1947.

Guzmán, P., Ecker, J.R. Exploiting the triple response of *Arabidopsis* to identify ethylene-related mutants. Plant Cell 2: 513–523; 1990.

Haring, V., Gray, J.E., McClure, B.A.,

Anderson, M.A., Clarke, A.E. Self-incompatibility: a self-recognition system in plants. Science 250: 937–941; 1990.

Harpham, N.V.J., Berry, A.W., Knee, E.M., Roveda-Hoyos, G., Raskin, I., Sanders, I.O., Smith, A.R., Wood, C.K., Hall, M.A. The effect of ethylene on the growth and development of wild-type and mutant *Arabidopsis thaliana* (L.) Heynh. Ann. Bot. 68: 55–61; 1991.

Haughn, G.W., Somerville, C.R. Genetic control of morphogenesis in *Arabidopsis*. Dev. Genet. 9: 73–89; 1988.

Heath, I.B., editor. Tip growth in plant and fungal cells. Academic Press, San Diego; 1990.

Hedge, I.C. A systematic and geographical survey of the old world Cruciferae. In: The biology and chemistry of the Cruciferae. J.G. Vaughan, A.J. Macleod, B.M.G. Jones, eds. Academic, London; 1976, pp. 1–45.

Heinen, W., Linskins, H.F. Enzymatic breakdown of stigmatic cuticle of flowers. Nature 191: 1416; 1961.

Heino, P., Snadman, G., Lång, V., Nordin, K., Palva, E.T. Abscisic acid deficiency prevents development of freezing tolerance in *Arabidopsis thaliana* (L.) Heynh. Theor. Appl. Genet. 79: 801–806; 1990.

Heslop-Harrison, J. Aspects of the structure, cytochemistry, and germination of rye (*Secale cereale*). Suppl. 1, Ann. Bot. 44: 1–47; 1979.

Heslop-Harrison, Y. Stigma characteristics and angiosperm taxonomy. Nord. J. Bot. 1; 401–420; 1981.

Hill, J.P., Lord, E.M. Dynamics of pollen tube growth in the wild radish *Raphanus raphanistrum* (Brassicaceae). II. Morphology, cytochemistry, and ultrastructure of transmitting tissues, and the path of pollen tube growth. Am. J. Bot. 74: 988–997; 1987.

Hill, J.P., Lord, E.M. Floral development in *Arabidopsis thaliana*: comparison of the wild type and the homeotic *pistillata* mutant. Can. J. Bot. 67: 2922–2936; 1989.

Hirono, Y., Rédei, G.P. Induced premeiotic exchange of linked markers in the angiosperm *Arabidopsis*. Genetics 51: 519–526; 1965.

Hogan, C.J. Microtubule patterns during meiosis in two higher plant species. Protoplasma 138: 126–136; 1987.

Hou, Y., von Arnim, A.G., Deng, X.-W. A new class of *Arabidopsis* constitutive photomorphogenic genes involved in regulating cotyledon development. Plant Cell 5: 329–339; 1993.

Huala, E., Sussex, I.M. *LEAFY* interacts with floral homeotic genes to regulate *Arabidopsis* floral development. Plant Cell 4: 901–913; 1992.

Huang, B.C. A comparative study of in vivo and in vitro embryogenesis in *Arabidopsis thaliana* L. Ph. D. Thesis, University of Edinburgh, Scotland; 1986.

Hugly, S., Kunst, L., Browse, J., Somerville, C. Enhanced thermal tolerance of photosynthesis and altered chloroplast ultrastructure in a mutant of *Arabidopsis* deficient in lipid desaturation. Plant Physiol. 90: 1134–1142; 1989.

Hugly, S., McCourt, P., Browse, J., Patterson, G.W., Somerville, C. A chilling sensitive mutant of *Arabidopsis* with altered steryl-ester metabolism. Plant Physiol. 93: 1053–1062; 1990.

Irish, V.F., Sussex, I.M. Function of the *APETALA1* gene during *Arabidopsis* floral development. Plant Cell 2: 741–753; 1990.

Irish, V.F., Sussex, I.M. A fate map of the *Arabidopsis* embryonic shoot apical meristem. Development 115: 745–753; 1992.

Izhar, S., Frankel, R. Mechanism of male sterility in *Petunia*: the relationship between pH, callase activity in anthers, and the breakdown of microsporogenesis. Theor. Appl. Genet. 41: 104–108; 1971.

Jack, T., Brockman, L.B., Meyerowitz, E.M. The homeotic gene *APETALA3* of *Arabidopsis thaliana* encodes a MADS Box and is expressed in petals and stamens. Cell 68: 683–697; 1992.

Janchen, E. Das system der Cruciferen. Oesterr. Bot. Z. 91: 1–28; 1942.

Johnson, H.B. Plant pubescence: an ecological perspective. Bot. Rev. 41: 233–258; 1975.

Jones, M.G.K. Host cell responses to endoparasitic nematode attack: structure and function of giant cells and syncytia. Ann. Appl. Biol. 97: 353–372; 1981.

Jongedijk, E. A rapid methyl salicylate clearing technique for routine phase-contrast observations on female meiosis in *Solanum*. J. Microscopy 146: 157–162; 1987.

References

Jürgens, G., Mayer, U., Torres Ruiz, R.A., Berleth, T., Miséra, S. Genetic analysis of pattern formation in the *Arabidopsis* embryo. Development Suppl. 1: 27–38; 1991.

Kapil, R.N., Tiwari, S.C. The integumentary tapetum. Bot. Rev. 44: 457–490; 1978.

Karlovská, V. Genotypic control of the speed of development in *Arabidopsis thaliana* (L.) Heynh. Lines obtained from natural populations. Biol. Plant. 16: 107–117; 1974.

Karssen, C.M., Brinkhorst-va der Swan, D.L.C., Breekland, A.E., Koornneef, M. Induction of dormancy during seed development by endogenous abscisic acid: studies on abscisic acid deficient genotypes of *Arabidopsis thaliana* (L.) Heynh. Planta 157: 158–165; 1983.

Kaul, M.L.H. Male sterility in higher plants. In: Monographs on Theoretical and Applied Genetics. R. Frankel, M. Grossman, H.F. Linskens, P. Maliga and R. Riley, eds. Springer-Verlag, Berlin; 1988.

Kausik, S.B. A contribution to the embryology of *Enalus acoroides* (l.fil.), Steud. Proc. Indian Acad. Sci. B. 11: 83–99; 1940.

Keijzer, C.J. The processes of anther dehiscence and pollen dispersal. I. The opening mechanism of longitudinally dehiscing anthers. New Phytol. 105: 487–498; 1987.

Khurana, J.P., Poff, K.L. Mutants of *Arabidopsis thaliana* with altered phototropism. Planta 178: 40–406; 1989.

Khurana, J.P., Ren, Z., Steinitz, B., Parks, B., Best, T.R., Poff, K.L. Mutants of *Arabidopsis thaliana* with decreased amplitude in their phototropic response. Plant Physiol. 91: 685–689; 1989.

Kiss, J.Z., Hertel, R., Sack, F.D. Amyloplasts are necessary for full gravitropic sensitivity in roots of *Arabidopsis thaliana*. Planta 177: 198–206; 1989.

Klee, H., Estelle, M. Molecular genetic approaches to plant hormone biology. Annu. Rev. Plant Physiol. Plant Mol. Biol. 42: 529–551; 1991.

Knox, R.B. Pollen–pistil interactions. In: H.F. Linskens and J. Heslop-Harrison, eds. Encyclopedia of Plant Physiology, New Series Vol. 17: Cellular Interactions; 1984, pp. 508–608.

Koch, E., Slusarenko, A.J. *Arabidopsis* is susceptible to infection by a downy mildew fungus. Plant Cell 2: 437–445; 1990.

Komaki, M.K., Okada, K., Nishino, E., Shimura, Y. Isolation and characterization of novel mutants of *Arabidopsis thaliana* defective in flower development. Development 104: 195–203; 1988.

Kroncz, C., Mayerhofer, R., Koncz-Kalman, Z., Nawrath, C., Rédei, G.P., Schell, J. Isolation of a gene encoding a novel chloroplast protein by T-DNA tagging in *Arabidopsis thaliana*. EMBO J. 9: 1337–1346; 1990.

Konjevi'c, R., Steinitz, B., Poff, K.L. Dependence of the phototropic response of *Arabidopsis thaliana* on fluence rate and wavelength. Proc. Natl. Acad. Sci. USA 86: 9876–9880, 1989.

Koornneef, M. The complex syndrome of *ttg* mutants. Arab. Inf. Serv. 18: 45–51; 1981.

Koornneef, M. Mutations affecting the testa colour in *Arabidopsis*. Arab. Inf. Serv. 27: 1–4; 1990a.

Koornneef, M. Linkage map of *Arabidopsis thaliana*. In: S.J. O'Brien, ed. Genetic Maps, 5th edn. Cold Spring Hargor Laboratory Press, Cold Spring Harbor; 1990b, pp. 6.94–6.97.

Koornneef, M., Dellaert, L.W.F., van der Veen, J.H. EMS- and radiation-induced mutation frequencies at individual loci in *Arabidopsis thaliana* (L.) Heynh. Mut. Res. 93: 109–123; 1982a.

Koornneef, M. Elgersma, A., Hanhart, C.J., van Loenen-Martinet, E.P., van Rijn, L., Zeevaart, J.A.D. A gibberellin insensitive mutant of *Arabidopsis thaliana*. Physiol. Plant. 65: 33–39; 1985.

Koornneef, M., Hanhart, C.J., Hilhorst, H.W.M., Karssen, C.M. In vivo inhibition of seed development and reserve protein accumulation in recombinants of abscisic acid biosynthesis and responsiveness mutants in *Arabidopsis thaliana*, Plant Physiol. 90: 463–469; 1989a.

Koornneef, M. Hanhart, C.J., Thiel, F.A genetic and phenotypic description of *eceriferum* (*cer*) mutants in *Arabidopsis thaliana*. J. Hered. 80: 118–122; 1989b.

Koornneef, M., Hanhart, C.J., van der Veen, J.H. A genetic and physiological analysis of late flowering mutants in *Arabidopsis thaliana*. Mol. Gen. Genet. 229: 57–66; 1991.

Koornneef, M., Jorna, M.L., Brinkhorst-van der Swan, D.L.C., Karssen, C.M.

The isolation of abscisic acid (ABA) deficient mutants by selection of induced revertants in non-germinating gibberellin sensitive lines of *Arabidopsis thaliana* (L.) Heynh. Theor. Appl. Genet. 61: 385–393; 1982b.

Koornneef, M., Luiten, W., de Vlaming, P., Schram, A.W. A gene controlling flavonoid-3'-hydroxylation in *Arabidopsis*. Arab. Inf Serv. 19: 113–115; 1982c.

Koornneef, M., Reuling, G., Karssen, C.M. The isolation and characterization of abscisic acid-insensitive mutants of *Arabidopsis thaliana*. Physiol. Plant. 61: 377–383; 1984.

Koornneef, M., Rolff, E., Spruit, C.J.P. Genetic control of light-inhibited hypocotyl elongation in *Arabidopsis thaliana* (L.) Heynh. Z. Planzenphysiol. Bd. 100S: 147–160; 1980.

Koornneef, M., Stam, P. Genetic analysis. In: C. Koncz, N.-H. Chua, and J. Schell, eds. Methods in Arabidopsis research. World Scientific, Singapore; 1992, pp. 83–99.

Koornneef, M., van der Veen, J.H. Induction and analysis of gibberellin sensitive mutants in *Arabidopsis thaliana* (L.) Heynh. Theor. Appl. Genet. 58: 257–263; 1980.

Koornneef, M., van Eden, J., Hanhart, C.J., Stan, P., Braaksma, F.J., Feenstra, W.J. Linkage map of *Arabidopsis thaliana*. J. Hered. 74: 265–272; 1983.

Kranz, A.R. Genetic and physiological damage induced by cosmic radiation on dry plant seeds during space flight. Adv. Space Res. 6: 135–138; 1986.

Kranz, A.R., Kirchheim, B. Genetic resources in *Arabidopsis*. Arab. Inf Serv. 24; 1987.

Krickhahn, D., Napp-Zinn, K. Fasciation studies in *Arabidopsis thaliana* (L.) HEYNH. Arab. Inf. Serv. 12: 9; 1975.

Kroh, M. An electron microscopic study of the behavior of Cruciferae pollen after pollination. In: H.F. Linskins, ed. Pollen physiology and fertilization. North-Holland Publishing Co., Amsterdam; 1964, pp. 221–224.

Kroh, M., Munting, A.J. Pollen germination and pollen tube growth in *Diplotaxis tenuifloia* after cross-pollination. Acta Bot. Neerl. 16: 182–187; 1967.

Kunst, L., Browse, J., Somerville, C. Altered chloroplast structure and function in a mutant of *Arabidopsis* deficient in plastid glycerol-3-phosphate acyl-

transferase activity. Plant Physiol. 90: 846–853; 1989a.

Kunst, L., Klenz, J.E., Martinez-Zapater, J., Haughn, G.W. *AP2* gene determines the identity of perianth organs in flowers of *Arabidopsis thaliana*. Plant Cell 1: 1195–1208; 1989b.

LaBrie, S.T., Wilkinson, J.Q., Tsay, Y.-F., Feldmann, K.A., Crawford, N.M. Identification of two tungstate-sensitive molybenum cofactor mutants, *chl2* and *chl7*, of *Arabidopsis thaliana*. Mol. Gen. Genet. 233: 169–176; 1992.

Laibach, F. Zur Frage nach der Individualität der Chromosomen im Pflanzenreich. Beih. Bot. Zentralbl. 1 Abt. 22: 191–210; 1907.

Laibach, F. *Arabidopsis thaliana* (L.) Heynh. als Objekt für genetische und entwicklungsphysiologische Untersuchungen. Bot. Archiv. 44: 439–455; 1943.

Laibach, F. Über sommer und winterannuelle Rasse von *Arabidopsis thaliana* (L.) Heynh. Ein Beitrag zur Atiologie der Blutenbildung. Beitr. Biol. Pflanz. 28: 173–210; 1951.

Lance-Nougarede, A. Comparison du fonctionnement reproducteur chez deux plants vivaces construisant des inflorescence engrappe indefinie sans fleur terminale *Teucrium scorodonia* L. (Labiees) et *Alyssum maritimum* Lamk. (Cruciferes). C.R. Acad. Sci. (Paris) 252: 924–926; 1961.

Langridge, J. Biochemical mutations in the crucifer *Arabidopsis thaliana* (L.) Heynh. Nature 176: 260–261; 1955.

Langridge, J. A hypothesis of developmental selection exemplified by lethal and semi-lethal mutants of *Arabidopsis*. Aust. J. Biol. Sci. 11: 58–68; 1958.

Lapushner, D., Frankel, R. Practical aspects, and the use of male sterility in the production of hybrid tomato seed. Euphytica 16: 300–310; 1967.

Laser, K.D., Lersten, N.R. Anatomy and cytology of microsporogenesis in cytoplasmic male sterile angiosperms. Bot. Rev. 38: 425–454; 1972.

Last, R.L., Bissinger, P.H., Mahoney, D.J., Radwanski, E.R., Fink, G.R. Tryptophan mutants in *Arabidopsis*: the consequences of duplicated tryptophan synthase β genes. Plant Cell 3: 345–358; 1991.

Last, R.L., Fink, G.R. Tryptophan-requiring mutants of the plant *Arabidopsis thaliana*. Science 240: 305–310; 1988.

Lawrence, G.H. Taxonomy of flowering plants. McMillan, New York and London; 1951, pp. 520–524.

Lawrence, M.J. Variations in natural populations of *Arabidopsis thaliana* (L.) Heynh. In: The biology and chemistry of the Cruciferae. J.G. Vaughan, A.J. Macleod, B.M.G. Jones, eds. Academic; London; 1976, pp. 167–190.

Lee, J.A., Bleecker, A., Amasino, R. Analysis of naturally occurring late flowering in *Arabidopsis thaliana*. Mol. Gen. Genet.: 237: 171–176; 1993.

Lee-Chen, S., Steinitz-Sears, L.M. The location of linkage groups in *Arabidopsis thaliana*. Can. J. Genet. Cytol. 9: 381–384; 1967.

Lemieux, B., Miquel, M., Somerville, C., Browse, J. Mutants of *Arabidopsis* with alterations in seed lipid fatty acid composition. Theor. Appl. Genet. 80: 234–240; 1990.

Leutwiler, L.S., Hough-Evans, B.R., Meyerowitz, E.M. The DNA of *Arabidopsis thaliana*. Mol. Gen. Genet. 194: 15–23; 1984.

Leyser, H.M.O., Furner, I.J. Characterisation of three shoot apical meristem mutants of *Arabidopsis thaliana*. Development 116: 397–403; 1992.

Li, S.L., Rédei, G.P. Estimation of mutation rates in autogamous diploids. Radiat. Bot. 9: 125–131; 1969a.

Li, S.L., Rédei, G.P. Thiamine mutants of the crucifer *Arabidopsis*. Biochem. Genet. 3: 163–170; 1969b.

Lin, T.-P., Caspar, T., Somerville, C., Preiss, J. Isolation and characterization of a starchless mutant of *Arabidopsis thaliana* (L.) Heynh. lacking ADP glucose pyrophosphorylase activity. Plant Physiol. 86: 1131–1135; 1988.

Lincoln, C. Britton, J.H., Estelle, M. Growth and development of the *axr1* mutants of *Arabidopsis*. Plant Cell 2: 1071–1080; 1990.

Lincoln, C., Estelle, M. The *axr1* mutation of *Arabidopsis* is expressed in both roots and shoots. J. Iowa Acad. Sci. 98: 68–71; 1990.

Lincoln, C., Turner, J., Estelle, M. Hormone-resistant mutants of *Arabidopsis* have an attenuated response to *Agrobacterium* strains. Plant Physiol. 98: 979–983; 1992.

Liscum, E., Hangarter, R.P. *Arabidopsis* mutants lacking blue light-dependent inhibition of hypocotyl elongation. Plant Cell 3: 685–694; 1991.

Lloyd, A.M., Walbot, V., Davis, R.W. *Arabidopsis* and *Nicotiana* anthocynain production activated by maize regulators *R* and *C1*. Science 258: 1773–1775; 1992.

Lolle, S.J., Cheung, A.Y. Promiscuous germination and growth of wild-type pollen from *Arabidopsis* and related species on the shoot of the *Arabidopsis* mutant, *fiddlehead*. Dev. Biol. 155: 250–258; 1993.

Lolle, S.J., Cheung, A.Y., Sussex, I.A. *fiddlehead*: an *Arabidopsis* mutant constitutively expressing an organ fusion program that involves interactions between epidermal cells. Dev. Biol. 152: 383–392; 1992.

Ludwig, S.R., Habera, L.F., Dellaporta, S.L., Wessler, S.R. *lc*, a member of the maize *R* gene family responsible for tissue-specific anthocyanin production, encodes a protein similar to transcriptional activators and contains the *myc*-homology region. Proc. Natl. Acad. Sci. USA 86: 7092–7096; 1989.

Lynch, M.A., Staehelin, L.A. Domain-specific and cell type-specific localization of two types of cell wall matrix polysaccharides in the clover root tip. J. Cell Biol. 118: 467–479; 1992.

Ma, H., Yanofsky, M.F., Meyerowitz, E.M. *AGL1-AGL6*, an *Arabidopsis* gene family with similarity to floral homeotic and transcription factor genes. Genes Dev. 5: 484–495; 1991.

Maher, E.P., Bell, C.J. Abnormal responses to gravity and auxin in mutants of *Arabidopsis thaliana*. Plant Sci. 66: 131–138; 1990.

Maher, E.P., Martindale, D.J.B. Mutants of *Arabidopsis thaliana* with altered responses to auxins and gravity. Biochem. Genet. 18: 1041–1043; 1980.

Maheshwari, P. An introduction to the embryology of angiosperms. McGraw Hill; New York; 1950.

Mandel, M.A., Bowman, J.L., Kempin, S.A., Ma, H., Meyerowitz, E.M., Yanofsky, M.F. Manipulation of flower structure in transgenic tobacco. Cell 71: 133–143; 1992a.

Mandel. M.A., Gustafson-Brown, C., Savidge, B., Yanofsky, M.F. Molecular characterization of the *Arabidopsis* floral homeotic gene *APETALA1*. Nature 360: 273–277; 1992b.

Mansfield, S.G., Briaty, L.G. Development of the free-nuclear endosperm in *Arabidopsis thaliana* (L.). Arab. Inf Serv. 27: 53–64; 1990a.

Mansfield, S.G., Briarty, L.G. Endosperm cellularization in *Arabidopsis thaliana* L. Arab. Inf. Serv. 27: 65–72; 1990b.

Mansfield, S.G., Briarty, L.G. Early embryogenesis in *Arabidopsis thaliana*. II. The developing embryo. Can. J. Bot. 69: 461–476; 1991.

Mansfield, S.G., Briarty, L.G. Cotyledon cell development in *Arabidopsis thaliana* during reserve deposition. Can. J. Bot. 70: 151–164; 1992.

Mansfield, S.G., Briarty, L.G., Erni, S. Early embryogenesis in *Arabidopsis thaliana*. I. The mature embryo sac. Can. J. Bot. 69: 447–460; 1991.

Marks, M.D., Esch, J.J. Trichome formation in *Arabidopsis* as a genetic model for studying cell expansion. Curr. Topics Plant Biochem. Physiol. 11: 131–142: 1992.

Marks, M.D., Esch, J., Herman, P., Sivakumaran, S., Oppenheimer, D.G. A model for cell-type determination and differentiation on plants. In: G.I. Jenkins and W. Schuch, eds. Molecular Biology of Plant Development, Symposium of the Society of Experimental Biology, XLV. The Company of Biologists Ltd., Cambridge; 1991, pp. 77–87.

Marks, M.D., Feldmann, K.A. Trichome development in *Arabidopsis thaliana*. I. T-DNA tagging of the *GLABROUS1* gene. Plant Cell 1: 1043–1050; 1989.

Marsden, M.P.F., Meinke, D.W. Abnormal development of the suspensor in an embryo-lethal mutant of *Arabidopsis thaliana*. Am. J. Bot. 72: 1801–1812; 1985.

Martinez, M.C., Jørgenson, J.-E., Lawton, M., Lamb, C.J., Doerner, P.W. Spatial pattern of *cdc2* expression in relation to meristem activity and cell proliferation during plant development. Proc. Natl. Acad. Sci. USA 89: 7360–7364; 1992.

Martínez-Zapater, J.M., Gil, P., Capel, J., Somerville, C.R. Mutations at the *Arabidopsis CHM* locus promote rearrangements of the mitochondrial genome. Plant Cell 4: 889–899; 1992.

Martínez-Zapater, J.M., Somerville, C.R. Effect of light quality and vernalization on late-flowering mutants of *Arabidopsis thaliana*. Plant Physiol. 92: 770–776; 1990.

Mayer, U., Büttner, G., Jürgens, G. Apical–basal pattern formation in the *Arabidopsis* embryo: studies on the role of the *gnom* gene. Development 117: 149–162; 1993.

Mayer, U., Torres Ruiz, R.A., Berleth, T., Miséra, S., Jürgens, G. Mutations affecting body orgnization in the *Arabidopsis* embryo. Nature 353: 402–407; 1991.

McKelvie, A.D. A list of mutant genes in *Arabidopsis thaliana* (L.) Heynh. Radiat. Bot. 1: 233–241; 1962.

Medford, J.I. Vegetative apical meristems. Plant Cell 4: 1029–1039; 1992.

Medford, J.I., Behringer, F.J., Callos, J.D., Feldmann, K.A. Normal and abnormal development in the *Arabidopsis* vegetative shot apex. Plant Cell 4: 631–643; 1992.

Meijer, G. The spectral dependence of flowering and elongation. Acta. Bot. Neer. 8: 189–246; 1959.

Meinke, D.W. Embryo-lethal mutants of *Arabidopsis thaliana*: evidence for gametophytic expression of the mutant genes. Theor. Appl. Genet. 63: 381–386; 1982.

Meinke, D.W. Embryo-lethal mutants of *Arabidopsis thaliana*: analysis of mutants with a wide range of lethal phases. Theor. Appl. Genet. 69: 543–552; 1985.

Meinke, D.W. Embryonic mutants of *Arabidopsis thaliana*. Dev. Genet. 12: 382–392; 1991a.

Meinke, D.W. Perspectives on genetic analysis of plant embryogenesis. Plant Cell 3: 857–866; 1991b.

Meinke, D.W. A homeotic mutant of *Arabidopsis thaliana* with leafy cotyledons. Science 258: 1647–1650; 1992.

Meinke, D.W., Sussex, I.M. Embryo-lethal mutants of *Arabidopsis thaliana*. A model system for genetic analysis of plant embryo development. Dev. Biol. 72: 50–61; 1979a.

Meinke, D.W., Sussex, I.M. Isolation and characterization of six embryo-lethal mutants of *Arabidopsis thaliana*. Dev. Biol. 72: 62–72; 1979b.

Meyerowitz, E.M. *Arabidopsis thaliana*. Annu. Rev. Genet. 21: 93–111; 1987.

Meyerowitz, E.M. *Arabidopsis*, a useful weed. Cell 56: 263–269; 1989.

Meyerowitz, E.M., Bowman, J.L., Brockman L.L., Drews, G.N., Jack, T., Sieburth, L.E., Weigel, D. A genetic and molecular model for flower development in *Arabidopsis*. Development Suppl. 1: 157–167; 1991.

Meyerowitz, E.M., Pruitt, R.E. *Arabidopsis thaliana* and plant molecular genetics. Science 229: 1214–1218; 1985.

Meyerowitz, E.M., Smyth, D.R., Bowman, J.L. Abnormal flowers and pattern formation in floral development. Development 106: 209–217; 1989.

Miksche, J.P., Brown, J.A.M. Development of vegetative and floral meristems of *Arabidopsis thaliana*. Am. J. Bot. 52: 533–537; 1965.

Miquel, M., Browse, J. *Arabidopsis* mutants deficient in polyunsaturated fatty acid synthesis. J. Biol. Chem. 267: 1502–1509; 1992.

Mirza, J.I. Seed coat retention and hypocotyl hook development in mutants of *Arabidopsis thaliana* L. Ann. Bot. 59: 35–39; 1987a.

Mirza, J.I. The effects of light and gravity on the horizontal curvature of roots of gravitropic and agravitropic *Arabidopsis thaliana* L. Plant Physiol. 83: 118–120; 1987b.

Mirza, J.I., Olsen, G.M., Iverson, T.-H., Maher, E.P. The growth and gravitropic responses of wild-type and auxin-resistant mutants of *Arabidopsis thaliana*. Physiol. Plant. 60: 516–522; 1984.

Misra, R.C. Contribution to the embryology of *Arabidopsis thalianum* (Gay & Monn.) Agra. Univ. J. Res. 11: 191–199; 1962.

Mizukami, Y., Ma, H. Ectopic expression of the floral homeotic gene *AGAMOUS* in transgenic *Arabidopsis* plants alters floral organ identity. Cell 71: 119–131; 1992.

Modrusan, Z., Feldmann, K., Haughn, G.W. Abnormal ovule development within *fruitless* (*fts*) mutant of *Arabidopsis thaliana*; XII International congress on sexual plant reproduction; Ohio State University; 1992, pp. 47–48.

Moore, R. Root graviresponsiveness and cellular differentiation in wild-type and a starchless mutant of *Arabidopsis thaliana*. Ann. Bot. 64: 271–277; 1989.

Morgan, T.H. The relation of genetics to physiology and medicine. (Nobel Prize lecture delivered June 4, 1934) In: Nobel Lectures, 1922–1941. Elsevier, Amsterdam; pp. 3–18.

Müller, A. Zur Charakterisierung der Blüten und Infloreszenzen von *Arabidopsis thaliana* (L.) Heynh. Kulturpflanze 9: 364–393, 1961.

Müller, A.J. Embryonentest zum Nachweis rezessiver Letalfaktoren bei *Arabidopsis thaliana*. Biol. Zentralbl. 82: 133–163; 1963.

Müller, A.J. The chimeral structure of M1 plants and its bearing on the determination of mutation frequencies in *Arabidopsis*. In: J. Veleminsky and T. Gichner, eds. Induction of Mutations and the Mutation Process. Czech. Acad. Sci., Prague; 1965; pp. 46–52.

Murray, D.L., Kohorn, B.D. Chloroplasts of *Arabidopsis thaliana* homozygous for the *ch1* locus lack chlorophyll *b*, lack stable LHCPII, and have stacked thylakoids. Plant Mol. Biol. 16: 71–79; 1991.

Nam, H.-G., Giraudat, J., den Boer, B., Moonan, F., Loos, W.D.B., Hauge, B.M., Goodman, H.M. Restriction fragment length polymorphism linkage map of *Arabidopsis thaliana*. Plant Cell 1: 699–705; 1989.

Nambara, E., Naito, S., McCourt, P. A mutant of *Arabidopsis* which is defective in seed development and storage protein accumulation is a new *abi3* allele. Plant J. 2: 435–441; 1992.

Napp-Zinn, K. On the genetic and developmental physiological basis of seasonal aspects of plant communities. Arb. Landwirt. Hochsch. Hohenheim 30: 33–49; 1964.

Napp-Zinn, K. *Arabidopsis thaliana* (L.) Heynh. In: L.T. Evans, ed. The induction of flowering: some case histories. Macmillan, Melbourne; 1969, pp. 291–304.

Napp-Zinn, K., *Arabidopsis thaliana*. In: A.H. Halevy, ed. CRC handbook of flowering, vol. 1. CRC Press, Boca Raton; 1985, pp. 492–503.

Nieuwhof, M. Effect of temperature on the expression of male sterility in brussels sprouts (*Brassica oleracea* L. var. *gemmifera* DC.) Euphytica 17: 265–273; 1968.

Norman, C., Runswick, M., Pollock, R., Treisman, R. Isolation and properties of cDNA clones encoding SRF, a transcription factor that binds to the c-*fos* serum response element. Cell 55: 989–1003; 1988.

Ockendon, D.J. Pollen tube growth and the site of the incompatibility reaction in *Brassica oleracea*. New Phytol. 71: 519–522; 1972.

Okada, K., Komaki, M.K., Shimura, Y. Mutational analysis of pistil structure and development of *Arabidopsis thaliana*. Cell Differ. Dev. 28: 27–38; 1989.

Okada, K., Shimura, Y. Mutational analysis of root gravitropism and phototropism of *Arabidopsis thaliana* seedlings. Aust. J. Plant Physiol. 19: 439–448; 1992a.

Okada, K., Shimura, Y. Aspects of recent developments in mutational studies of plant signaling pathways. Cell 70: 369–372; 1992b.

Okada, K., Shimura, Y. Reversible root tip rotation in *Arabidopsis* seedlings induced by obstacle-touching stimulus. Science 250: 274–276; 1990.

Okada, K., Ueda, J., Komaki, M.K., Bell, C.J., Shimura, Y. Requirement of the auxin polar transport system in early stages of *Arabidopsis* floral bud formation. Plant Cell 3: 677–684; 1991.

Okamuro, J., den Boer, B., van Montagu, M., Jofuku, D. *Apetala2*: characterization of a novel plant homeotic regulatory gene from *Arabidopsis*. J. Cell. Biochem. Suppl. 17B: Abstract D113, p. 16; 1993.

Olsen, G.M., Mirza, J.I., Maher, E.P., Iverson, T.-H. Ultrastructure and movements of cell organelles in the root cap of agravitropic mutants and normal seedlings of *Arabidopsis thaliana*. Physiol. Plant. 60: 523–531; 1984.

Oppenheimer, D.G., Esch, J.J., Marks, M.D. Molecular genetics of *Arabidopsis* trichome development. In: D.P.S. Verma, ed. Control of Plant Gene Expression. CRC Press, Inc., Boca Raton, Florida; 1993, pp. 275–286.

Oppenheimer, D.G., Herman P.L., Sivakumaran, S., Esch, J., Marks, M.D. A *myb* gene required for leaf trichome differentaition in *Arabidopsis* is expressed in stipules. Cell 67: 483–493; 1991.

Orr, A.R. Inflorescence development in *Brassica campestris* L. Am. J. Bot. 65: 466–470; 1978.

Palmer, R.G., Winger, C.L., Albertsen, M.C. Four independent mutations at the ms_1 locus in soybeans. Crop Sci. 18: 727–729; 1978.

Palmer, R.G., Johns, C.W., Muir, P.S.

Genetics and cytology of the ms_3 male-sterile soybean. J. Hered. 71: 343–348; 1980.

Parks, B.M., Quail, P.H. *hy8*, a new class of *Arabidopsis* long hypocotyl mutants deficient in functional phytochrome A. Plant Cell 5: 39–48; 1993.

Parks, B.M., Shanklin, J., Koornneef, M., Kendrick, R.E., Quail, P.H. Immunochemically detectable phytochrome is present at normal levels but is photochemically nonfunctional in the *hy1* and *hy2* long hypocotyl mutants of *Arabidopsis*. Plant Mol. Biol. 12: 425–437; 1989.

Passmore, S., Maine, G.T., Elble, R., Christ, C., Tye, B.-K. A *Saccharomyces cerevisiae* protein involved in plasmid maintenance is necessary for mating of MATa cells. J. Mol. Biol. 204: 593–606; 1988.

Patton, D.A., Franzmann, L.H., Meinke, D.W. Mapping genes essential for embryo development in *Arabidopsis thaliana*. Mol. Gen. Genet. 227: 337–347; 1991.

Patton, D.A., Meinke, D.W. Ultrastructure of arrested embryos from lethal mutants of *Arabidopsis thaliana*. Am. J. Bot. 77: 653–661; 1990.

Payer, J.-B. Traité d'organogénie comparé de la fleur. Libraire de Victor Masson, Paris; 1857.

Peng, J., Harberd, N.P. Derivative alleles of the *Arabidopsis gibberellin-insensitive* (*gai*) mutation confer a wild-type phenotype. Plant Cell 5: 351–360; 1993.

Pepper, A., Delaney, T.P., Chory, J. Genetic interactions in plant photomorphogenesis. Sem. Dev. Biol. 4: 15–22; 1993.

Peterson, R.L. Differentiation and maturation of primary tissues in white mustard root tips. Can. J. Bot. 45: 319–331; 1967.

Pickert, M. *In vitro* germination and storage of trinucleate *Arabidopsis thaliana* (L.) pollen grains. Arab. Inf Serv. 26: 39–42; 1988.

Pickett, F.B., Wislon, A.K., Estelle, M. The *aux1* mutation of *Arabidopsis* confers both auxin and ethylene resistance. Plant Physiol. 94: 1462–1466; 1990.

Polowick, P.L., Sawhney, V.K. A scanning electron microscope study on the initiation and development of floral organs of *Brassica napus* (cv. Westar). Am. J. Bot. 73: 254–263; 1986.

Polyakova, T.F. Development of male

and female gametophytes of *Arabidopsis thaliana* (L.) Heynh. Issledov. Genet. USSR 2: 125–133; 1964 (in Russian with English summary).

Pruitt, R.E., Chang, C., Pang, P.P.-Y., Meyerowitz, E.M. Molecular genetics and development of *Arabidopsis*. In: W. Loomis, ed. Genetic regulation of development, 45th Symp. Soc. Dev. Biol., Liss, New York; 1987, pp. 327–338.

Pruitt, R.E., Horejsi, T.F., Pierskalla, B.K., Ploense, S.E. Genetic analysis of cellular interactions during fertilization of *Arabidopsis thaliana*. J. Cell. Biochem. Suppl. 15A: Abstract A629, p. 137; 1991.

Pruitt, R.E., Meyerowitz, E.M. Characterization of the genome of *Arabidopsis thaliana*. J. Mol. Biol. 187: 169–183; 1986.

Pyke, K.A., Leech, R.M. Chloroplast division and expansion is radically altered by nuclear mutations in *Arabidopsis thaliana*. Plant Physiol. 99: 1005–1008; 1992.

Pyke, K.A., Marrison, J.L., Leech, R.M. Temporal and spatial development of the cells of the expanding first leaf of *Arabidopsis thaliana* (L.) Heynh. J. Exp. Bot. 42: 1407–1416; 1991.

Rédei, G.P. Supervital mutants of *Arabidopsis*. Genetics 47: 443–460; 1962.

Rédei, G.P. Non-mendelian megagametogenesis in *Arabidopsis*. Genetics 51: 857–872; 1965a.

Rédei, G.P. Genetic blocks in the thiamine synthesis of the angiosperm *Arabidopsis*. Am. J. Bot. 52: 834–841; 1965b.

Rédei, G.P. Biochemical aspects of a genetically determined variagation in *Arabidopsis*. Genetics 56: 431–443; 1967.

Rédei, G.P. *Arabidopsis thaliana* (L.) Heynh. A review of the genetics and biology. Bibliog. Genet. 20: 1–151; 1970.

Rédei, G.P. Hereditary structural alterations of plastids induced by a nuclear mutator gene in *Arabidopsis*. Protoplasma 77: 361–380; 1973a.

Rédei, G.P. Extra-chromosomal mutability determined by a nuclear locus in *Arabidopsis*. Mut. Res. 18: 149–162; 1973b.

Rédei, G.P. *Arabidopsis* as a genetic tool. Annu. Rev. Genet. 9: 111–127; 1975.

Rédei, G.P. A heuristic glance at the past of *Arabidopsis* genetics. In: C. Koncz, N.-H. Chua, and J. Schell, eds. Methods in Arabidopsis research. World Scientific, Singapore; 1992; pp. 1–15.

Rédei, G.P., Acedo, G., Gavazzi, G. Flower differentiation in *Arabidopsis*. Stadler Symp. 6: 135–168; 1974.

Rédei, G.P., Hirono, Y. Linkage studies. Arab. Inf. Serv. 1: 9–10; 1964.

Rédei, G.P., Koncz, C. Classical mutagenesis. In: C. Koncz, N.-H. Chua, and J. Schell, eds. Methods in Arabidopsis research. World Scientific, Singapore; 1992; pp. 16–82.

Reed, J.W., Nagpal, P., Poole, D.S., Furuya, M., Chory, J. Mutations in the gene for the red/far-red light receptor phytochrome B alter cell elongation and physiological responses throughout *Arabidopsis* development. Plant Cell 5: 147–157; 1993.

Regan, S.M., Moffatt, B.A. Cytochemical analysis of pollen development in wild-type *Arabidopsis* and a male-sterile mutant. Plant Cell 2: 877–889; 1990.

Reinholz, E. Beeinflussung der Morphogenese embryonaler Organe durch ionisierende Strahlungen. I. Keimlingsanomalien durch Röntgenbestrahlung von *Arabidopsis thaliana*-Embryonen in verschiedenen Entwicklungsstadien. Strahlentherapie 190: 537–553; 1959.

Reiser, L., Fischer, L.R. Identification of a potential mutant in *Arabidopsis thaliana* with abnormal ovule development. Abstract and Program for the 3rd International Congress on Plant Molecular Biology, Tucson (no. 652); 1991.

Reiter, R.S., Williams, J.G.K., Feldmann, K.A., Rafalski, J.A., Tingey, S.V., Scolnik, P.A. Global and local genome mapping in *Arabidopsis thaliana* by using recombinant inbred lines and random amplified polymorphic DNAs. Proc. Natl. Acad. Sci. USA 89: 1477–1481; 1992.

Relichová, J. Some new mutants. Arab. Inf Serv. 13: 25–28; 1976.

Relichová, J. Causes of the spontaneous outcrossing in *Arabidopsis*. Arab. Inf Serv. 15: 59–63; 1978.

Richmond, T.R., Kohel, R.J. Analysis of a completely male sterile character in american upland cotton. Crop Sci. 1: 397–401; 1961.

Rick, C.M. Genetics and development of

nine male-sterile tomato mutants. Hilgardia 18: 599–633; 1948.

Rick, C.M., Butler, L. Cytogenetics of the tomato. Adv. Genet. 8: 267–282; 1956.

Röbbelen, G. Über heterophyllie bei *Arabidopsis thaliana* (l.) Heynh. Ber. dt. bot. Ges. 70: 39–44; 1957a.

Röbbelen, G. Untersuchungen an strahleninduzierten Blattforbmutanten von *Arabidopsis thaliana* (L.) Heynh. Z. Ind. Abst. Vererb.-Lehre 88: 189–252; 1957b.

Röbbelen, G. Flower malformations in mutants as a means of partitioning the developmental process. Arab. Inf Serv. 2: 12–13; 1965a.

Röbbelen, G. The Laibach standard collection of natural races. Arab. Inf Serv. 2: 36–47; 1965b.

Röbbelen, G. Genbedingte rotlicht-empfindlichkeit der chloroplastendifferenzierung bei *Arabidopsis*. Planta 80: 237–254; 1968.

Robinson-Beers, K., Pruitt, R.E., Gasser, C.S. Ovule development in wild-type *Arabidopsis* and two female-sterile mutants. Plant Cell 4: 1237–1249; 1992.

Roggen, H.P.J.R. Scanning electron microscopical observations on compatible and incompatible pollen–stigma interactions in *Brassica*. Euphytica 21: 1–10; 1972.

Rose, A.B., Casselman, A.L., Last, R.L. A phosphoribosylanthranilate transferase gene is defective in blue fluorescent *Arabidopsis thaliana* tryptophan mutants. Plant Physiol. 100: 582–592; 1992.

Ruth, J., Klekowski, E.J., Jr., Stein, O.L. Impermanant initials of the shoot apex and diplontic selection in a juniper chimera. Am. J. Bot. 72: 1127–1135; 1985.

Sachs, T. Pattern formation in plant tissues. Cambridge University Press, New York; 1991.

Sæther, N., Iverson, T.-H. Gravitropism and starch statoliths in an *Arabidopsis* mutant. Planta 184: 491–497; 1991.

Sassen, M.M.A. The stylar transmitting tissue. Acta. Bot. Neerl. 23: 99–108; 1974.

Satina, S., Blakeslee, A.F. Periclinal chimeras in *Datura stramonium* in relation to development of the leaf and flower. Am. J. Bot. 28: 862–871; 1941.

Satina, S., Blakeslee, A.F., Avery, A.G. Demonstration of the three germ layers in the shoot apex of *Datura* by means of induced polyploidy in periclinal chimeras. Am. J. Bot. 27: 895–905; 1940.

Sawhney, V.K., Bhadula, S.K. Microsporogenesis in the normal and the male-sterile *stamenless2* mutant of tomato (*Lycopersicon esculentum*). Can. J. Bot. 66: 2013–2021; 1988.

Schiefelbein, J.W., Shipley, A., Rowse, P. Calcium influx at the tip of growing root-hair cells of *Arabidopsis thaliana*. Planta 187: 455–459; 1992.

Schiefelbein, J.W., Somerville, C. Genetic control of root-hair development in *Arabidopsis*. Plant Cell 2: 235–243; 1990.

Schmid, R. Filament histology and anther dehiscence. Bot. J. Linn. Soc. 73: 303–315; 1976.

Schmidt, A. Histologische Studien an phanerogamen Vegetationspunkten. Bot. Arch. 8: 345–404; 1924.

Schnall, J.A., Quatrano, R.S. Abscisic acid elicits the water-stress response in root hairs of *Arabidopsis thaliana*. Plant Physiol. 100: 216–218; 1992.

Schneider, T., Dinkins, R., Robinson, K., Shellhammer, J., Meinke, D.W. An embryo-lethal mutant of *Arabidopsis thaliana* is a biotin auxotroph. Dev. Biol. 131: 161–167; 1989.

Schröter, H., Mueller, C.G.F., Meese, K., Nordheim, A. Synergism in ternary complex formation between the dimeric glycoprotein p67^{SRF}, polypeptide p62^{TCF}, and the *c-fos* serum response element. EMBO J. 9: 1123–1130; 1990.

Schultz, E.A., Haughn, G.W. *LEAFY*, a homeotic gene that regulates inflorescence development in *Arabidopsis*. Plant Cell 3: 771–781; 1991.

Schultz, E.A., Pickett, F.B., Haughn, G.W. The *FLO10* gene product regulates the expression domain of homeotic genes *AP3* and *PI* in *Arabidopsis* flowers. Plant Cell 3: 1221–1237; 1991.

Schulz, O.E. Cruciferae. In: A. Engler and H. Harms, ed. Das natürlichen Pflanzenfamilien. Engelmann, Leipzig; 1936, pp. 17b: 227–658.

Schulz, P., Jensen, W.A. *Capsella* embryogenesis: the development of the free-nuclear endosperm. Protoplasma 80: 183–205; 1974.

References

Schulz, S.R., Jensen, W.A. *Capsella* embryogenesis: the egg, zygote, and young embryo. Am. J. Bot. 55: 807–819; 1968a.

Schulz, S.R., Jensen, W.A. *Capsella* embryogenesis: the early embryo. J. Ultrastruct. Res. 22: 376–392; 1968b.

Schwarz-Sommer, Z., Huijser, P., Nacken, W., Saedler, H., Sommer, H. Genetic control of flower development by homeotic genes in *Antirhinum majus*. Science 250: 931–936; 1990.

Schweizer, D., Ambros, P., Gründler, P., Varga, F. Attempts to relate cytological and molecular chromosome data of *Arabidopsis thaliana* to its linkage map. Arab. Inf. Serv. 25: 27–34; 1988.

Sears, L.M.S., Lee-Chen, S. Cytogenetic studies in *Arabidopsis thaliana*. Can. J. Genet. Cytol. 12: 217–223; 1970.

Sessions, A., Feldmann, K., Zambriski, P. Mutations affecting gynoecium development in *Arabidopsis thaliana*. J. Cell. Biochem. Suppl. 17B: Abstract D224, p. 23; 1993.

Shannon, S., Meeks-Wagner, D.R. A mutation in the *Arabidopsis TFL1* gene affects inflorescence meristem development. Plant Cell 3: 877–892; 1991.

Shirley, B.W., Hanley, S., Goodman, H.M. Effects of ionizing radiation on a plant genome: analysis of two *Arabidopsis transparent testa* Mutations. Plant Cell 4: 333–347; 1992.

Sijmons, P.C., Grundler, F.M.W., von Mende, N., Burrows, P.R., Wyss, U. *Arabidopsis thaliana* as a new model host for plant–parasitic nematodes. Plant J. 1: 245–254; 1991.

Singh, S.P., Rhodes, A.M. A morphological and cytological study of male sterility in *Curcurbita maxima*. Am. J. Soc. Hort. Sci. 78: 375–378; 1961.

Skorupska, H., Palmer, R.G. Genetic and cytology of the *ms6* male-sterile soybean. J. Hered. 80: 304–310; 1989.

Smyth, D.R., Bowman, J.L., Meyerowitz, E.M. Early flower development in *Arabidopsis*. Plant Cell 2: 755–767; 1990.

Snape, J.W., Lawrence, M.J. The breeding system of *Arabidopsis thaliana*. Heredity 27: 299–302; 1971.

Somers, D.E., Sharrock, R.A., Tepperman, J.T., Quail, P.H. The *hy3* long hypocotyl mutant of *Arabidopsis* is deficient in phytochrome B. Plant Cell 3: 1263–1274; 1991.

Somerville, C.R. Analysis of photosynthesis with mutants of higher plants and algae. Annu. Rev. Plant Physiol. 37: 467–507; 1986.

Somerville, C., Browse, J. Plant lipids: Metabolism, mutants, and membranes. Science 252: 80–87; 1991.

Somerville, C.R., Portis, Jr., A.R., Ogren, W.L. A mutant of *Arabidopsis thaliana* which lacks activation of RuBP carboxylase in vivo. Plant Physiol. 70: 381–387; 1982.

Somerville, S.C., Ogren, W.L. An *Arabidopsis thaliana* mutant defective in chloroplast dicarboxylate transport. Proc. Natl. Acad. Sci. USA 80: 1290–1294; 1983.

Sparrow, A.H., Price, H.J., Underbrink, A.G. A survey of DNA content per cell and per chromosome of prokaryotic and eukaryotic organisms: some evolutionary considerations. Brookhaven Symp. Biol. 23: 451–494; 1972.

Steeves, T.A., Sussex, I.M. Patterns in plant development. Cambridge University Press, New York; 1989.

Steinert, P.M., Roop, D.R. Molecular and cellular biology of intermediate filaments. Annu. Rev. Biochem. 57: 593–625; 1988.

Steinheil, A. Considérations sur l'usage qu'on peut faire des rapports de position qui existent entre la bractée et les parties de chaque verticille floral, dans la détermination du plan normal sur lequel les différentes fleurs sont construites. Ann. Sci. Nat. Sér. 2, 12: 169–361; 1839.

Steinitz-Sears, L.M. Chromosome studies in *Arabidopsis thaliana*. Genetics 48: 483–490; 1963.

Stettler, R.F., Ager, A.A. Mentor effects in pollen interactions. In: H.F. Linskens and J. Heslop-Harrison, eds. Encylopedia of Plant Physiol., New Series Vol. 17: Cellular Interactions. Springer-Verlag, Berlin; 1984, pp. 609–623.

Su, W., Howell, S.H. A single genetic locus, *ckr1*, defines *Arabidopsis* mutants in which root growth is resistant to low concentrations of cytokinin. Plant Physiol. 99: 1569–1574; 1992.

Sun, T., Goodman, H.M., Ausubel, F.M. Cloning the *Arabidopsis GA1* locus by genomic subtraction. Plant Cell 4: 119–128; 1992.

Sung, Z.R., Belachew, A., Shunong, B., Bertrand-Garcia, R. *EMF*, an *Arabidopsis* gene required for vegetative shoot development. Science 258: 1645–1647; 1992.

Susek, R.E., Ausabel., F., Chory, J. Intracellular signal transduction mutants of *Arabidopsis* uncouple nuclear and chloroplast gene expression. *Cell* 74: 787–799; 1993.

Sussex, I.M. Developmental programming of the shoot meristem. Cell 56: 225–229; 1989

Talon, M., Koornneef, M., Zeevaart, J.A.D. Endogenous gibberellins in *Arabidopsis thaliana* and possible steps blocked in the biosynthetic pathways of the semidwarf *ga4* and *ga5* mutants. Proc. Natl. Acad. Sci. USA: 7983–7987; 1990a.

Talon, M., Koornneef, M., Zeevaart, J.A.D. Accumulation of C_{19}-gibbereliins in the gibberellin-insensitive dwarf mutant *gai* of *Arabidopsis thaliana* (L.) Heynh. Planta 182: 501–505; 1990b.

Theis, R., Röbbelen, G. Anther and microspore development in different male-sterile lines of oilseed rape (*Brassica napus* L.). Angew. Botanik 64: 419–434; 1990.

Tilton, V.R., Horner, H.T. Stigma, style, and obturator of *Ornithogalum caudatum* (Liliaceae) and their function in the reproductive process. Am. J. Bot. 67: 1113–1131; 1980.

Timpte, C.S., Wilson, A.K., Estelle, M. Effects of the *axr2* mutation of *Arabidopsis* on cell shape in hypocotyl and inflorescence. Planta 188: 271–278; 1992.

Usmanov, P.D. A spontaneous mutant of *Arabidopsis thaliana* with an altered type of inflorescence. Arab. Inf. Serv. 7: 32; 1970.

Vandendries, R. Contribution à l'histoire du développement des crucifères. Cellule 25: 412–459; 1909.

van der Veen, J.H., Blankestijn-de Vries, H. Double reduction in tetraploid *Arabidopsis thaliana*, studied by means of a chlorophyll mutant with a distinct simplex phenotype. Arab. Inf. Serv. 10: 11–12; 1973.

van der Veen, J.H., Wirtz, P. EMS-induced genetic male sterility in *Arabidopsis thaliana*: a model selection experiment. Euphytica 17: 371–377; 1968.

van Montagu, M., Dean, C., Flavell, R., Goodman, H., Koornneef, M., Meyerowitz, E., Peacock, J., Shimura, Y., Somervilee, C. The multinational coordinated *Arabidopsis thaliana* genome research project. Progress report: Year two. Publ. 92–112. National Science Foundation, Washington, D.C.; 1992.

Vaughan, J.G. Structure of the angiosperm apex. Nature 169: 458–459; 1952.

Vaughan, J.G. The morphology and growth of the vegetative and reproductive apices of *Arabidopsis thaliana* (L.) Heynh., *Capsella bursa-pastoris* (L.) Medic. and *Anagallis arvensis*, L.J. Linn. Soc. Lond. Bot. 55: 279–301; 1955.

Webb, M.C. Aspects of embryo sac development, fertilization and proembryogenesis in *Arabidopsis thaliana*, with emphasis on the microtubular cytoskeleton. Ph.D. Thesis, The Australian National University, Canberra; 1991.

Webb, M.C., Gunning, B.E.S. Embryo sac development in *Arabidopsis thaliana*. I. Megasporogenesis, including the microtubular cytoskeleton. Sex. Plant Reprod. 3: 244–256; 1990.

Webb, M.C., Gunning, B.E.S. The microtubular cytoskeleton during development of the zygote, proembryo, and free-nuclear endosperm in *Arabidopsis thaliana* (L.) Heynh. Planta 184: 187–195; 1991.

Webb, M.C., Gunning, B.E.S. Cell biology of embryo sac development in *Arabidopsis*. In: E.G. Williams, R.B. Knox, A.E. Clarke, eds. Genetic control of self-incompatibility and reproductive development in flowering plants. Kluwer Academic Publishers, Dordrecht; 1993, in press.

Wei, N., Deng, X.-W. *COP9*: A new genetic locus involved in light-regulated development and gene expression in *Arabidopsis*. Plant Cell 4: 1507–1518; 1992.

Weigel, D., Alvarez, J., Smyth, D.R., Yanofsky, M.F., Meyerowitz, E.M. *LEAFY* controls floral meristem identity in *Arabidopsis*. Cell 69: 843–859; 1992.

Weigel, D., Meyerowitz, E.M. Genetic hierarchy controlling flower development. In: M. Bernfield, ed. Molecular basis of morphogenesis. Wiley-Liss, New York; 1993, pp. 91–105.

References

Westerman, J.M., Lawrence, M.J. Genotype-environment interaction and developmental regulation in *Arabidopsis thaliana*. Heredity 25: 609–627; 1971.

Whalen, M.C., Innes, R.W., Bent, A.F., Staskawicz, B.J. Identification of *Pesudomonas syringae* pathogens of *Arabidopsis* and a bacterial locus determining avirulence on both *Arabidopsis* and soybean. Plant Cell 3: 49–59; 1991.

Wilkinson, J.Q., Crawford, N.M. Identification of the *Arabidopsis CHL3* gene as the nitrate reductase structural gene *NIA2*. Plant Cell 3: 461–471; 1991.

Williams, E.G., Maheswaran, G. Somatic embryogenesis: factors influencing coordinated behavior of cells as an embryogenic group. Ann. Bot. 57: 443–462; 1986.

Wilson, A.K., Pickett, F.B., Turner, J.C., Estelle, M.A. Dominant mutation in *Arabidopsis* confers resistance to auxin, ethylene, and abscisic acid. Mol. Gen. Genet. 222: 377–383; 1990.

Wilson, R.N., Heckma, J.W., Somerville, C.R. Gibberellin is required for flowering in *Arabidopsis thaliana* under short days. Plant Physiol. 100: 403–408; 1992.

Worrall, D., Hird, D.L., Hodge, R.,

Wyatt, P., Draper, J., Scott, R. Premature dissolution of the microsporocyte callose wall causes male sterilty in transgenic tobacco. Plant Cell 4: 759–771; 1992.

Wyss, U. Observations on the feeding behavior of *Heterodera schachtii* throughout its development, including events during moulting. Fund. Appl. Nematol. 15: 75–89; 1992.

Yakovlev, M.S., Alimova, G.K. Embryogenesis in *Arabidopsis thaliana* (L.) Heynh. (Cruciferae). Bot. Zh. 61: 12–24; 1976 (in Russian with English summary).

Yakubova, M.M., Nazarova, A.A., Krendeleva, T.E. Structural and functional peculiarities of the photosynthetic apparatus of *Arabidopsis thaliana* (L.) Heynh. mutants. Biokhimiya 45: 684–872; 1980.

Yanofsky, M.F., Ma, H., Bowman, J.L., Drews, G.N., Feldmann, K.A., Meyerowitz, E.M. The protein encoded by the *Arabidopsis* homeotic gene *AGAMOUS* resembles transcription factors. Nature 346: 35–39; 1990.

Zagotta, M.T., Shannon, S., Jacobs, C., Meeks-Wagner, D.R. Early-flowering mutants of *Arabidopsis thaliana*. Aust. J. Plant Physiol. 19: 411–418; 1992.

Subject Index

Flowering (cont.)
  mutants
    early, 50–51, 141, 178
    late, 50–51, 141, 178–179
  transition to, 12, 135–136, 141, 178
  vernalization, effect of, 136, 178
Flowers, 135–143, **133–273**
  development, 139–141, 148–161
    models of, 143–144
    stages of
      definition, 139–141
      stage 1, 139–140, 148–149, 152–153
      stage 2, 139–140, 148–149, 152–153
      stage 3, 139–140, 148–149, 152–153
      stage 4, 140, 152–153
      stage 5, 140, 152–153, 156–157
      stage 6, 140, 156–157
      stage 7, 140–141, 156–157, 277
      stage 8, 140, 156–157, 277
      stage 9, 140–141, 160–161, 277, 300
      stage 10, 140, 160–161, 168–169, 277, 300
      stage 11, 140, 160–161, 168–169, 277, 300
      stage 12, 140, 160–161, 168–169, 277, 300
      stage 13+, 140, 277, 300
  meristem, 136–137, 139, 146–151, 254–257
  mutants, 141–143
    floral organ, 143, 258–273
      carpel, 264–274
      other, 258, 263
    floral primordia formation, 182–199
    homeotic, 142
    meristem identity, 142, 200–213
    meristic, 142
    organ identity, 142–143, 200–203, 216–249, 252–253
      carpel, 216–225
      petal, 200–205, 226–243
      sepal, 200–205, 226–233
      stamen, 216–225, 234–243
    organ number, 143, 250–251, 254–257
    whorl number, 143, 250–251
  primordia
    mutant, 182–199
    wild-type, 136–137, 146–151
  structure, 137–139
    floral diagram, 139
    related species, 138
  whorl, definition of, 137–138

Fruit. *See* Silique
Funiculus. *See* Ovules

G

Gene for gene hypothesis, 405
Germination
  pollen, 88, 288–289, 336
  seed, 12–13
  See also Seed, dormancy
Gibberellins, 8, 74, 84, 136, 178
Gravitropism, 8, 9, 48–49, 78, 80, 118–121
Gynoecium
  development, 156–161
    growth rate, 160
    initiation, 158–159
    primoridia, 156–159
  epidermis, 170–171
  ovary, 158–159
  ovary wall, 170–171
  placentae, 299, 306–307, 342–345
  septum, 158–159, 170–171, 299, 302–303, 342–345
  stigma, 160–161, 168–171, 336, 338–345
  structure, 299
  style, 170–171, 338, 339, 342–345
  transmitting tissue, 299, 336, 342–345

H

Homeosis, 142
Hormone mutants, 8–9
  ABA, 76–77
  auxin, 78–81
  gibberellin, 74–76
Hours after flowering (HAF), 320
Hypersensitive response, 408–411
Hypocotyl, 12–13, 44–45, 50–53, 80–85, 359–360, 374–377
Hypophysis, 93, 368–369, 374–375

I

Indole, 9, 10
Inflorescence
  development, 141, 146–147
  meristem, 6, 135, 136–137, 141, 146, 147, 148–153
    central initiation zone, 136–137, 148, 151
    flank meristem, 136–137, 148, 151
    rib (file) meristem, 136–137, 148, 151
    mutants, 141, 180–199, 214–215, 254–257
  phyllotaxy, 135, 148–149, 254–257

Integumentary tapetum. *See* Ovules
Integuments, 355–356, 362–363, 370–371, 376–377, 386–389
  See also Ovules

L

L1, L2, L3. *See* Meristem
Leaf
  cauline, 4–5, 135, 146–147
  development, 5–6, 28–39
  epidermis, 38–43
  heteroblasty, 4–5, 88
  lamina, 6, 32–33
  mesophyll, 5–6, 28–33, 38–39, 54–55
  mutants, 6–7, 88–89
  petiole, 4–5
  phyllotaxy, 4–5, 16–17
  primordia, 3, 5, 12–21, 34–35
  rosette, 4–5, 135, 178–179
  shape, 4–5, 6–7, 72–73, 88–89
  stipules, 5, 14–15, 16–17, 34–37, 146–147
  vasculature, 32–33, 38–39
  See also Flowers; Trichome
Light regulated development, 7–8, 44–47
  blue, 52–53
  far-red, 52–53
  mutants, 7, 48–55, 82–83
  phototropism, 7–8
  red, 52–53
Lignin, 38–39
Lipids, 10, 40–41, 86–87, 170–171
Locule. *See* Stamen

M

MADS box, 144, 200, 216
Male sterility, 278–279, 284–295, 335
Megagametogenesis, 300, 302, 310–311
Megaspore. *See* Embryo sac
Megasporogenesis, 299, 300, 302, 306, 308–309, 314–315
Meiosis. *See* Ovules; Pollen
Meristem
  apical, 12–19, 135, 136–137, 138, 146, 147, 148–153
  L1, L2, L3, 3, 5, 16–17, 136–137, 148–152, 154–156, 158–159, 236–239, 302
  mutants, 6, 18–27, 214–217
  See also Flowers, meristem; Inflorescence, meristem; Roots, meristem; *Tunica-corpus*; Vegetative meristem
Mesophyll. *See* Leaf

Sepal
  development
    differentiation, 152–153
    growth rate, 160
    initiation, 152–155
    primordia, 152–155
  epidermis, 162–163
  lateral versus medial, 152–153,
    154
Septum. *See* Gynoecium
Sieve elements, 420–423
Silique, 86–87, 88–89, 346–347
Stamen
  anther, 156–157, 166–167
    endothecium, 218, 277, 278, 282–
      283
    locule, 156–157, 277, 282–283
    middle layer, 277, 278, 282–283
    parietal tissue, 277–278
    pollen sac, 156, 277, 278, 282–283
    septum, 282–283
    sporogenous tissue, 156
    stomium, 282–283, 288–289
    tapetum, 160, 277, 280–285
  connective, 218, 278
  development
    differentiation, 156–157, 160–161
    growth rate, 160
    initiation, 154–157
    primordia, 152–153, 156–157
  epidermis, 164–167, 282–283
  filament, 156–157, 160–161, 166–167
  lateral versus medial, 156–157

Starch, 9, 50–51, 174–175, 322–325,
  328–329, 362–363, 378–379
Stem, 3, 80–81, 84–88, 168–169
Stigma, 160–161, 168–171, 336, 338–
  345
  See also Gynoecium
Stipules. *See* Leaf
Stomata, 38–41, 168–171, 174–175,
  224–225, 414–417
Style. *See* Gynoecium
Suspensor, 93, 312–313, 352, 357, 366–
  373, 388–389, 394–395
Synergid. *See* Embryo sac
Systemic acquired resistance, 405

T

Tapetum. *See* Stamen
Temperature sensitivity, 10, 76, 106–
  111, 226–231, 240–241
Testa. *See* Seed, coat
Thiamine, 357
Thylakoid membrane, 44–47, 50–51
Transition to flowering. *See* Flowering
Transmitting tissue. *See* Gynoecium
Trichome
  branching, 58–59, 62–63
  development, 16–17, 28–29, 42–43,
    56–57, 60–61
  mutants, 62–73, 398, 401
Tryptophan, 9, 10
Tunica-corpus, 3, 148, 151

V

Vasculature
  cotyledon, 12–13
  leaf, 32, 38–39, 78–79
  ovule, 320–321
  root, 102, 103
  seedling, 358
Vegetative development, 3–11, **1–
  89**
Vegetative meristem
  mutant, 6, 18–23
  wild-type, 3–5, 14–19
    adult, 3, 4–5, 16–17
    central initiation zone, 3, 4
    flank meristem, 3
    juvenile, 4–5, 14–17
    rib (file) meristem, 3–4, 5

W

Whorl. *See* Flowers

X

Xylem, 102–103, 418–421

Z

Zygote, 312–315, 318–319, 351, 353,
  362–365, 386–387

Gene Index

449